Geomagnetism and Palaeomagnetism

NATO ASI Series

Advanced Science Institutes Series

A Series presenting the results of activities sponsored by the NATO Science Committee, which aims at the dissemination of advanced scientific and technological knowledge, with a view to strengthening links between scientific communities.

The Series is published by an international board of publishers in conjunction with the NATO Scientific Affairs Division

A	Life Sciences	Plenum Publishing Corporation
B	Physics	London and New York
C	Mathematical and Physical Sciences	Kluwer Academic Publishers
		Dordrecht, Boston and London
D	Behavioural and Social Sciences	
E	Applied Sciences	
F	Computer and Systems Sciences	Springer-Verlag
G	Ecological Sciences	Berlin, Heidelberg, New York, London,
H	Cell Biology	Paris and Tokyo

Series C: Mathematical and Physical Sciences - Vol. 261

Geomagnetism and Palaeomagnetism

edited by

F. J. Lowes

D. W. Collinson

J. H. Parry

S. K. Runcorn

D. C. Tozer

Physics Department,
University of Newcastle upon Tyne, U.K.

and

A. Soward

Mathematics Department,
University of Newcastle upon Tyne, U.K.

Kluwer Academic Publishers

Dordrecht / Boston / London

Published in cooperation with NATO Scientific Affairs Division

Proceedings of the NATO Advanced Study Institute on
Geomagnetism and Palaeomagnetism
Newcastle upon Tyne, U.K.
11–22 April, 1988

Library of Congress Cataloging in Publication Data

Geomagnetism and palaeomagnetism.

(NATO ASI series. Series C, Mathematical and physical
sciences ; vol. 261)
"Lectures given at the NATO advanced study institute,
held 11-22 April 1988 at Newcastle upon Tyne, England"--
Pref.
Includes index.
1. Magnetism, Terrestrial--Secular variation--
Congresses. 2. Paleomagnetism--Congresses. I. Lowes,
F. J. II. Series: NATO ASI series. Series C, Mathe-
matical and physical sciences ; no. 261.
QC828.G35 1989   538'.72   88-27364

ISBN-13: 978-94-010-6896-3   e-ISBN-13: 978-94-009-0905-2
DOI: 10.1007/978-94-009-0905-2

Published by Kluwer Academic Publishers,
P.O. Box 17, 3300 AA Dordrecht, The Netherlands.

Kluwer Academic Publishers incorporates the publishing programmes of
D. Reidel, Martinus Nijhoff, Dr W. Junk, and MTP Press.

Sold and distributed in the U.S.A. and Canada
by Kluwer Academic Publishers,
101 Philip Drive, Norwell, MA 02061, U.S.A.

In all other countries, sold and distributed
by Kluwer Academic Publishers Group,
P.O. Box 322, 3300 AH Dordrecht, The Netherlands.

TABLE OF CONTENTS

PREFACE

This volume presents lectures given at the NATO Advanced Study Institute held 11-22 April 1988 at Newcastle upon Tyne, England.

The aim of the Institute was to improve the interaction between workers in observational geomagnetism (using historical data) and archaeo- and palaeo-magnetism (using the remanent magnetization of man-made artefacts and of natural sediments and rocks) and those trying to interpret the data in terms of mechanisms inside or outside the Earth, particularly those developing dynamo theories of the field.

The material discussed ranged from magnetic bacteria swimming round a circle in a few seconds, the effect of El Niño, through secular variation with time scales of tens to thousands of years and the mechanics of individual field reversals and excursions (aborted reversals?) to possible modulation of average reversal frequency on the hundred million year time scale.

Many members of the Physics Department helped with the organization, and we are most grateful to them, and in particular to Anne Codling for her very many contributions. We also gratefully acknowledge the painstaking work of Aileen Dryburgh and Lynn Whiteford in so carefully typing the manuscript.

F.J. Lowes   Newcastle upon Tyne
D.W. Collinson   21 September 1988
J.H. Parry
S.K. Runcorn
A. Soward
D.C. Tozer

Institute Director: Professor S.K. Runcorn

                                       Department of Physics
                                       University of Newcastle upon Tyne

Organizing Committee: Professor R. Hide

                                       Meteorological Office
                                       Bracknell, Berks

                                       Professor K.M. Creer

                                       Department of Geophysics
                                       University of Edinburgh

                                       Professor A. Soward

                                       Department of Mathematics
                                       University of Newcastle upon Tyne

We are grateful to the Scientific Affairs Division of NATO for
sponsoring this Institute.

A SPHERICAL CAP HARMONIC MODEL OF THE CRUSTAL MAGNETIC ANOMALY FIELD
IN EUROPE OBSERVED BY MAGSAT

A. De Santis*, D.J. Kerridge and D.R. Barraclough
British Geological Survey
Murchison House
West Mains Road
Edinburgh, UK, EH9 3LA

*Istituto Nazionale di Geofisica
Via di Villa Ricotti 42
00161 Roma, Italy

ABSTRACT. The geomagnetic field observed in current-free regions
above the Earth's surface may be expressed as the gradient of a scalar
potential satisfying Laplace's equation. Spherical cap harmonic
analysis enables solution of Laplace's equation, subject to boundary
conditions appropriate to geomagnetic field analysis, in a region
bounded by a spherical cap. Magsat data within a spherical cap of
half-angle 35° centred on latitude 45°N, longitude 10°E have been
analysed for their crustal content. The resulting estimates of the
crustal vector field have been used to derive a spherical cap harmonic
model of the crustal scalar potential. The model contains 256
parameters and portrays wavelengths of 1000 km and above. Vector
anomaly maps derived from the model show several prominent features of
which the largest is that in the Kursk region of the USSR. The model
has been used to correct both the vector and total intensity data on
to a 2° by 2° grid at an altitude of 400 km. Anomaly maps produced by
contouring the grid averages are in good agreement with those derived
from the model. The major difference is for the vertical component of
the anomaly field over the Kursk region of the USSR. This is a
high-amplitude short-wavelength feature which the model smooths.

1. INTRODUCTION

In current-free regions the geomagnetic field may be expressed as the
gradient of a scalar potential which satisfies Laplace's equation.
This follows from the physical requirement that the curl and
divergence of the field are both zero. The solution of Laplace's
equation in spherical polar co-ordinates, using spherical harmonic
functions, subject to boundary conditions appropriate to global
geomagnetic field analysis, is well-known and forms the basis of
models of the main field, such as the International Geomagnetic
Reference Field (IGRF) (Barraclough 1987). The potential is expressed

1

F. J. Lowes et al. (eds.), Geomagnetism and Palaeomagnetism, 1–17.

in terms of a finite number of model coefficients which are estimated using geomagnetic field data, and the model can be used to synthesize field values, satisfying the potential field constraints, in any source-free region. This includes upward and downward continuation beyond the boundaries defined by the data distribution Errors in synthesised field values can be estimated if the uncertainties in the model coefficients are known.

For regional geomagnetic field analyses, where a model applicable to a limited portion of the Earth's surface is required, the spherical harmonics used in global analyses are no longer suitable. The global spherical harmonics are not orthogonal over the restricted area and the techniques for determining the model coefficients, generally based on the method of least-squares, become numerically unstable. One purpose of a regional model is to show smaller scale structure than is revealed by a global model. The shortest wavelength that a conventional spherical harmonic model can represent at the Earth's surface is equal to the circumference of the Earth divided by the maximum degree of the spherical harmonic expansion (Bullard 1967). Even if the technique could be applied without numerical difficulties, it would be highly inefficient.

A common technique in regional field analysis is to express the field as a polynomial in latitude and longitude, ignoring altitude variations in the data, or using some approximation to correct the data to the Earth's surface. Where more than one geomagnetic field element is modelled in this way it is possible to constrain the polynomials to partially obey the potential constraints (Tsubokawa 1952, Fougere and McClay 1957), but the resulting model cannot be upward or downward continued. Often, e.g. Molina et al. (1984), the polynomials for different elements are derived independently; no constraints are applied. The method of rectangular harmonic analysis has been used by Alldredge (1981). This allows regional data to be modelled in terms of a solution of Laplace's equation using Cartesian co-ordinates (x,y,z) at the origin of which x and y are horizontal and z is vertical. The solution is periodic in x and y and decays exponentially with z. The method can be applied successfully over small areas, but numerical difficulties arise because of the wide range of values the exponential term assumes when the data spans an appreciable range in z.

In an elegant extension of conventional spherical harmonic analysis Haines (1985a) has shown how a solution to Laplace's equation may be found, subject to boundary conditions appropriate to geomagnetic field analysis, for a spherical cap. The technique leads to models of the field which can be upward and downward continued and, in deriving a model, proper account can be taken of altitude variations in the data. The spherical cap model is expressed in terms of associated Legendre functions of integer order but of (in general) non-integer degree. For a model comprising a given number of terms the minimum wavelength which the model can portray is determined by the maximum degree which is in turn determined by the half-angle of the spherical cap. The minimum wavelength is always less than that of a global model with the same number of terms; hence short wavelengths are modelled more efficiently than with global spherical harmonics. The theory developed by Haines is summarized in Section 2.

A geomagnetic field observation is the resultant of fields due to currents flowing in the Earth's core, the crust, the ionosphere and the magnetosphere, and of fields due to magnetization of crustal rocks. Any of these fields can be modelled using spherical cap harmonic analysis, if the required signal is first isolated. The technique has been used by Haines (1985b) to model crustal fields, by Haines (1985c) to model secular variation in Canada, and by Haines and Newitt (1986) to define a Canadian geomagnetic reference field for which both the main field and the secular variation have been modelled.

In this paper spherical cap harmonic analysis has been used to model crustal field data over Europe. The data set was derived from the vector component data collected by the Magsat spacecraft. The Magsat mission is outlined in Section 3 and the data selection criteria and the methods of isolating the crustal signal are described in Section 4. The analysis carried out is described in Section 5 and the results of the analysis, in the form of crustal anomaly maps are presented in Section 6.

2. THEORY

The standard form of the solution of Laplace's equation for internal sources used in geomagnetism is the following:

$$V = a \sum_{n=1}^{\infty} (a/r)^{n+1} \sum_{m=0}^{n} (g_n^m \cos m\phi + h_n^m \sin m\phi)\, P_n^m(\cos\theta). \qquad (2.1)$$

In equation (2.1) r is the geocentric distance, θ is colatitude, ϕ is longitude, a is the radius of the reference sphere, and $P_n^m(\cos\theta)$ is the (Schmidt normalised) associated Legendre polynomial of degree n and order m. The constants g_h^m and h_n^m, called Gauss coefficients, are the model coefficients. (In actual analysis the expansion in n is truncated at some finite degree $n = n^*$.) This represents a superposition of solutions of the form

$$V = \sum_{n=1}^{\infty} \sum_{m=0}^{n} V_n^m ,$$

with

$$V_n^m = a(a/r)^{n+1}(g_n^m \cos m\phi + h_n^m \sin m\phi)\, P_n^m(\cos\theta).$$

The solutions V_n^m are obtained using the technique of separation of variables to reduce the problem to solving three coupled ordinary differential equations. The three equations are coupled by eigenvalues m^2 and $n(n+1)$ and the permissible values of m and n are determined by the boundary conditions. Continuity of the potential in longitude dictates that m is real and integer, and the conditions of regularity of the potential at $\theta = 0°$ and at $\theta = 180°$ restrict n to real and integer values. With integer values of m and n the solutions to the ordinary differential equation in θ are finite polynomials, the associated Legendre polynomials, $P_n^m(\cos\theta)$.

For a spherical cap the boundary conditions in longitude are identical to those in the global problem and so m is again real and integer. The boundary conditions at $\theta = 0°$ are also the same as in global analysis, but those at the boundary of the spherical cap at $\theta = \alpha$ are new. The potential and its θ-derivative must satisfy

$$V(r,\alpha,\phi) = f(r,\phi) \tag{2.2}$$

and

$$\partial V(r,\alpha,\phi)/\partial\theta = g(r,\phi), \tag{2.3}$$

where the functions f and g are arbitrary functions satisfying identical conditions in r and ϕ as V and $\partial V/\partial\theta$ respectively. Haines (1985a) shows that equation (2.2) may be satisfied by choosing functions V_n^m such that

$$\partial V_n^m(r,\alpha,\phi)/\partial\theta = 0.$$

This is achieved by determining the roots of

$$dP_n^m(\cos\alpha)/d\theta = 0, \tag{2.4}$$

regarded as an equation in n, for $m = 0,1,2 \ldots \infty$. Similarly, equation (2.3) is satisfied by functions V_n^m which are zero at the boundary of the spherical cap. These are found by determination of the values of n for which

$$P_n^m(\cos\alpha) = 0. \tag{2.5}$$

The values of n which enable equations (2.4) and (2.5) to hold are, in general, non-integer. For a given value of m if the values of n determined by solving (2.4) and (2.5) are written down in increasing order they alternate, the smallest value being a solution to (2.4). This property is convenient for labelling the solutions at each value of m. As m begins at zero Haines (1985a) defines an index k, also starting at zero, so k=0 labels the smallest root n at each m. The solutions of (2.4) are then characterised by k-m=even, those of (2.5) by k-m=odd. The two sets of solutions form two sets of basis functions. Within each set the functions are orthogonal over the spherical cap in the same way that the Legendre polynomials in conventional spherical harmonic analysis are orthogonal over the interval $\theta=0°$ to $\theta=180°$. However functions with k-m=even are not orthogonal to those with k-m=odd.

The potential may then be written as

$$V = a \sum_{k=0}^{\infty} (a/r)^{n+1} \sum_{m=0}^{k} (g_k^m \cos m\phi + h_k^m \sin m\phi) \, P_n^m(\cos \theta), \qquad (2.6)$$

which is similar to (2.1), the expression for conventional spherical harmonic analysis, but the summation over (integer) n has been replaced by a summation over the new index k which is also used to index the Gauss coefficients. If the expansion is truncated at a maximum index $k=k_{max}$ then, if the k=0 term is included, there are $(k_{max}+1)^2$ terms in the expansion.

Given values of n and m the $P_n^m(\cos \theta)$ can be computed using the following formula.

$$P_n^m(\cos \theta) = \sum_{j=0}^{\infty} A_j(m,n) \left(\frac{1-\cos \theta}{2} \right)^j , \qquad (2.7)$$

where $A_o(m,n) = K_n^m \sin^m \theta,$

and where other values of A_j can be calculated using the recurrence relation

$$A_j(m,n) = \left[\frac{(j+m-1)(j+m) - n(n+1)}{j(j+m)} \right] A_{j-1}(m,n). \qquad (2.8)$$

The K_n are given by

$$K_n^m = 1 \quad \text{for } m=0,$$
$$K_n^m = \frac{2^{\frac{1}{2}}}{m! \ 2^m} \left(\frac{\Gamma(n+m+1)}{\Gamma(n-m+1)} \right)^{\frac{1}{2}} .$$

The expression for $P_n^m(\cos \theta)$ given in equation (2.7) is an infinite series which will truncate to a finite polynomial only when n is integer. In practice a tolerance for convergence is prescribed and the summation is performed until this is achieved. From equation (2.8) it is clear that numerical precision is important. For example for small m, large n the A_j will initially increase rapidly in magnitude and alternate in sign.

Haines (1985a) uses Stirling's formula to derive the following approximation for the K_n^m:

$$K_n^m = \frac{2^{-m}}{\sqrt{(m\pi)}} \left(\frac{n+m}{n-m} \right)^{\frac{1}{2}n+\frac{1}{4}} p^{\frac{1}{2}m} \exp(e_1 + e_2),$$

where $p = (n/m)^2 - 1,$

$e_1 = -(1/12m)(1 + 1/p)$

and $e_2 = (1/360m^3)(1 + 3/p^2 + 4/p^3).$

From equation (2.7) it is easily shown that

$$\frac{dP_n^m}{d\theta} = \sum_{j=1}^{\infty} A_j(m,n) \frac{j}{2} \sin\theta \left(\frac{1-\cos\theta}{2}\right)^n \qquad \text{for } m=0, \quad (2.9a)$$

$$= \frac{m\cos\theta}{\sin\theta} P_n^m + \sum_{j=1}^{\infty} A_j(m,n) \frac{j}{2} \sin\theta \left(\frac{1-\cos\theta}{2}\right)^j \quad \text{for } m>0. \quad (2.9b)$$

Equations (2.7) and (2.9) enable computation of the associated Legendre functions and their θ-derivatives and so provide a means for solving equations (2.5) and (2.4) respectively for the appropriate values of n using numerical methods.

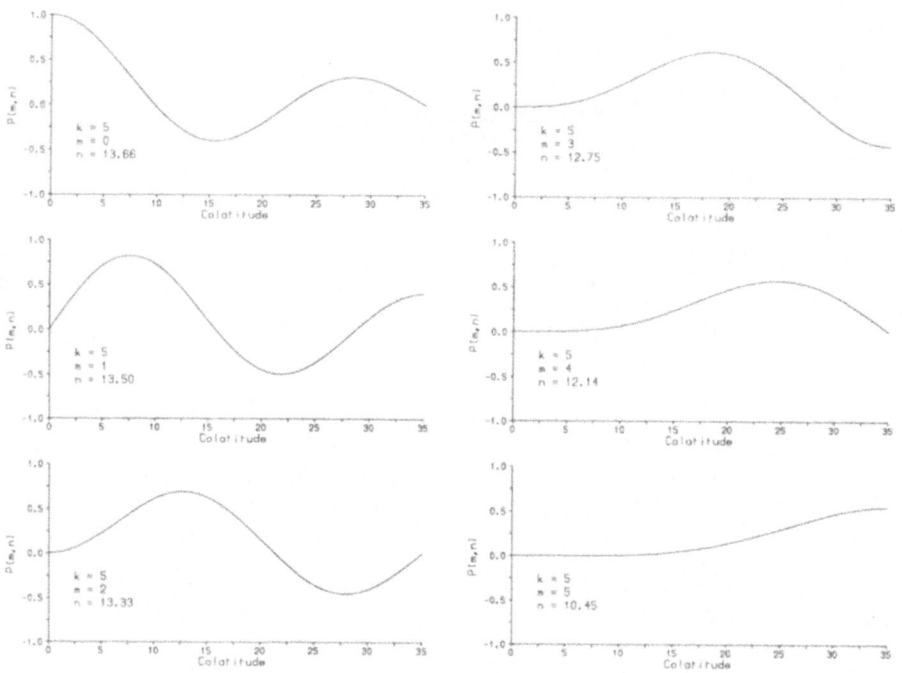

Figure 1. Associated Legendre functions for a spherical cap of half-angle 35°, for k = 5, m = 0 to 5.

For illustration, in Fig. 1 the $P_n^m(\cos\theta)$ for a spherical cap of half-angle 35°, for k=5, m=0,1,...5 are shown. Note that the functions with k-m=even are zero at the boundary of the cap and that those with k-m=odd have zero gradient at the boundary as the conditions (2.5) and (2.6), respectively, require.

3. THE MAGSAT MISSION

The Magsat spacecraft was launched in October 1979 into a twilight
sun-synchronous orbit of inclination 96.76°, apogee 561 km, perigee
352 km. The orbital period was approximately 96 minutes. The
spacecraft carried a three-component ring-core fluxgate magnetometer
(Acuña et al., 1978) with a measurement accuracy of ±3 nT, and a
caesium vapour magnetometer capable of measuring total intensity to an
accuracy of ±1.5 nT. Vector measurements were made at a rate of 16
samples per second, scalar measurements at 8 samples per second. The
scalar magnetometer developed a fault soon after launch and
subsequently functioned intermittently. Sufficient scalar data were
obtained, however, to enable in-flight calibration of the fluxgate
magnetometer until shortly before the end of the mission in June 1980.
Spacecraft fields, position and time uncertainties, and, for the
vector data, attitude uncertainties, contributed errors to the
measured data. Vector data obtained when the spacecraft attitude was
known with an accuracy of ±20 seconds of arc are termed
'fine-attitude' data. The various sources of error are discussed in
detail by Langel et al. (1981). Langel et al. (1982) quote the
overall error in the scalar data as ±2 nT and the error in the
fine-attitude vector component data as ±6 nT.

A subset of the data collected by Magsat is available on the
Investigator-B tapes (Langel et al. 1981). These contain spot
measurements of the geocentric X (north), Y (east) and Z (radially
inwards) components of the geomagnetic field and the total intensity,
together with the geocentric latitude, longitude and radial distance,
at intervals of approximately 5 seconds (37 km) along track. The
total intensity data on the tapes are derived from the fluxgate vector
measurements. Mean values computed by averaging over the 40
measurements before and after the spot values are also supplied.

The planetary K-index, K_p, a three-hour measure of geomagnetic
field activity at mid- and sub-auroral latitudes, and the disturbance
storm time index, D_{st}, a one-hour measure of the disturbance of the
horizontal component of the geomagnetic field near the equator due to
the storm time ring current, are supplied for each pass (orbit). The
pass number and the modified Julian day number (January 1st 1980 is
day number 44239) are also recorded.

4. DATA SELECTION

The object of the study was to model long-wavelength (>1000 km)
features of the crustal vector field observed by Magsat over Europe.
The area of interest was defined as that contained within a spherical
cap of half-angle 35° centred on latitude 45°N, longitude 10°E, (in
northern Italy). This is an area (at the Earth's surface) of
approximately 46.10^6 km^2. The accuracy of a spherical cap harmonic
model deteriorates close to the boundary of the spherical cap so it is
advisable to extend the boundary beyond the area where high model

accuracy is important. However, for a given number of model
parameters, the shortest wavelength the model portrays increases as
the half-angle of the spherical cap is increased. These
considerations were taken into account in defining the spherical cap.

The spot values of the vector and scalar field recorded on the
Investigator-B tapes were the data used in the study. The spot values
were preferred to the averages because although the averageing reduces
noise, the averaged value may not be representative of the field at
the mean position if field gradients (due to any source) are
significantly non-linear over the 37 km of the average. In order to
reduce the effects of magnetic disturbance data from passes which
crossed the spherical cap during magnetically quiet intervals (K_p 2)
were selected.

Langel and Estes (1985) used a selection of Magsat data to derive
spherical harmonic models of the main field (degree 13), secular
variation (degree 10) and the external (magnetospheric) field (degree
1), and designated these models GSFC(12/83). The external terms were
written as linear functions of D_{st}. The effect of induced fields, an
internal source, was included by modifying the axial dipole term of
the main field model with a term proportional to the axial term in the
external field model. For this study the modelled contributions of
the main-field, the magnetospheric field and the induction field
calculated from GSFC(12/83) were subtracted from the observed data.
The residual data were then plotted and scrutinized pass by pass.
These residual data, given adequate modelling of the main-field,
magnetospheric fields and induction fields, would represent the
crustal field and any unmodelled ionospheric fields. The twilight
orbit of Magsat was chosen to minimize the effects of ionospheric
currents, but these were not always negligible. For instance, Langel
and Estes attributed differences in main-field models derived
independently from dawn and dusk data to the presence of the
equatorial electrojet (an eastward flowing current in the ionosphere)
at dusk but not at dawn. Maeda et al. (1982) found evidence for
meridianal currents, also present at dusk only.

A common feature of the plots was disturbance of the X and Y
components at high latitudes due to field-aligned currents, which, as
the geomagnetic field is near-vertical at high latitudes, generate
horizontal magnetic fields. The most common editing decision was to
reject X and Y data at high latitudes, often retaining the Z
component. Occasional offsets in the vector data were seen. Mayhew
et al. (1985) noted this effect in the raw Magsat data and attributed
it to changes in the instruments used in the attitude determination.
Subjective decisions were made about passes in which this effect was
observed; they were either wholly retained or wholly rejected.

In an earlier study using Magsat data over Europe, Kerridge et
al. (1985) noted that plots of residuals over the same geographic
regions but at different times displayed inconsistent offsets or
trends. This effect has been noted by many authors and is believed to
result from inadequate modelling of external fields (e.g. Coles 1985).
Mayhew (1979) found that removing quadratic fits to residual scalar
data collected by the POGO satellites improved pass to pass

consistency. This is a somewhat arbitrary procedure, but it has been found effective by many authors and Kerridge et al. (1985) found that it improved consistency between data from Magsat passes over Europe. This procedure was applied to the edited data in each pass to generate the final data set which comprised measurements made at 88 897 points from a total of 665 passes. Of these points 85 308 included all three components. The geocentric distance of the data ranged from 6665 km to 6891 km with a mean of 6774 km. The data density is one set of measured values (i.e. at least one component) to approximately 520 km^2 at the Earth's surface.

5. ANALYSIS

The minimum wavelength that a truncated spherical cap harmonic model can portray is determined by the maximum degree n in the spherical harmonic expansion. This, in turn, is dependent on the half-angle (α) of the spherical cap and the maximum value of the index k. For a given value of k the maximum value of n occurs for m = 0. In Table 1 the values of n for k = 0 to 25, m = 0 are listed for α = 35°. As the Earth's circumference is approximately 40 000 km a value of n = 40 corresponds to a wavelength of 1000 km at the Earth's surface. From Table I it can be seen that n = 39.36 at k = 15. This determined our choice of k_{max} = 15 for the model, which has 256 coefficients.

Table I. Values of the spherical harmonic degree n(k,0) for a spherical cap of half-angle α = 35°.

k	m	n(k,m)	k	m	n(k,m)
0	0	0.0000	13	0	34.2227
1	0	3.4260	14	0	36.7622
2	0	5.7930	15	0	39.3645
3	0	8.5318	16	0	41.9079
4	0	10.9958	17	0	44.5065
5	0	13.6633	18	0	47.0530
6	0	16.1619	19	0	49.6487
7	0	18.8008	20	0	52.1977
8	0	21.3171	21	0	54.7910
9	0	23.9405	22	0	57.3420
10	0	26.4675	23	0	59.9334
11	0	29.0813	24	0	62.4861
12	0	31.6156	25	0	65.0759

The first step in the analysis was to convert the latitudes and longitudes of the data from 'old' values in the normal geographic system to 'new' values relative to the new pole. Colatitude and longitude in the conventional system defined in relation to the geographic north pole and the Greenwich meridian will be denoted θ and ϕ, the corresponding coordinates relative to the new pole will be written as θ' and ϕ'. The $\phi' = 0°$ meridian is defined by analogy with the definition of geomagnetic coordinates (e.g. Chapman and Bartels 1940) to be the great circle passing through the new north pole and the south geographic pole. If the coordinates of the new pole in the old system are θ_o and ϕ_o then the new coordinates of a point $P(\theta,\phi)$ may be calculated using the following equations which are derived from standard formulae of spherical trigonometry:

$$\cos \theta' = \cos \theta_o \cos \theta + \sin \theta_o \sin \theta \cos (\phi - \phi_o)$$

$$\tan(\pi - \phi') = \sin \theta \sin(\phi - \phi_o)/(\sin \phi_o \cos \theta -$$

$$- \cos \theta_o \sin \theta \cos (\phi - \phi_o))$$

(The solution for ϕ' is indeterminate when the point P is at the new pole.) The magnetic field components (which will be assumed to be geocentric rather than geodetic values) must be rotated into the new coordinate system. The transformation from (X,Y,Z) to (X′,Y′,Z′) is performed as follows:

$$X' = X \cos \beta - Y \sin \beta \qquad\qquad (5.1a)$$

$$Y' = X \sin \beta + Y \cos \beta \qquad\qquad (5.1b)$$

$$Z' = Z \qquad\qquad (5.1c)$$

where

$$\sin \beta = \sin \theta_o \sin(\phi - \phi_o)/\sin \theta' \text{ provided } \theta' \neq 0.$$

It was originally thought that because of the convergence of the Magsat tracks to the north of the region the data density would be greater in the north than in the south. The (approximately) equal area grid comprising the 3730 cells shown in Fig. 2a was used to index each data value according to the cell into which it fell. (The cells were defined by bands 1° wide, the number of cells in each band being chosen so that the area of each cell would approximate to an area of 1° by 1° at the equator.) The data set was then ordered by cell number which facilitated analysis of the data density. The average density was 24 data per cell; the maximum number of data in a cell was 70 and 56 cells were empty. It was found that overall there was no marked imbalance between the data density in the north and in the south. The editing process had resulted in data being rejected predominantly at high latitudes which tended to compensate for the effect of the convergence of the tracks.

From the ordered data set, cell averages of the data and of position were determined. This averaged data set was used to derive a preliminary spherical cap harmonic model (k_{max} = 8) of the crustal anomaly field by the method of least squares using the algorithm of Malin et al. (1982) to form and solve the normal equations. The residuals of the full data set from the preliminary model were

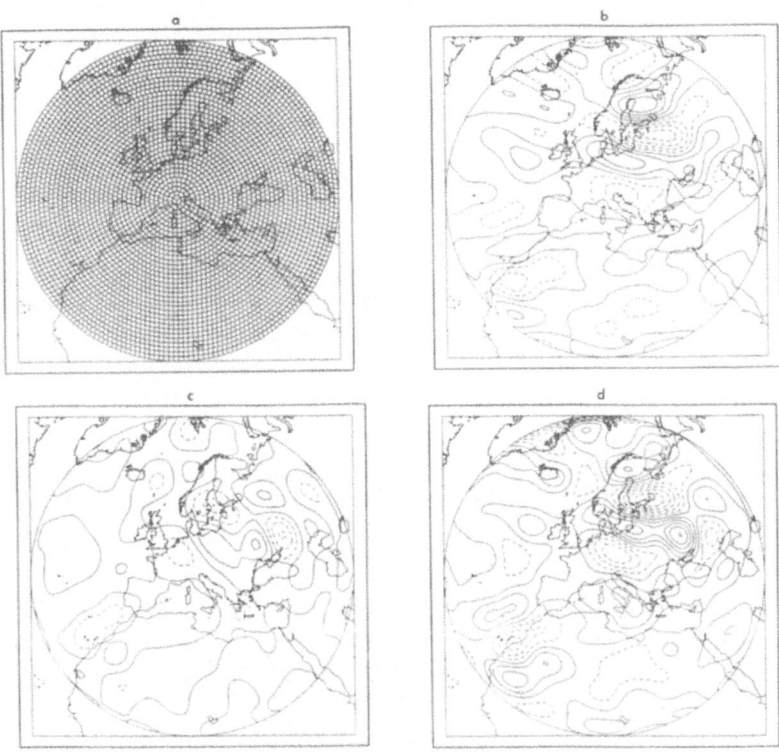

Figure 2. a. The (approximately) equal-area grid used to index the
               Magsat data.
           b,c,d. Anomaly maps at 400 km derived from the spherical
               cap harmonic model. In each map the contour interval
               is 2 nT, the zero and positive contours are shown as
               full lines, the negative contours as dashed lines.
           b, X component; c, Y component; d, Z component.

computed to screen the data set for gross outliers, but none were
found. Using cell averages is unsatisfactory for a method which is
capable of taking position into account properly, but, provided the
range of the field values within a cell is small compared with the
uncertainties in the measurements, noise reduction will be achieved by
the averaging. The advantage of using averages is that the data set
is reduced (in this case) to 3730 sets of three-component data and so
the least-squares determination is relatively inexpensive in
computational terms.

Table II The rms values, in nT, of the anomaly component data in each 'colatitude' band (subscript o), and the rms residual of the data from the model (subscript res). N is the number of sets of three-component data in each band, n is the number of sets with one or two components only. The final line summarizes the statistics over the whole spherical cap.

θ	X_o	X_{res}	Y_o	Y_{res}	Z_o	Z_{res}	N	n
1	3.93	2.79	3.98	2.06	1.89	1.74	101	7
2	3.60	2.93	3.91	2.58	2.41	1.92	298	10
3	3.80	2.89	3.61	2.46	3.14	1.85	408	5
4	3.49	2.90	3.58	2.70	3.89	1.85	530	17
5	3.08	2.60	3.94	2.86	4.36	1.83	663	14
6	2.52	2.45	3.82	2.80	4.70	1.87	908	19
7	2.90	2.69	3.34	2.67	4.81	1.79	1143	29
8	3.54	2.88	3.11	2.65	4.32	1.93	1326	46
9	3.81	2.60	3.14	2.83	3.56	2.01	1414	90
10	4.10	2.76	3.08	2.83	2.86	2.05	1686	77
11	3.68	2.74	3.07	2.93	3.14	2.11	1867	70
12	3.16	2.89	3.04	2.87	3.44	2.11	2128	44
13	3.18	2.92	3.11	2.95	3.40	2.19	2161	51
14	3.52	2.84	3.01	2.81	3.11	2.18	2284	69
15	3.94	2.92	2.90	2.76	2.84	2.24	2298	83
16	4.05	2.87	3.13	2.88	2.93	2.23	2609	113
17	3.74	2.77	3.32	2.86	3.20	2.23	2849	143
18	3.56	2.93	3.54	3.02	4.47	2.55	2768	174
19	3.27	2.93	3.86	3.04	4.55	2.56	2655	171
20	3.47	2.95	3.68	3.02	4.61	2.82	2865	198
21	3.95	2.95	3.82	2.90	4.59	2.73	2921	192
22	3.68	2.84	3.77	3.00	4.13	2.77	3064	152
23	3.54	2.91	3.73	3.06	3.90	2.77	3183	148
24	3.29	2.90	3.69	3.11	3.71	2.72	3608	162
25	3.18	2.95	3.61	3.04	3.45	2.58	3475	155
26	3.11	2.88	3.61	3.09	3.59	2.55	3549	137
27	3.09	2.92	3.59	3.06	3.38	2.49	3728	144
28	2.90	2.69	3.41	3.05	3.17	2.49	3646	173
29	2.88	2.66	3.26	2.97	3.10	2.60	3993	112
30	2.71	2.51	3.11	2.88	3.05	2.57	3654	137
31	2.56	2.43	3.08	2.79	3.00	2.48	3517	114
32	2.55	2.42	2.93	2.72	2.97	2.52	3685	101
33	2.56	2.40	2.91	2.75	2.76	2.43	3629	105
34	2.55	2.40	2.92	2.79	2.87	2.61	3404	121
35	2.91	2.77	3.32	3.20	3.27	2.98	3291	206
	3.25	2.76	3.36	2.93	3.54	2.48	85308	3589

However, to take full advantage of the method the entire data set should be used, providing the data density has no geographical bias, with each item of data used to form an equation of condition. The final model, with k = 15, was generated using the full data set, with all data weighted equally. The rms values of the input data, X', Y' and Z', in each 'colatitude' band together with the rms residuals from the final model are shown in Table II. The final line in Table II shows the overall values for the spherical cap.

6. CRUSTAL ANOMALY MAPS

The spherical cap harmonic model was used to compute values of the X', Y' and Z' components of the anomaly field at an altitude of 400 km on a 1° by 1° grid in the new coordinate system. The components were rotated into X, Y and Z using expressions obtained by rearrangement of equations (5.1). The grid values of X, Y and Z were then contoured to produce the anomaly maps shown in Figs. 2b-2d. The projection is azimuthal equidistant, the contour interval is 2 nT, the zero and positive contours are shown as full lines, the negative contours as dashed lines.

The Kursk region is well-known as an area with very large magnetic anomalies due to iron ore deposits and its signature in Magsat data has been studied in detail by Taylor and Frawley (1987). To test the ability of the spherical cap harmonic model to portray this large-amplitude anomaly, data from the strip of latitude 49.75°N to 50.25°N and for longitude 0° to 45°E were selected from the full data set, (the centre of the Kursk anomaly is at approximately 50°N, 37°E). In Fig. 3a the altitudes of the Z-component anomaly data selected are plotted; these range from 316.7 km to 514.6 km. In Fig. 3b differences between the modelled values at the positions of the input data and modelled values at 400 km altitude, 50°N latitude and the nearest integer value of longitude are shown. Fig. 3c shows the original Z-anomaly data and Fig. 3d shows the data (plotted against original longitude) with the corrections from Fig. 3b applied, with the modelled profile along latitude 50°N at 400 km superimposed.

To portray the Z-component data over the Kursk region more accurately the model would clearly have to extend to a value of k greater than 15 to include shorter wavelengths. However the general fit of the model to the data along the profile is good. Using the model to correct the data to constant altitude will be a legitimate procedure where there are no wavelengths in the data shorter than the minimum wavelength in the model. In the profile in Fig. 3 this is not the case only over the Kursk region. In practice for this data set the corrections in Fig. 3b are mostly smaller than the measurement errors, so there is no clear noise reduction from Fig. 3c to Fig. 3d.

To examine the fit of the model to the data over the entire region further contour maps were produced by gridding the data onto a 2° by 2° grid, using the spherical cap harmonic model to correct the data to the nearest grid point at 400 km altitude. The standard deviations of the grid averages were improved over the averages with

14

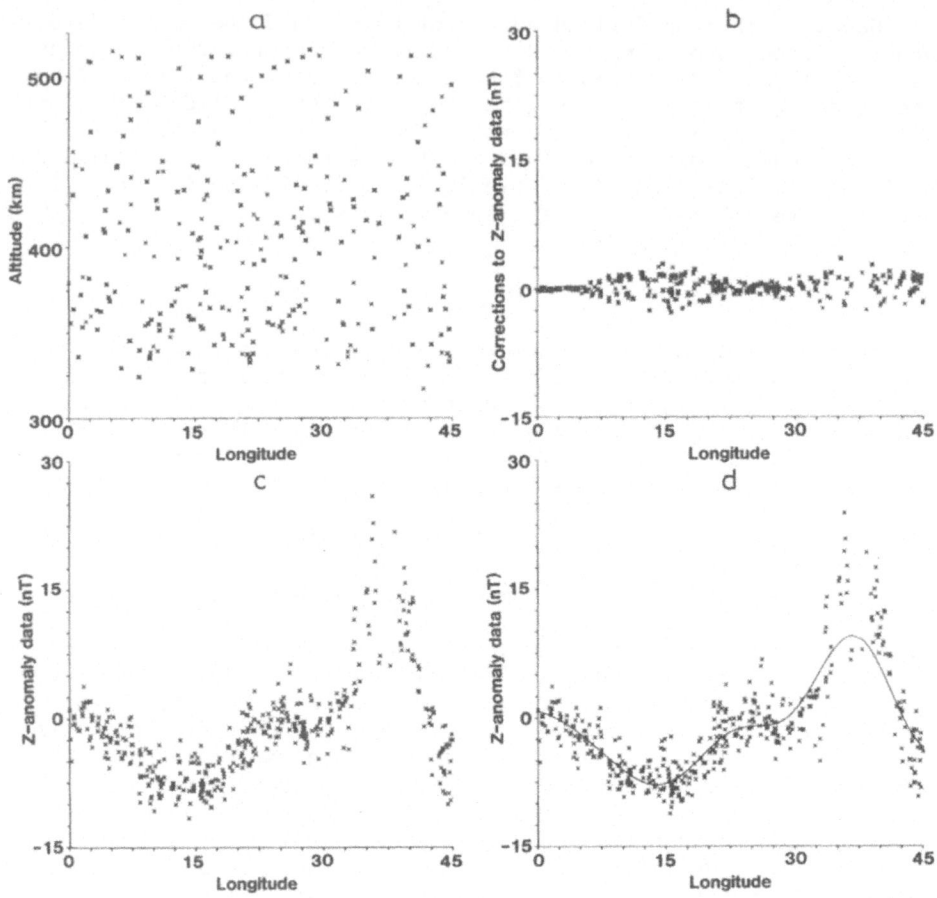

Figure 3. Z-component data in the (conventional) latitude range
49.75° to 50.25° and longitude range 0° to 45°E.
a. The altitudes (above 6371.2 km) of the data.
b. The differences between the modelled values at the data
positions and those at latitude 50°N, altitude 400 km,
and nearest integer value of longitude.
c. The original Z-component data.
d. The Z-component data plotted against original longitude
but with the corrections in Fig. 3b applied. The full
line shows the model values along the profile.

no correction applied, but only marginally because of the relative
sizes of the corrections and the data uncertainties. The grid points
with no data, (principally close to the pole of the spherical cap)
were assigned model values. Anomalies in the total field (F) were
computed using

$$FF_m = XX_m + YY_m + ZZ_m$$

where the subscript m refers to main-field values which were computed using the main-field model of GSFC(12/83). The maps of X, Y, Z and F obtained by contouring the grid averages are shown in Fig. 4.

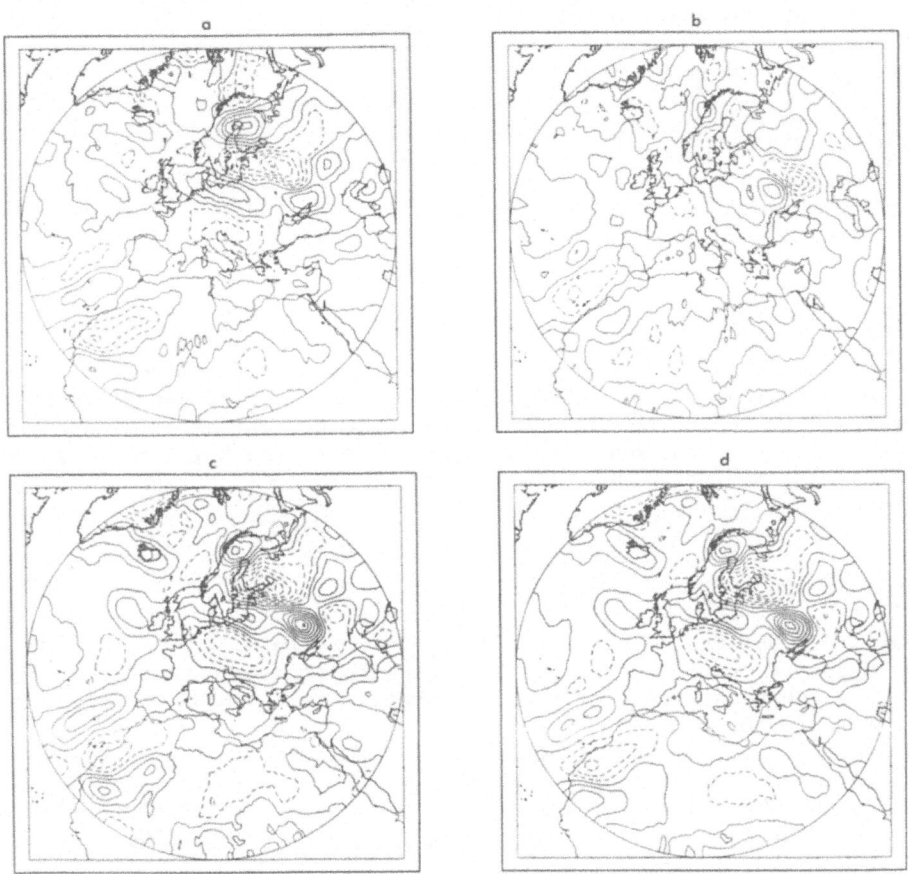

Figure 4.　Anomaly maps derived by using the spherical cap harmonic model to correct the original data onto a 2° by 2° grid at 400 km and then contouring the grid averages. The contour interval is 2 nT, the zero and positive contours are shown as full lines, the negative contours as dashed lines.
　　　　　a. The X-component.
　　　　　b. The Y-component.
　　　　　c. The Z-component.
　　　　　d. The scalar anomaly field.

The maps in Fig. 4 show more shorter wavelength detail than those produced from the model shown in Fig. 2. However, much of this detail is spurious, reflecting noise in the grid averages rather than real short-wavelength features in the anomaly field. The exception is the Z-component anomaly in the Kursk region which is, as indicated by Fig. 3d, over-smoothed by the model. Otherwise the positions and amplitudes of the anomalies are in good agreement.

The ability of spherical cap harmonic analysis to model accurately the anomaly field over a spherical cap centred in Europe, using a model with 256 parameters, has been demonstrated. The model, because it is the appropriate solution of Laplace's equation, allows upward and downward continuation, subject to errors due to uncertainties in the model coefficients. The method could be applied to any geomagnetic data set over a spherical cap and so is a powerful tool for regional field analysis.

ACKNOWLEDGEMENTS

This paper is published with the permission of the Director, British Geological Survey (NERC). The work was carried out while Dr. De Santis visited the Geomagnetism Research Group at the British Geological Survey, supported by an award made by the Royal Society and the Accademia Nazionale dei Lincei under the European Science Exchange Programme.

REFERENCES

Acuña, M.H., Scearce, J.B., Seek, J.B. and Scheifele, J., 1978. The Magsat vector magnetometer - A precision fluxgate magnetometer for the measurement of the geomagnetic field. NASA/GSFC Tech. Mem. 79656.

Alldredge, L.R., 1981. Rectangular harmonic analysis applied to the geomagnetic field. J. Geophys. Res., 86, 3021-3026.

Barraclough, D.R., 1987. International Geomagnetic Reference Field: the fourth generation. Phys. Earth Planet. Inter., 48, 279-292.

Bullard, E.C., 1967. The removal of trend from magnetic surveys. Earth Planet. Sci. Letts., 2, 293-300.

Chapman, S. and Bartels, J., 1940. Geomagnetism. Oxford University Press.

Coles, R.L., 1985. Magsat scalar magnetic anomalies at northern high latitudes. J. Geophys. Res.. 90, 2576-2582.

Fougere, P. and McClay, J., 1957. Preparation of mutually consistent magnetic charts. Geophys. Res. Pap. 55, Air Force Cambridge Res. Cent., Bedford, Mass., 39pp.

Haines, G.V., 1985a. Spherical cap harmonic analysis. J. Geophys. Res., 90, 2583-2592.

Haines, G.V., 1985b. Magsat vertical field anomalies above 40°N from spherical cap harmonic analysis. J. Geophys. Res., **90**, 2593–2598.

Haines, G.V., 1985c. Spherical cap harmonic analysis of geomagnetic secular variation over Canada 1960–1983. J. Geophys. Res., **90**, 563–12574.

Haines, G.V. and Newitt, L.R., 1986. Canadian Geomagnetic Reference Field 1985. J. Geomag. Geoelectr., **38**, 895–921.

Kerridge, D.J., Parr, R.S. and Barraclough, D.R., 1985. Scalar and vector magnetic anomaly maps for Europe derived from Magsat data. British Geological Survey, Geomagnetism Research Group Report No. 85/20.

Langel, R.A., Berbert, J., Jennings, T. and Horner, R., 1981. Magsat data processing: an interim report for investigators. NASA Tech. Mem. 82160.

Langel, R.A., Ousley, G. and Berbert, J., 1982. The Magsat mission. Geophys. Res. Lett., **9**, 243–245.

Langel, R.A. and Estes, R.H., 1985. The near-Earth magnetic field at 1980 determined from Magsat data. J. Geophys. Res., **90**, 2495–2509.

Malin, S.R.C., Barraclough, D.R. and Hodder, B.M., 1982. A compact algorithm for the formation and solution of normal equations. Computers and Geosciences, **8**, 355–358.

Maeda, H., Iyemari, T., Araki, T. and Kamei, T., 1982. New Evidence of a meridional current system in the equatorial ionosphere. Geophys. Res. Lett., **9**, 337–340.

Mayhew, M.A. Inversion of satellite magnetic anomaly data. J. Geophys., **45**, 119–128.

Mayhew, M.A., Johnson, B.D. and Wasilewski, P.J., 1985. A review of problems and progress in studies of satellite magnetic anomalies. J. Geophys. Res., **90**, 2511–2522.

Molina, F., Meloni, A., Battelli, O. and De Santis, A., 1984. Comparison of geomagnetic planetary reference fields over Italy. Phys. Earth Planet. Inter., **37**, 35–45.

Taylor, P.T. and Frawley, J.J., 1987. Magsat anomaly data over the Kursk region, USSR. Phys. Earth Planet. Inter., **45**, 255–265.

Tsubokawa, I., 1952. Reduction of the results obtained by the magnetic survey of Japan (1948–51) to the epoch 1950.0 and deduction of the empirical formulae expressing the magnetic elements. Bull. Geogr. Surv. Inst., **3**, 1–29.

GEOMAGNETIC SECULAR VARIATION

Malcolm G. McLeod
Naval Ocean Research and Development Activity
NORDA, Code 352
Bay St. Louis, MS 39529-5004
United States

ABSTRACT. The spatial and temporal variation of the geomagnetic field component annual means has been analyzed. The data used in this study consist of the annual means of the vector magnetic field components measured at nearly one hundred magnetic observatories widely distributed about the Earth. The data that have been analyzed are for the time interval 1961–1977.

Spherical harmonic models for each year within the indicated time interval have been computed for filtered first, second, and third time derivatives of the field using filtered first, second, and third time differences of the annual means as input data. A set of 48 magnetic observatories selected to minimize noise was used for the analysis. Coefficients corresponding to both internal and external sources were included in the models.

The geomagnetic jerk of 1969 was found to be due to internal sources. External current systems were found to be responsible for a significant portion of the first, second and third time differences of the annual means of the geomagnetic field. While most of the effect of the external current systems can be modeled by degree one spherical harmonics, higher degree spherical harmonics are also important for modeling the field due to external sources. The time variation of the external sources included harmonics of the sunspot cycle.

1. INTRODUCTION

The temporal variation of the geomagnetic field is a subject of importance to scientific theories of the Earth. This subject also has a number of practical applications. Increased understanding of the temporal (or "secular") variation of the geomagnetic field is expected to lead to increased understanding of the geomagnetic dynamo, of motion of the core fluid, and of the electrical conductivity of the Earth's mantle. The relationship of the secular variation of the geomagnetic field to these topics has been discussed by Backus (1983), Ducruix et al. (1980), Gubbins (1984), Bloxham and Gubbins (1985), and others. Increased understanding of the geomagnetic secular variation

F. J. Lowes et al. (eds.), Geomagnetism and Palaeomagnetism, 19–30.
© 1989 by Kluwer Academic Publishers.

is important also to a number of practical applications that involve determination of the orientation of an object relative to the Earth. These applications include navigation of ships and aircraft, alignment of communications antennas, and determination of the orientation of satellites in near-Earth orbit. These applications require models or charts of the geomagnetic field. Because the geomagnetic field changes with time, models or charts of the secular variation are also needed to update the field models.

The directional properties of the compass have been used for navigational purposes for centuries. The first really scientific theory of the origin of Earth's magnetism was proposed by Gilbert (1600) who thought the Earth was permanently magnetized. Halley (1683, 1692) published papers on the secular variation of the geomagnetic field. Navigators' logs were the source of much of his information. Further discussion of Halley's work can be found in an article by Evans (1988). This is the same Halley for whom the Comet Halley is named.

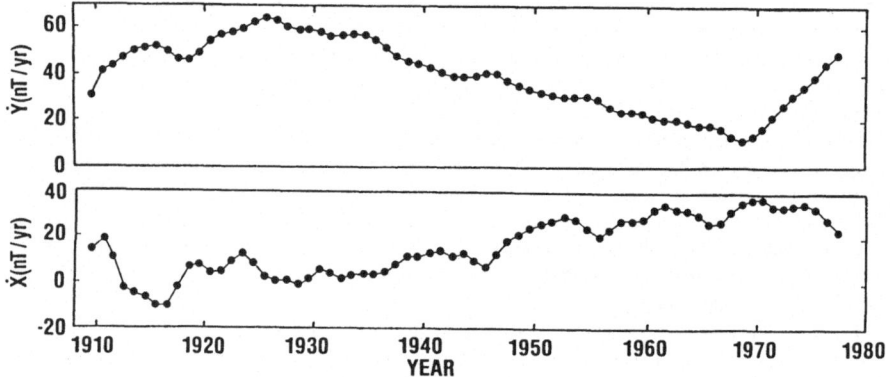

Figure 1. Smoothed first differences of annual means of X and Y components at Eskdalemuir, Scotland. From Alldredge (1984).

The data on which the research described in this paper is based consist of a set of magnetic observatory annual means for the vector geomagnetic field components. Graphs of the smoothed first differences of the annual means of two field components vs. time are shown in Figure 1. The means are averages of the field components at Eskdalemuir, Scotland. The averages are over all times of the day and all days of the year. The smoothing is done by a symmetric digital filter with weights (1/4, 1/2, 1/4). This figure is taken from a paper by Alldredge (1984). It can be seen that the slope of the East-West field derivative is nearly constant from 1925 to 1970 and again from 1970 to the end of the data in 1978. The change of slope near 1970 has been termed a "geomagnetic jerk" or "secular variation impulse" and has been discussed by a number of authors including Courtillot et al. (1978), Malin and Hodder (1982), Alldredge (1984, 1985), Gubbins (1984), McLeod (1985), and Backus et al. (1987). A review article on this subject has been published by Courtillot and Le

Mouel (1984). Two news articles on this phenomenon by Kerr (1984) and Weisburd (1985) have also appeared in print. There is still some controversy in the scientific community concerning the jerk, especially concerning the time duration of the "impulse" and whether this phenomenon is worldwide or local.

In addition to the features previously noted, the Y-component (East-West) first differences exhibit relatively small oscillations of about 11 year period. The X-component (North-South) first differences are similar in character to those of the Y-component except that the oscillations of approximate 11 year period are larger. These oscillations are of the same period as the solar cycle and are due primarily to current systems external to the Earth which are influenced by the solar wind and solar flares. The oscillations due to external currents are smaller on the Y-axis (East-West) than on the X-axis (North-South) partly because some of the currents are in planes approximately parallel to the equatorial plane. Moreover, the external current systems do not rotate with the Earth so they tend not to produce average field components in the East-West direction if the average is over an entire day or an integral number of days.

Some of the scientific controversy concerning the jerk is perhaps a result of the difficulty in distinguishing between fields of internal and external origin in the data as well as difficulty in distinguishing between these fields and noise. Noise is defined as errors due to instrument drift and malfunction, calibration errors, manmade magnetic signals, and possible errors in the data processing that produces annual means. The principal objective of the research reported in this paper is the separation of fields of internal from external origin and an evaluation of the noise or errors in the data set. This research has not yet been completed; however, some significant new results are reported.

2.   DATA SET SELECTION

The data used for this research are magnetic observatory annual means. The data set and some of the basic programs used for this research were supplied by Robert Langel and some of his colleagues at the NASA Goddard Space Flight Center. The data set and programs are essentially the same as were used for the research reported by Langel et al. (1982). These data are not the most recent currently available; however, it was felt that they are adequate for initiating the present research.

The initial data set chosen for study consisted of the annual means of the vector field components from 88 observatories having vector data for every year in the interval from 1961.5 to 1977.5. (The annual mean for calendar year 1961 is associated with the time 1961.5 corresponding to the middle of the year.) This time period was chosen as a compromise between the length of the time interval and the number of observatories with continuous data for the interval. If the interval were extended to include 1961.5 to 1980.5, there would be only 75 observatories. A shorter interval from 1970.5 to 1977.5

contained 121 observatories.

As is well known, the magnetic observatories are not very nearly uniformly distributed about the Earth; of the initial 88 chosen, only 18 were south of the geomagnetic equator. It was subsequently found that 4 of these 88 initial observatories did not have 3 component vector data for all years in the interval. Thus only 84 observatories remained in the initial data set including only 17 south of the geomagnetic equator.

The slope of the first differences of the field components shown in Figure 1 is nearly constant except for the changes in slope at approximately 1925 and 1970 and except for fluctuations apparently due to external current systems. Therefore, it was felt that RMS values of the third differences of the field components at the various observatories might be a convenient indicator of the data errors for these observatories. The third differences (unfiltered) for each field component at each observatory were computed for each year from 1963.0 to 1976.0. Each difference was associated with the time midway between the two times used to compute the difference. The vector field components were converted to geomagnetic coordinates, and the RMS values for each geomagnetic vector field component at each observatory were computed. These values are shown in Table I together with the observatory name and location in geomagnetic coordinates. The observatories are listed in order of their geomagnetic latitude. Also shown in Table I are the RMS residuals after subtracting the best fitting (in the least squares sense) degree one external zonal harmonic from the third difference annual means. The best fit was determined by including all degree one internal terms and the degree one zonal external term in a spherical harmonic analysis. Only the fields corresponding to the external term were subtracted from the third difference annual means to compute the residuals.

It can be seen that the signals on the X-axis (N-S) and Z-axis (vertical) are generally larger than the signal on the Y-axis (E-W) as would be expected if the signals on the X-axis and Z-axis are largely due to external current systems. Also, some of the values appear to be anomalously large. It was decided to select a subset of these 84 observatories by deleting those observatories with an RMS third derivative greater than 8.0 units for the Y-axis and greater than 25.0 units for the X-axis or Z-axis. There were 32 observatories with the Y-component greater than 8 units; of these 32 observatories, 22 had Z-components greater than 25.0 units. There is a high correlation between large values on one axis and large values on another axis as would be expected for "noise" but not for long spatial wavelength geomagnetic signals. Four additional observatories were deleted because they had values of RMS third derivatives large compared with corresponding RMS third derivatives at nearby observatories. A total of 36 observatories were deleted leaving 48 selected observatories including 7 south of the geomagnetic equator.

Table II lists the 48 selected observatories with both their geographic and geomagnetic coordinates. The RMS third derivatives are given as well as the residuals after subtracting the best fitting first degree zonal external harmonic and the residuals after

Table I. Initial 84 Observatories. (RMS) values of (unfiltered) third differences, unit nT/yr^3, field components in geomagnetic coordinates. (For RESIDUAL see text.)

OBSERVATORY	GEOMAGNETIC LAT	LONG	THIRD DERIVATIVE X	Y	Z	RESIDUAL X	Y	Z
THULE II	88.56	13.93	7.5	3.7	24.4	7.6	3.7	18.9
RESOLUTE BAY	83.12	-65.05	7.8	13.2	27.7	8.4	13.2	26.4
BJORNOYA	70.89	124.78	7.8	9.8	18.2	8.7	9.8	18.5
LEIRVOGUR	69.66	72.08	13.3	5.5	14.6	12.0	5.5	8.4
TROMSO	66.84	117.26	14.5	8.4	22.9	12.2	8.4	21.8
CHELYUSKIN II	66.47	177.34	10.6	14.8	40.9	9.4	14.8	37.1
COLLEGE	64.89	-101.12	11.8	3.2	11.1	10.0	3.2	6.4
SODANKYLA	63.49	120.69	9.6	3.8	10.4	7.0	3.8	7.4
LOPARSKOYE	63.25	126.48	11.0	11.4	25.3	9.9	11.4	26.7
DIKSON II	63.09	162.41	13.4	14.3	62.3	13.8	14.3	55.9
UELEN	62.11	-120.91	15.6	71.9	30.8	15.3	71.9	28.3
LERWICK	62.03	89.60	7.1	5.6	8.0	3.8	5.6	7.3
DOMBAS II	61.83	101.00	11.1	5.6	5.4	7.9	5.6	8.5
MEANOOK	61.78	-56.32	11.7	5.7	16.4	11.0	5.7	11.0
SITKA	60.13	-82.20	10.6	3.1	10.4	6.4	3.1	7.1
ESKDALEMUIR	57.92	84.08	8.0	3.2	7.8	4.5	3.2	5.7
LOVO	57.66	106.76	7.5	1.8	8.7	4.1	1.8	4.2
NURMIJARVI	57.50	113.53	8.3	3.3	7.0	5.5	3.3	3.6
VALENTIA	56.05	74.79	7.5	2.5	8.1	4.4	2.5	7.5
VOYEYKOVO	55.92	118.33	10.6	10.1	17.5	12.2	10.1	11.9
RUDE SKOV	55.38	99.62	11.6	3.0	10.0	7.5	3.0	10.0
VICTORIA	54.20	-64.63	9.9	4.2	11.6	6.1	4.2	9.9
HARTLAND	54.06	80.29	8.4	3.9	7.4	4.3	3.9	3.0
WINGST	54.05	95.21	7.5	2.7	6.9	2.1	2.7	4.0
WITTEVEEN	53.63	92.38	24.3	6.0	10.1	21.7	6.0	5.0
HEL	53.01	104.80	10.3	5.5	10.6	5.7	5.5	9.3
NIEMEGK	51.77	97.75	8.3	3.0	10.8	2.9	3.0	11.7
DOURBES	51.45	88.97	8.4	3.9	10.4	3.8	3.9	8.3
PLESHENITZI	51.24	112.91	11.5	14.4	12.6	9.8	14.4	8.2
YAKUTSK	51.19	-164.74	22.4	15.5	39.5	21.2	15.5	39.1
KRASNAYA PAKHRA	50.55	121.59	9.6	4.7	12.8	7.9	4.7	11.7
CHAMBON FORET	49.92	85.69	12.4	6.9	11.7	9.1	6.9	12.6
FREDERICKSBURG	49.12	-7.97	10.5	3.9	12.5	8.3	3.9	9.4
FURSTNFELDBRUCK	48.31	94.59	10.0	2.7	6.1	3.7	2.7	3.8
LVOV	47.60	107.07	17.0	7.1	21.6	19.3	7.1	25.9
WIEN KOBENZL	47.43	99.45	10.0	2.9	8.4	4.6	2.9	3.8
NAGYCENK	46.75	99.56	18.9	10.8	9.1	17.3	10.8	8.8
HURBANOVO	46.70	101.05	14.5	6.3	12.5	11.1	6.3	9.8
TIHANY	45.82	100.37	20.0	11.9	13.3	16.5	11.9	14.6
SAN MIGUEL III	44.98	52.53	17.7	114.3	24.8	15.6	114.3	21.8
COIMBRA	44.43	71.73	9.5	4.4	11.2	6.6	4.4	12.3
EBRO	43.32	81.07	11.5	4.1	10.9	6.8	4.1	11.9
STEPANOVKA	43.29	112.37	12.7	6.0	13.1	9.0	6.0	12.5
TOLEDO	43.28	76.14	11.7	5.8	7.2	8.4	5.8	5.1
AQUILA	42.35	94.30	10.9	3.8	6.5	5.3	3.8	5.5
SURLARI	42.11	107.37	39.4	12.6	94.9	43.6	12.6	91.5
PATRONY	40.82	176.10	12.7	6.0	15.0	6.7	6.0	13.0
PANAGYURISHTE	40.40	104.69	11.9	3.3	10.7	5.3	3.3	9.9
SAN FERNANDO	40.37	72.83	22.0	20.8	118.1	20.8	20.8	116.9
TUCSON	40.30	-45.67	10.6	3.0	9.7	4.5	3.0	6.9
ALMERIA	40.05	76.78	17.0	6.9	15.0	13.3	6.9	12.9
ISTANBL KNDILLI	38.09	108.79	11.5	10.6	13.0	6.8	10.6	10.4
DUSHETI	36.37	123.43	12.2	5.1	10.1	5.8	5.1	7.1
MEMAMBETSU	34.26	-149.93	14.7	3.1	12.6	8.2	3.1	9.9
GORNOTAYEZHNAYA	33.05	-160.30	12.9	7.1	35.6	9.2	7.1	36.4
KAKIOKA	26.27	-152.37	11.6	1.6	3.4	3.8	1.6	3.4
SIMOSATO	23.28	-155.90	17.7	4.4	7.3	9.9	4.4	7.3
HONOLULU IV	21.30	-91.61	14.4	3.8	8.6	9.0	3.8	8.5
KANOYA	20.77	-160.29	11.8	4.3	2.5	4.8	4.3	3.9
M BOUR	20.70	56.71	11.7	8.9	10.9	8.5	8.9	11.6
SHESHAN	19.95	-169.11	11.8	8.4	16.7	7.1	8.4	16.6
FUQUENE	16.61	-3.09	16.2	9.0	45.0	21.5	9.0	43.8
CHA PA	11.07	174.90	24.7	20.5	43.2	23.2	20.5	42.3
ALIBAG	9.40	145.25	21.5	18.1	31.0	25.0	18.1	30.2
GUAM	4.33	-145.38	22.2	6.3	10.6	19.2	6.3	10.6
MUNTINLUPA	3.32	-168.58	18.6	18.4	8.7	11.7	18.4	8.4
ANNAMALAINAGAR	1.52	151.06	14.9	41.5	53.7	13.4	41.5	53.7
HUANCAYO	-0.81	-4.48	19.4	7.0	11.2	12.2	7.0	11.1
TRIVANDRUM	-1.06	148.09	21.2	36.8	28.0	19.2	36.8	28.0
LUANDA BELAS	-7.51	82.32	21.9	9.4	62.8	16.6	9.4	63.4
VASSOURAS	-12.23	25.56	20.6	2.3	16.7	15.4	2.3	15.9
APIA III	-15.59	-98.12	16.1	8.9	13.9	10.0	8.9	13.1
PORT MORESBY	-18.09	-140.37	10.4	3.9	24.5	7.9	3.9	24.7
MAPUTO	-27.82	97.76	19.0	7.9	28.5	20.6	7.9	27.8
HERMANUS	-33.50	82.46	12.1	2.9	4.2	5.9	2.9	7.0
GNANGARA	-42.72	-172.22	10.3	4.2	8.7	5.9	4.2	7.5
TOOLANGI	-46.02	-137.42	7.1	8.0	6.3	5.4	8.0	8.1
AMBERLEY II	-47.05	-106.04	9.4	4.0	13.0	7.8	4.0	10.8
ARGENTINE ISLND	-53.96	4.65	24.6	7.8	10.7	21.7	7.8	10.1
KERGUELEN	-57.22	130.47	16.9	14.5	8.5	15.4	14.5	6.9
MACQUARIE ISLND	-60.46	-115.51	12.4	24.9	27.6	10.5	24.9	24.9
NOVOLAZAREVSKAY	-66.53	55.59	24.3	34.2	88.6	24.3	34.2	92.1
MAWSON	-73.22	106.11	20.2	10.1	45.2	20.9	10.1	46.7
MIRNYY	-76.80	150.68	25.5	9.8	66.0	23.8	9.8	64.3

Table II. Selected 48 Observatories. RMS values of (unfiltered) third differences, unit nT/yr³, field components in geomagnetic coordinates. (For RESIDUALS see text.)

OBSERVATORY	GEOGRAPHIC LAT	LONG	GEOMAGNETIC LAT	LONG	THIRD DERIVATIVE X	Y	Z	RESIDUAL 1 X	Y	Z	RESIDUAL 2 X	Y	Z
KANOYA	31.42	130.88	20.77	-160.29	11.8	4.3	2.5	4.7	4.3	4.5	4.9	4.3	5.1
GUAM	13.58	144.87	4.33	-145.38	22.2	6.3	10.6	20.1	6.3	10.6	20.1	6.1	11.6
HUANCAYO	-12.04	-75.34	-0.81	-4.48	19.4	7.0	11.1	10.9	7.0	11.1	10.6	7.0	11.3
VASSOURAS	-22.40	-43.65	-12.23	25.56	20.6	2.3	16.7	13.8	2.3	15.6	10.6	2.3	15.5
PORT MORESBY	-9.41	147.15	-18.09	-140.37	12.1	2.9	24.5	8.2	2.9	24.8	13.3	2.9	21.8
HERMANUS	-34.43	19.23	-33.50	-82.46	12.1	4.2	4.2	4.7	2.9	8.0	8.2	3.9	8.0
GNANGARA	-31.78	115.95	-42.72	-172.22	10.3	4.2	8.7	5.2	4.2	8.1	5.2	4.2	8.1
AMBERLEY II	-43.15	172.72	-47.05	-106.04	9.4	4.0	13.0	8.5	4.0	10.1	5.2	4.0	9.5
ARGENTINE ISLND	-65.24	-64.26	-53.96	4.65	24.6	7.8	10.7	21.1	7.8	10.8	20.7	7.8	10.4
THULE II	77.48	-69.17	88.56	13.93	7.5	3.7	24.4	7.6	3.7	17.8	7.4	3.7	11.7
LEIRVOGUR	64.18	-21.70	69.66	72.08	13.3	5.5	14.6	12.4	5.5	8.1	9.7	5.2	8.2
COLLEGE	64.86	-147.84	64.89	-101.12	11.1	3.8	10.4	10.1	3.8	5.9	8.1	3.2	5.6
SODANKYLA	67.37	26.63	63.49	120.69	9.6	3.8	10.4	7.1	3.8	7.7	5.8	3.8	7.3
LERWICK	60.13	-1.18	62.03	89.60	7.1	5.6	8.0	3.3	5.6	7.4	5.7	5.6	6.6
DOMBAS II	62.07	-9.12	61.83	101.00	11.1	5.6	5.6	7.4	5.6	9.4	6.7	5.6	8.7
MEANOOK	54.62	-113.33	61.78	-56.32	11.7	5.7	16.4	11.1	5.7	9.7	11.5	5.7	9.7
SITKA	57.06	-135.32	60.13	-82.20	10.6	3.1	10.4	5.5	3.1	6.3	5.0	3.1	5.6
ESKDALEMUIR	55.32	-3.20	57.92	84.08	8.0	3.2	7.8	4.6	3.2	6.3	3.2	3.2	6.0
LOVO	59.34	17.83	57.66	106.76	7.5	1.8	8.7	3.8	1.8	4.4	4.6	1.8	3.6
NURMIJARVI	60.51	24.66	57.50	113.53	8.3	3.3	7.0	5.9	3.3	4.4	6.1	3.3	4.0
VALENTIA	51.93	-10.25	56.05	74.79	7.5	2.5	8.1	4.6	2.5	8.5	5.0	2.5	4.0
RUDE SKOV	55.84	12.46	55.38	99.62	11.6	3.0	10.0	6.5	3.0	11.5	6.0	3.0	11.2
VICTORIA	48.52	-123.42	54.28	-64.63	9.9	4.2	11.6	6.5	4.2	9.3	5.8	4.2	8.1
HARTLAND	50.99	-4.48	54.06	80.29	8.4	3.9	7.4	4.5	3.9	3.5	4.2	3.9	3.5
WINGST	53.74	9.07	54.05	95.21	7.5	2.7	6.9	2.0	2.7	4.5	1.8	2.7	4.5
HEL	54.61	18.82	53.01	104.80	10.3	5.5	6.6	5.3	5.5	9.7	4.9	5.5	9.3
NIEMEGK	52.07	12.68	51.77	97.75	8.3	3.0	10.8	2.2	3.0	12.0	2.3	3.0	11.5
DOURBES	50.10	4.59	51.45	88.97	8.4	3.9	10.4	3.0	3.9	8.4	3.2	3.9	8.1
KRASNAYA PAKHRA	55.48	37.31	50.55	121.59	9.6	4.7	12.8	8.6	4.7	11.4	8.6	4.7	11.1
CHAMBON FORET	48.02	2.26	49.92	85.69	12.4	6.9	11.8	8.0	6.9	13.1	7.5	6.9	11.4
FREDERICKSBURG	38.21	-77.37	49.12	-7.97	10.5	3.9	12.5	8.0	3.9	8.3	7.7	3.9	7.7
FURSTNFELDBRUCK	48.17	11.28	48.31	94.59	10.0	2.7	6.1	2.3	2.7	4.7	2.7	2.7	4.2
WIEN KOBENZL	48.26	16.32	47.43	99.45	10.0	2.9	8.4	3.5	2.9	4.0	3.4	2.9	4.2
HURBANOVO	47.87	18.19	46.70	101.05	9.5	4.4	12.5	9.7	6.3	9.2	9.2	6.3	8.7
COIMBRA	40.22	-8.42	43.32	71.73	9.5	4.1	12.5	6.6	4.4	13.0	6.6	4.4	12.2
EBRO	40.82	0.49	43.32	81.07	11.5	4.0	10.9	6.6	4.1	12.7	5.8	4.1	12.0
STEPANOVKA	46.78	30.88	43.28	112.37	12.7	5.0	7.2	8.6	6.0	13.0	9.3	6.0	12.0
TOLEDO	39.88	-4.05	43.28	76.14	11.7	5.8	5.8	8.4	5.8	5.5	8.1	5.8	5.7
AQUILA	42.38	13.32	42.35	94.30	11.9	6.0	10.0	5.0	3.8	6.7	5.8	3.8	6.0
PATRONY	52.17	104.45	40.82	176.10	12.7	6.0	15.0	6.9	6.0	13.0	7.5	6.0	13.4
PANAGYURISHTE	42.51	24.18	40.40	104.69	11.9	3.0	10.7	3.9	3.0	13.0	4.4	3.0	10.5
TUCSON	32.25	-110.83	40.30	-45.67	10.6	3.3	9.7	3.5	3.0	6.4	4.2	3.0	6.6
ALMERIA	36.85	-2.46	40.05	76.78	10.6	6.9	15.0	12.2	6.9	11.9	12.1	6.9	12.8
DUSHETI	42.09	44.71	36.37	123.43	12.2	5.1	7.2	4.8	5.1	7.2	4.9	5.1	6.8
MEMAMBETSU	43.91	144.19	34.26	-149.93	14.7	3.1	12.6	7.9	3.1	10.1	7.9	3.1	9.7
KAKIOKA	36.23	140.19	26.27	-152.37	14.7	1.6	3.4	2.7	1.6	4.0	3.8	1.6	4.2
SIMOSATO	33.58	135.94	23.28	-155.90	17.7	4.4	7.3	7.6	4.4	7.8	7.7	4.4	5.6
HONOLULU IV	21.32	-158.00	21.30	-91.61	14.4	3.8	8.6	8.5	3.8	8.8	8.4	3.8	7.7

subtracting the set of best fitting first, third, fifth, and seventh degree zonal external harmonics. The best fitting external harmonics were computed as for Table I. The degree one internal terms were included in the spherical harmonic expansion but only the external fields were subtracted from the third difference annual means to compute the residuals. The residuals from subtracting the degree one zonal external field are somewhat smaller in general than they are in Table I, since this zonal harmonic is fit to a smaller and presumably more accurate set of observatory data. In general, this one external term accounts for over half the power in the RMS third differences. The residuals are further reduced when additional external harmonics are included in the spherical harmonic analysis. For the deleted observatories, Table I shows that the residuals are not very appreciably smaller than the original third differences, which suggests that for these deleted observatories the third differences are largely noise.

3. MAGNETIC OBSERVATORY THIRD DERIVATIVE FIELD

As discussed in the preceding section, the third differences of the magnetic observatory annual means contain considerable "noise", even though the noise is reduced by deleting those observatories with the most noise from the data set. In order to further reduce noise, the observatory third differences were filtered by a symmetric filter with weights (1/16, 4/16, 6/16, 1/16) and interpolated to yield time series values every 1/2 year. This filter is equivalent to applying a filter that forms the running average of two adjacent values four times in succession. This lowpass filter has an impulse response half width of 1.3 years and a half power frequency of 0.125 cycles/year. Because the filter is symmetric in time, there is no phase shift. This filter was used in an attempt to reduce the noise relative to the jerk of 1969 so that the jerk might be better distinguished in the data.

Figure 2 shows the RMS values of the filtered third difference field components for the two sets of observatories – the selected 48 observatories and the initial 84 observatories. The field components are in geomagnetic coordinates. The geomagnetic jerk of 1969 is evident in the Y-component (E-W) for the selected 48 observatories. The peaks on the X-axis and Z-axis at 1967 and 1973 are presumably due to external current systems since they do not appear on the Y-axis. The initial data set appears to be much noisier than the selected data set, and the geomagnetic jerk of 1969 can just barely be observed on the Y-axis for the initial data set.

Figure 3 shows the RMS values of the filtered third difference field components for the selected 48 observatories. Two graphs are shown, one for which the field components are in geomagnetic coordinates and one for which the field components are in geographic coordinates. The two graphs are very similar except for peaks in the RMS geographic Y-component at 1967 and 1973 corresponding to similar peaks in the X-component and Z-component that are presumably due to external current systems. These graphs indicate that the external

current systems produce smaller mean fields in the Y-direction in
geomagnetic coordinates than in geographic coordinates. These graphs
also suggest that the mean fields due to external currents can be
modeled by only zonal terms for a spherical harmonic model in
geomagnetic coordinates.

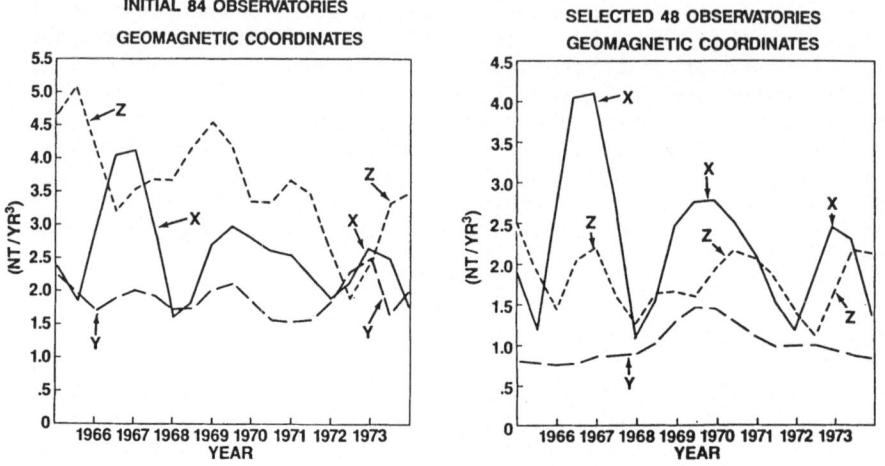

Figure 2. RMS values (over the set of observatories) of the filtered
and interpolated third differences. Field components in geomagnetic
coordinates.

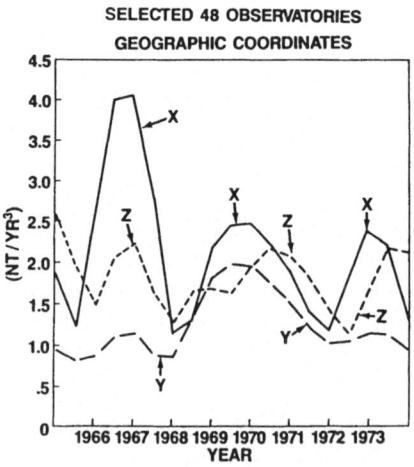

Figure 3. RMS values (over the set of observatories) of the filtered
and interpolated third differences, for field components in
geomagnetic and in geographic coordinates.

4. SPHERICAL HARMONIC ANALYSES

Spherical harmonic models were computed from the filtered and interpolated first, second, and third differences of the field components for the selected 48 observatories. The field components were in geomagnetic coordinates and only odd degree zonal terms were used to model the external sources. Fifteen internal terms were included in the models corresponding to all first, second, and third degree internal terms. Only four external terms were included corresponding to the first, third, fifth, and seventh zonal external terms. Except for the five second degree internal terms, all terms were of odd degree. Because there were only 7 observatories below the geomagnetic equator in the selected data set, it was felt that the number of terms of even degree should be kept small to avoid unstable solutions to the least squares equations.

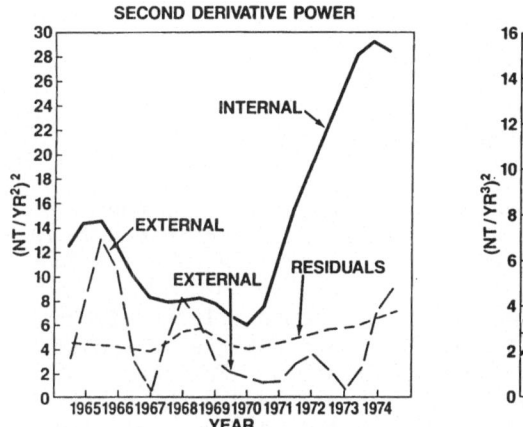 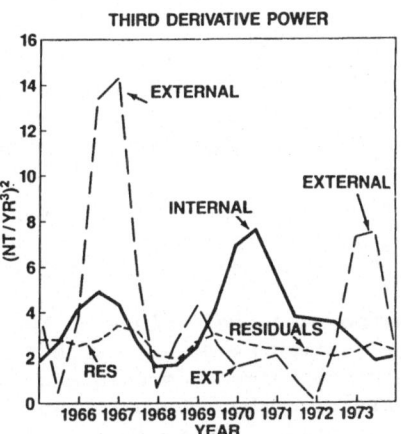

Figure 4. Mean square values, over the Earth's surface, of the internal and external parts of the 'fields' given by the spherical harmonic models of the second and third time derivatives.

Figure 4 shows the mean squared field derivatives associated with the internal and external terms of the spherical harmonic models. These mean squared values are plotted against time for both the second time derivative and the third time derivative of the filtered field. It can be seen that the peaks in the third derivative power at 1967 and 1973 appear to be due to external current systems as suggested previously in this paper. There is a peak in the internal power at about 1970 corresponding to the geomagnetic jerk and there are smaller peaks in the external power at about 1969 and 1971. The second derivative power shows a minimum in the internal power at about 1970 corresponding to the geomagnetic jerk. Maxima in the second derivative external power correspond to minima in the third derivative external power, and minima in the second derivative external power correspond to maxima in the third derivative external power. The internal power is larger relative to the external power and residuals for the second derivative field than it is for the third derivative field. The minor peak in the third derivative internal power at about 1967 is possibly due to aliasing of unmodeled external power.

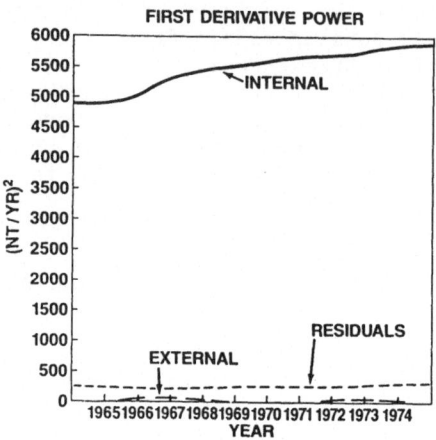

Figure 5. Mean square values, over the Earth's surface, of the internal and external parts given by the spherical harmonic models of the first and second time derivatives.

The mean squared values of the first and second filtered field time derivatives associated with the internal and external terms of the spherical harmonic models are shown in Figure 5. For the first time derivative the internal power is larger relative to the external power and residuals than it is for the second time derivative. The change in relative power for the internal terms, external terms, and residuals as the degree of the derivative is varied is due to different temporal power spectra for these different terms.

The internal and external third derivative power has been previously computed as a function of time for approximately the same time period considered here by Alldredge (1984). Alldredge included all internal and external terms through degree four in the spherical harmonic analysis for a total of 48 parameters, and used data from 83 observatories. This was nearly the same set of observatories as the initial data set described in this paper. Alldredge did not find a significant difference between the internal and external power plotted against time, in contrast to the results reported here. Perhaps part or all of the explanation for these differences may be the presumably noisier data set used by Alldredge and the larger number of parameters that were determined by him.

5. CONCLUSIONS

External current systems are responsible for a significant portion of
the time derivatives of the magnetic observatory annual means. The
relative power due to external sources increases as the degree of the
time derivative increases. The degree one zonal coefficient
(geomagnetic coordinates) is the most significant external term in a
spherical harmonic analysis and accounts for over half of the total
power in the third derivative for a select set of observatories.
Other zonal external terms are also significant, and the time
variation of the external coefficients include harmonics of the
sunspot cycle. The geomagnetic jerk of 1969 is evident in a plot of
internal power of the third time derivative against time, but other
phenomena than the jerk are also present for this time period.
Perhaps some or all of these other phenomena are simply "noise". In
any case these phenomena will be responsible for errors in
computations of the jerk spherical harmonic coefficients.

6. ACKNOWLEDGEMENT

This research was supported by the Office of Naval Research and by
Chief of Naval Operations of the U.S. Navy.

7. REFERENCES

Alldredge, L.R., 1985. 'More on the alleged 1970 geomagnetic jerk',
     Phys. Earth and Planet. Inter., 39, 255-264.
Alldredge, L.R., 1984. 'A discussion of impulses and jerks in the
     geomagnetic field', J. Geophys. Res., 89, 4403-4412.
Backus, G.E., R.H. Estes, D. Chinn, and R.A. Langel, 1987. 'Comparing
     the jerk with other global models of the geomagnetic field from
     1960 to 1978', J. Geophys. Res., 92, 3615-3622.
Backus, G.E., 1983. 'Application of mantle filter theory to the
     magnetic jerk of 1969', Geophys. J.R. Astron. Soc., 74,
     713-746.
Bloxham, J. and D. Gubbins, 1985. 'The secular variation of Earth's
     magnetic field', Nature, 317, 777.
Courtillot, V. and J.-L. Mouel, 1984. 'Geomagnetic secular variation
     impulses', Nature, 311, 709-716.
Courtillot, V., J. Ducruix and J.-L. Le Mouel, 1978. 'Sur une
     accélération récente de la variation séculaire du champ
     magnétique terrestre, C.R. Hebd. Seances Acad. Sci. Ser. D, 287,
     1095-1098.
Courtillot, V. and J.-L. Le Mouel, 1976. 'On the long-period
     variations of the earth's magnetic field from 2 months to 20
     years', J. Geophys. Res., 81, 2941-2950.

30

Ducruix, J., V. Courtillot and J.-L. Le Mouel, 1980. 'The late 1960's secular variation impulse, the eleven year magnetic variation and the electrical conductivity of the deep mantle', Geophys. J. R. Astron. Soc., 61, 73-79.

Evans, M., Edmund Halley, 1988. 'Geophysicist', Physics Today, 41, No. 2, 41-45.

Gilbert, W., 1958. 'De Magnete', 1600, reprinted by Dover, New York.

Gubbins, D., 1984. 'Geomagnetic field analysis, II, Secular variation consistent with a perfectly conducting core', Geophys. J.R. Astron. Soc., 77, 753-766.

Halley, E., 1692. 'An account of the cause of the change of the variation of the magnetical needle with a hypothesis of the structure of the internal parts of the Earth', Philos. Trans. R. Soc. London, 17, 563.

Halley, E., 1683. 'A Theory of the Variation of the Magnetical Compass', Philos. Trans. R. Soc. London, 13, 208.

Kerr, R.A., 1984. 'Magnetic "jerk" gaining wider acceptance', Science, 225, 1135-1136.

Langel, R.A., R.H. Estes and G.D. Mead, 1982. 'Some new methods in geomagnetic field modeling applied to the 1960-1980 epoch', J. Geomagn. Geoelectr., 34, 327-349.

Malin, S.R.C. and B.M. Hodder, 1982. 'Was the 1970 geomagnetic jerk of internal or external origin?', Nature, 296, 726-728.

McLeod, M.G., 1985. 'On the geomagnetic jerk of 1969', J. Geophys. Res., 90, 4597-4610.

Weisburd, S., 1985. 'The Earth's magnetic hiccup', Science News, 128, 218-219.

HISTORICAL SECULAR VARIATION AND GEOMAGNETIC THEORY

David Gubbins
Department of Earth Sciences
Bullard Laboratories
Madingley Road
Cambridge   CB3 0EZ

ABSTRACT. The historical record of magnetic observations from AD 1695 has been analyzed to give maps of the magnetic field at the core-mantle boundary for the last 300 yr. These maps have been interpreted qualitatively in terms of a stationary dynamo field with hemispheric symmetry and a pattern of secular variation linked to the base of the solid mantle by lateral variations of temperature. In consequence the palaeomagnetic time-averaged field is expected to resemble the historical average and geomagnetic reversals to be related to secular variation features currently active on the core-mantle boundary beneath the South Atlantic.

1. INTRODUCTION

In the discussion at the Advanced Study Institute I described work done on historical magnetic observations that has led to maps of the Earth's magnetic field at the core-mantle boundary (CMB), and gave speculations arising from inspection of these maps on the main dynamo, the secular variation, and the mechanism of polarity reversal. Also at the Institute, Professor V. Courtillot suggested that I publish maps of the field at the Earth's surface, as this would be useful for workers in palaeomagnetism and recent secular variation. Since the data analysis and theory are both available in the published literature, I decided to take this opportunity to publish maps of declination, inclination anomaly, and total intensity, at the Earth's surface for five epochs from AD 1715.0 to AD 1969.5 at approximately 60 year intervals (see Figures 2-4), and restrict the text to a brief guide to the published papers.

2. DATA ANALYSIS

The magnetic field is represented in terms of a magnetic potential assuming the Earth's mantle and entire region of observation, which includes the atmosphere to a height of several hundred kilometres for

31

F. J. Lowes et al. (eds.), Geomagnetism and Palaeomagnetism, 31–43.
© 1989 by Kluwer Academic Publishers.

satellite data, is an electrical insulator. The model fit to the
observations is a least squares criterion and the inversion is
stabilised by requiring the magnetic field to be smooth at the CMB.
Methods were evolved for linear data (i.e. magnetic components that
are linearly related to the core field (X,Y,Z but not D,I,H,F) by
Whaler and Gubbins (1981), Shure et al. (1982), Parker and Shure
(1982), Shure et al. (1983), and Gubbins (1983, 1984), with mainly
applications to secular variation rather than main field. The methods
of Shure et al. (1982) and Gubbins (1983) are different but produce
the same field model; the latter used a Bayesian argument to derive
error estimates for the model. Although both methods are
basis-independent, they make use of spherical harmonic functions and
are in many respects technically similar to conventional least squares
spherical harmonic fits. The smoothness criterion at the CMB replaces
the truncation of the series. The models should be superior to those
based on truncated spherical harmonic series because they contain the
additional information that the field originate in the core.
     The method was extended to nonlinear data by Gubbins and Bloxham
(1984), who applied it to main field models for epochs 1980, 1969.5,
and 1960.0. They provide a somewhat different derivation of the
method of Gubbins (1983) based on maximum likelihood methods. The
method was also extended to allow calculation of models which fit the
frozen-flux conditions required if the core is a perfect electrical
conductor. The conditions on secular variation are linear and
consequently easy to apply (Gubbins 1984), while the conditions on
main field are nonlinear and require an iterative fit (Gubbins and
Bloxham 1984).
     MAGSAT data supplied by Dr. R. Langel was used for the 1980
model. POGO total intensity data and magnetic observatory annual
means were used for epoch 1969.5. Land, air, and marine survey data
from a magnetic tapes supplied by World Data Center Cl at Edinburgh
compiled originally by Hendricks and Cain (1963), combined with
observatory annual means, were used for 1960. Observatory data and
the survey tapes were used for the rest of the 20th century to produce
models for every decade from 1905.5 to 1955.5 and POGO total intensity
data for 1966.0. No two field models contained the same observation,
except for 1955.5 and 1960 which have some overlap. The 1960 model is
not regarded as one of the final set. The analysis of these models
will be described in Bloxham et al. (1988).
     Data from 1860-1900 was uncatalogued and has been researched
recently to produce a model for epoch 1880.0. Details are described
in Bloxham et al. (1988). Data windows were chosen in earlier times
to produce models at epochs 1715.0, 1777.5 and 1842.5. Details appear
in Bloxham (1986a) and Bloxham and Gubbins (1985). Models prior to
1842.5 contain no intensity information, and are therefore ambiguous
by at least an arbitrary multiplicative constant. There is unlikely
to be further ambiguity (Proctor and Gubbins 1988). The model of 1715
required extensive navigation corrections for the longitudes, made by
reading ship's logs and checking landfalls. This model contains
rather few inclination measurements, but those that are present run
from north to south down the Atlantic Ocean and there is good reason

to suppose that this configuration, combined with good declination coverage, will be sufficient (Gubbins 1986). An analysis is currently underway of the 17th century, and a field model will eventually be forthcoming. This will complete the analysis of the historical record. Nearly 200,000 original measurements have been analyzed; at the start of the work the earliest field model based on original measurements was for 1965.

3. RESULTS

The maps of radial component of magnetic field at the CMB appear to be accurate to 10-100 μT. They show some stationary features (1-9 in Figure 1), some westward-drifting features (10 and most others in Figure 1), and a point beneath Indonesia (B in Figure 1) where the field pattern changes without moving (Bloxham and Gubbins 1985; Bloxham et al. 1988). There are no eastward-drifting features, except in the 19th century beneath Alaska where stationary feature 2 (Figure 1) appears to split in two with one half going west and the other east, the probable consequence of a westward-drifting patch interacting with a stationary patch.

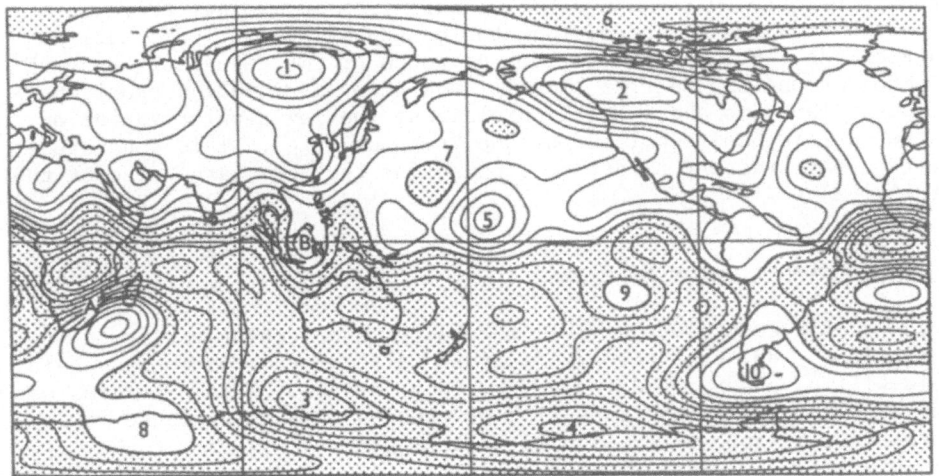

Figure 1. Radial component of field at the CMB in 1980. Contour interval 100 μT. Numbers and letters indicate features discussed in the text. Cylindrical equidistant projection (Plate Carrée).

The main constituents of the field at the CMB are four stationary highs centred on about 60°N and S and 120°E and W (1-4 in Figure 1). Surprisingly (or perhaps not if we take Cowling's theorem into account!), the flux is nearly zero at both geographical poles (6 and 8 in Figure 1). The coincidence of longitudes of the main highs is

evidence of a dipole-type symmetry, which is a fundamental feature of
the dynamo equations; absence of flux near the poles may be an effect
of the inner core (Gubbins and Bloxham 1987). This aspect of the CMB
field is believed to be the main dynamo-generated field.

The secular variation consists mainly of westward movement of
flux patches from Indonesia to the west coast of the American
continent, oscillations near Indonesia, and a large growth in a patch
of "reverse flux", i.e. where field lines point the opposite way to
that of a dipole field, near A in Figure 1. This has been shown to be
a continuation of the flux changes found for 1960-1980 by Bloxham and
Gubbins (1986) (see Bloxham and Gubbins 1985, Bloxham et al. 1988);
the violation of the frozen-flux conditions is therefore now well
established, casting some doubt on calculations of core flow based on
the hypothesis. Secular variation is notably absent from a region
enclosing most of the Pacific Ocean, except for the southwest corner.

The existence of both stationary and drifting patches is
explained by Bloxham and Gubbins (1987) by a theory of thermal
interaction with the mantle: fluid downwelling beneath cold mantle
causes flux concentration while upwelling beneath hot mantle causes
flux dispersion or, in the Indian Ocean patch at A in Figure 1, flux
expulsion (Bloxham 1986b). Indonesia is a potential site for a bump
on the CMB, where in situ oscillations persist. There is some support
for a correlation with mantle temperature from seismology (Dziewonski
1984) and from mantle convection studies (Hager et al. 1985, Gubbins
and Richards 1986). There is some evidence of CMB topography beneath
trenches around the Pacific (Morelli and Dziewonski 1987), providing
an Indonesian bump and possibly explaining the low secular variation
in the Pacific basin enclosed by the "ring of fire". This conjecture
awaits a satisfactory theory to account for the very large size of the
effect.

4. PALAEOMAGNETIC CONSEQUENCES

If the mantle controls the field morphology and secular variation, as
it appears to do, then the same pattern of secular variation and main
field will have persisted for a very long period of time - perhaps as
long as 100 Myr, the overturn time of the lower mantle. There is
therefore some hope of explaining palaeomagnetic results for the
time-averaged field and secular variation, provided the historical
record is long enough to allow a satisfactory time average. In fact
this is not quite the case: the historical record is too short for a
proper average, but an "eyeball" average can be done using maps of
declination and inclination at the Earth's surface (see Figures 2, 3)
and comparing their behaviour with the CMB field.

In this way Gubbins (1988) argued that a time-average of the
historical record produces "far-sidedness" (Wilson 1970), or negative
inclination anomalies for normal polarity, in most areas. The
exceptions are the south Atlantic, extreme north and south Pacific,
and southeast Pacific. Analysis of virtual geomagnetic pole
positions, collected into latitude bins by the two hemispheres

("Atlantic" and "Pacific") confirm the predominance of far-sidedness and near-sidedness in the regions predicted, except for the southeast Pacific. This is taken as support for the theory of thermal core-mantle interaction, with the proviso that the historical record is atypical for the southeast Pacific.

Gubbins (1987) showed the current fall in the dipole moment is due to the growth and southward movement of the reverse-flux features in the southern hemisphere. Past movement of these features can explain the long-term fall in the field intensity observed by archaeomagnetism, and the appearance of another such patch over the next 500 years could produce a continued fall in dipole moment to half its peak value. Such oscillations may or may not be a typical feature of secular variation. If they are, then occasionally they may spawn a full polarity reversal. The reverse-flux features resemble sunspots, and the instability may procede in similar fashion to the period reversal of field on the sun.

If oscillations in dipole moment and reverse flux features are linked to reversal mechanism, they may control the frequency of reversals. This has increased almost monotonically since the Cretaceous quiet zone (Lowrie and Kent 1983). The change could arise from increased activity of the instability producing the reverse-flux features, such as a change in temperature at the CMB beneath the Indian Ocean zone of flux expulsion. Reducing this temperature, or moving the patch to equatorial or polar regions where toroidal fields are small, would change the reversal frequency. The timescale for change in reversal frequency is about right for overturn in the lower mantle.

5. CONCLUSIONS

This report has described a number of speculative suggestions, based on the historical record of secular variation. Perhaps the most encouraging result is that dynamo theory may now be driven by the observations, rather than by the need to understand basic processes as has been the case up to now. These speculative ideas need confirmation by theoretical studies, which is the next stage of this work. No doubt some of them will prove naïve, but the observations are firm enough that some such explanation is necessary.

The historical record is now exhausted. The complete dataset will be left with the World Data Center for distribution to anyone wishing to make use of it. The next stage in the data analysis is to examine palaeomagnetic data, starting with lake sediments which record the next most recent times. This data is quite different in character and will require quite different methods of analysis. Hopefully we can develop the theory of secular variation to provide a bridge between the historical secular variation and this longer-term record.

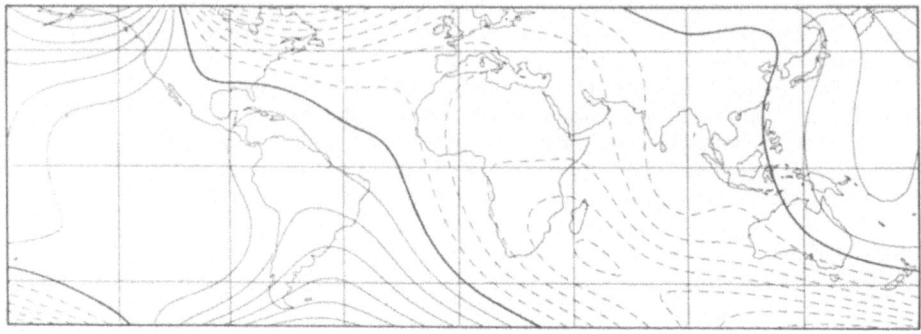

1715.0

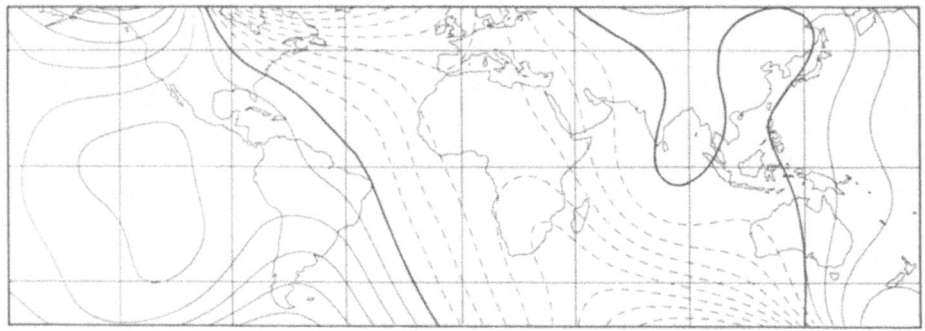

1775.5

Figure 2. Declination at the Earth's surface. Contour interval 5°.
Epochs 1715.0, 1775.5, 1842.5, 1905.5 and 1969.5. Cylindrical
equidistant projection (Plate Carreé) to ±60°.

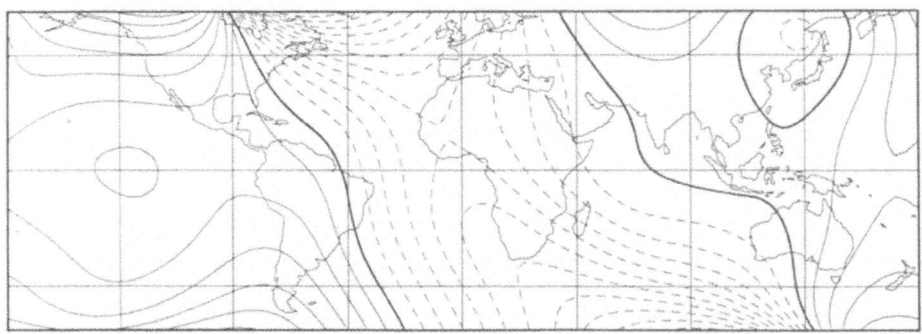

1842.5

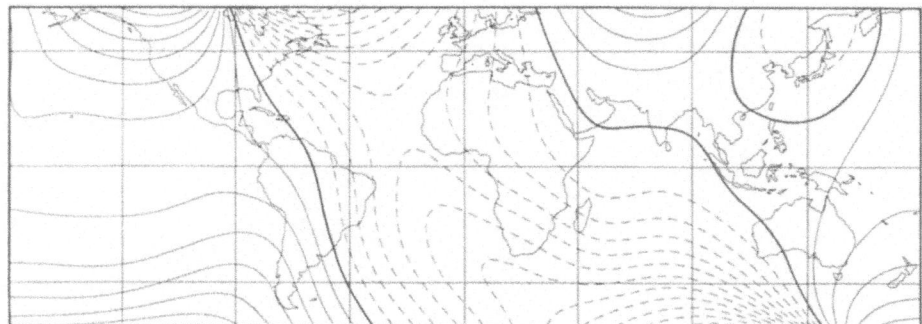

1905.5

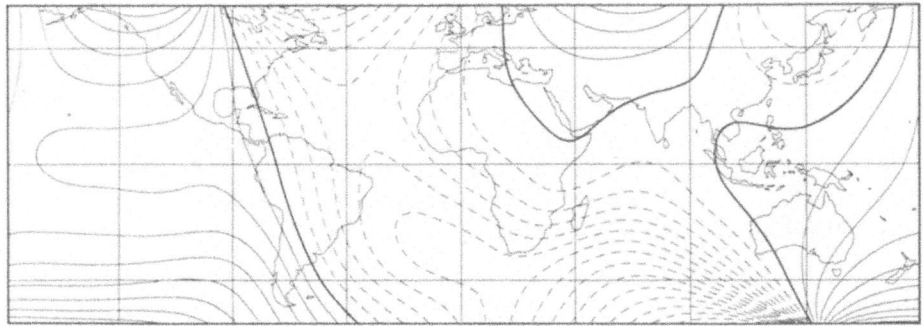

1969.5

38

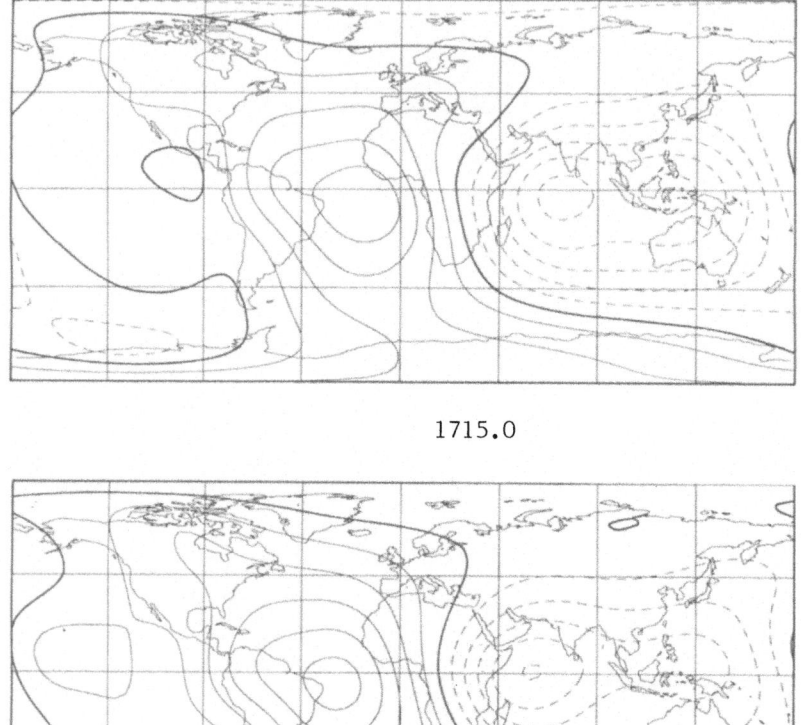

1715.0

1775.5

Figure 3. Inclination anomaly (inclination minus that for an axial
dipole) at the Earth's surface. Contour interval 5°. Epochs 1715.0,
1775.5, 1842.5, 1905.5 and 1969.5. Cylindrical equidistant projection
(Plate Carreé).

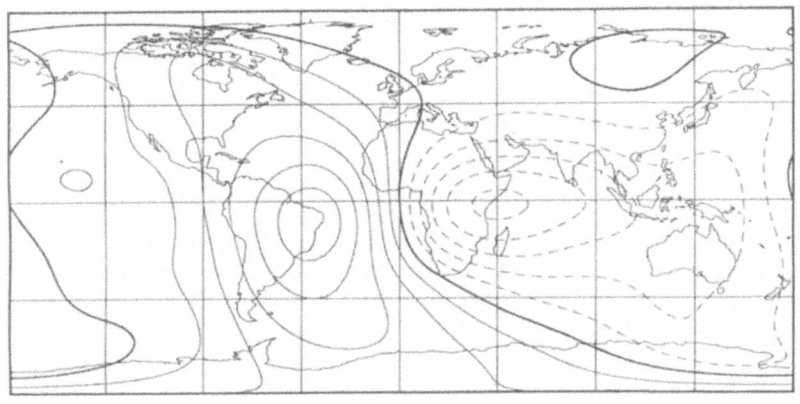

1842.5

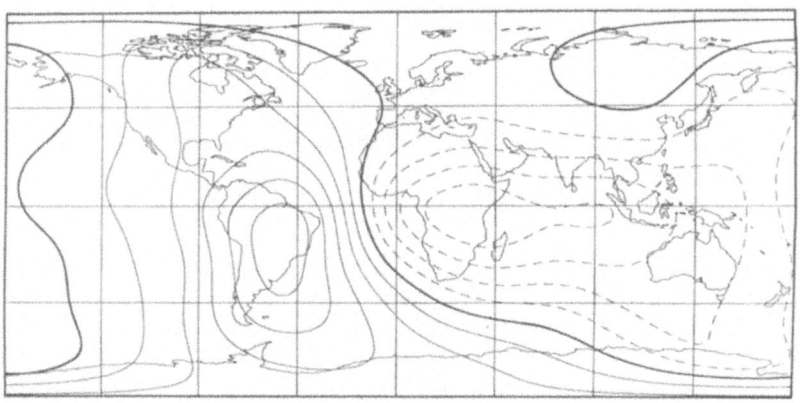

1905.5

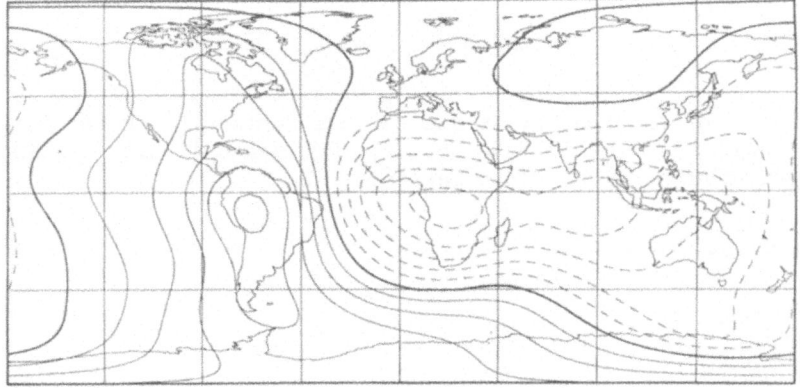

1969.5

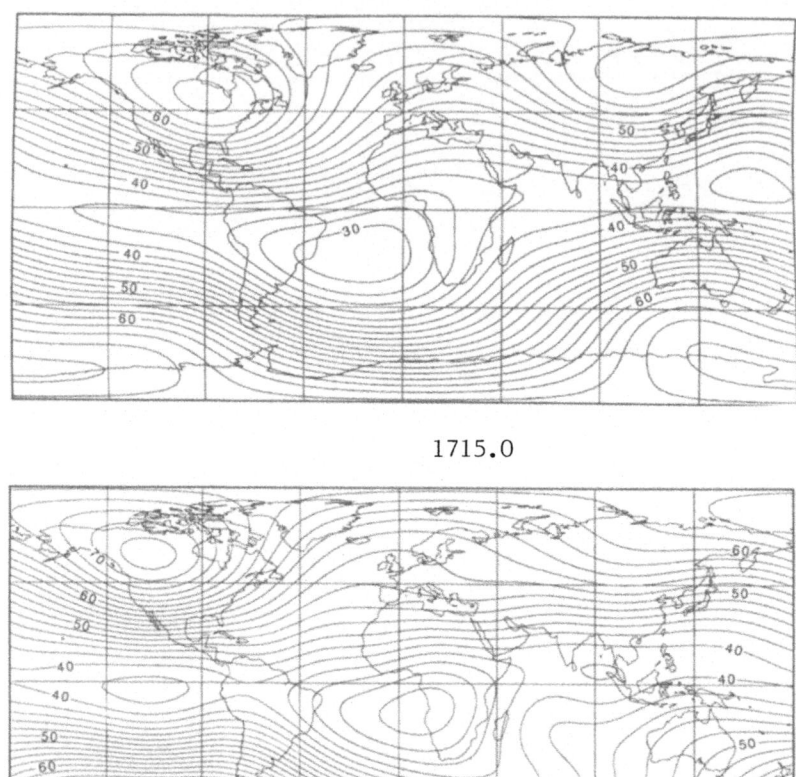

1715.0

1775.0

Figure 4. Total intensity at the Earth's surface. Contour interval 1 μT. Epochs 1715.0, 1775.5, 1842.5, 1905.5 and 1969.5. Cylindrical equidistant projection (Plate Carreé).

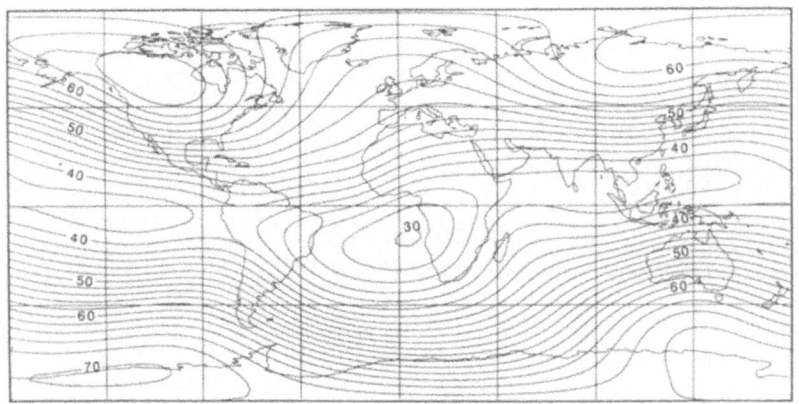

1842.5

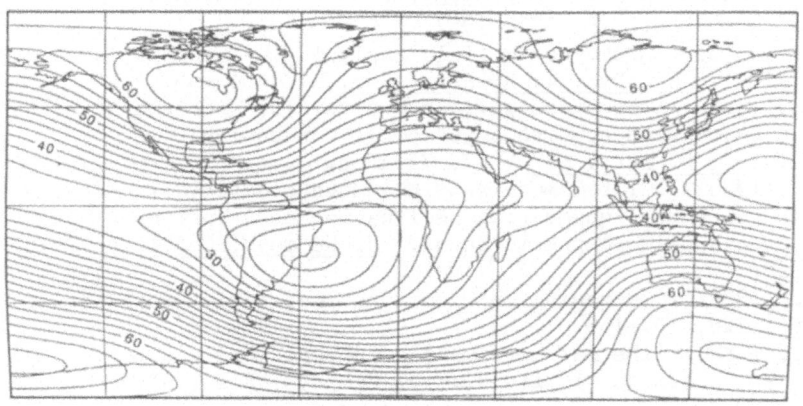

1905.5

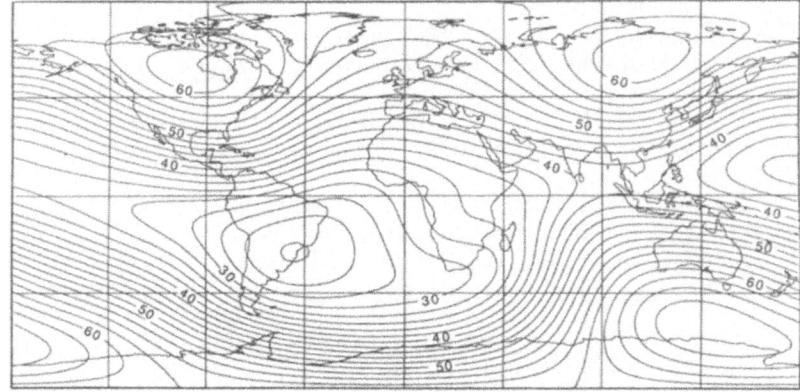

1969.5

42

6. REFERENCES

Bloxham, J., 1986a. 'Models of the Magnetic Field at the Core-Mantle Boundary for 1715, 1777, and 1842'. J. Geophys. Res., 91, 13954-13966.

Bloxham, J., 1986b. 'The expulsion of magnetic flux from the Earth's core'. Geophys. J.R. Astr. Soc., 87, 669-678.

Bloxham, J. and Gubbins, D., 1985. 'The secular variation of the Earth's magnetic field'. Nature, 317, 777-781.

Bloxham, J. and Gubbins, D., 1986. 'Geomagnetic field analysis IV - Testing the frozen-flux hypothesis'. Geophys. J. R. Astr. Soc., 84, 139-152.

Bloxham, J. and Gubbins, D., 1987. 'Thermal core-mantle interactions'. Nature, 325, 511-513.

Bloxham, J., Gubbins, D. and Jackson, A., 1988. 'Geomagnetic Secular Variation', submitted to Phil. Trans. R. Soc. Lond.

Dziewonski, A.M., 1984. 'Mapping the lower mantle: determination of lateral hererogeneity in P velocity up to degree and order 6', J. Geophys. Res., 89, 5929-5952.

Gubbins, D., 1983. 'Geomagnetic field analysis I - Stochastic Inversion'. Geophys. J. R. Astr. Soc., 73, 641-652.

Gubbins, D., 1984. 'Geomagnetic field analysis II - Secular Variation consistent with a perfectly conducting core'. Geophys. J. R. Astr. Soc., 77, 753-766.

Gubbins, D., 1986. 'Global models of the magnetic field in historical times: augmenting declination observations with archeo- and paleo-magnetic data'. J. Geomagn. Geoelectr., 38, 715-720.

Gubbins, D., 1987. 'Mechanism for geomagnetic polarity reversals'. Nature, 326, 167-169.

Gubbins, D., 1988. 'Thermal core-mantle interactions and time-averaged paleomagnetic field'. J. Geophys. Res., 93, 3413-3420.

Gubbins, D. and Bloxham, J., 1984. 'Geomagnetic field analysis III - Magnetic fields on the core-mantle boundary'. Geophys. J. R. Astr. Soc., 80, 696-713.

Gubbins, D. and Bloxham, J., 1987. 'Morphology of the geomagnetic field and implications for the geodynamo'. Nature, 325, 509-511.

Gubbins, D. and Richards, M., 1986. 'Coupling of the core dynamo and mantle: thermal or topographic?'. Geophys. Res. Lett., 13, 1521-1524.

Hager, B.H., Clayton, R.W., Richards, M.A., Comer, R.P. and Dziewonski, A.M., 1985. Nature, 313, 541-545.

Hendricks, S.J. and Cain, J.C., 1963. 'World Magnetic Survey Data', NASA Goddard Space Flight Centre report X-611-63-178, August 1963.

Lowrie, W. and Kent, D.V., 'Geomagnetic reversal frequency since the late Cretaceous'. Earth Planet. Sci. Lett., 62, 305-313.

Morelli, A. and Dziewonski, A.M., 1987. 'Topography of the core-mantle boundary and lateral homogeneity of the liquid core'. Nature, 325, 678-683.

Parker, R.L. and Shure, L., 1982. 'Efficient modelling of the Earth's magnetic field with harmonic splines'. Geophys. Res. Lett., 9, 812-815.

Proctor, M.R.E. and Gubbins, D., 1988. 'Ambiguity of magnetic field models based solely on directional data', in preparation for J. Geomagn. Geoelect.

Shure, L., Parker, R.L. and Backus, G.E., 1982. 'Harmonic splines for geomagnetic modelling'. Phys. Earth Planet. Interiors, 28, 215-229.

Shure, L., Whaler, K.A., Gubbins, D. and Hobbs, B., 1983. 'Physical constraints for the analysis of the geomagnetic secular variation'. Phys. Earth Planet. Interiors, 32, 114-131.

Whaler, K.A. and Gubbins, D., 1981. 'Spherical harmonic analysis of the geomagnetic field: an example of a linear inverse problem'. Geophys. J. R. Astr. Soc., 65, 645-693.

Wilson, R.L., 1970. 'Permanent aspects of the Earth's non-dipole magnetic field over upper Tertiary times'. Geophys. J. R. Astr. Soc., 19, 417-437.

CHANGES IN THE EARTH'S RATE OF ROTATION ON AN EL NIÑO TO CENTURY BASIS

Nils-Axel Mörner
Paleogeophysics & Geodynamics
Geological Institute
S-106 91 Stockholm
Sweden

ABSTRACT. Accelerations and decelerations in the Earth's rate of rotation of more than one year's duration have generally been assumed to be balanced by corresponding core motions. It is shown, however, that changes of duration from El Niño events to 50-150 years are primarily balanced by, or rather driven by, hydrospheric motions, i.e. oceanic circulation changes. Two (or three) of the major LOD changes during the last 350 years are linked to geomagnetic "jerks". Whilst the corresponding interchange of angular momentum primarily seems to have taken place between the "solid" Earth and the hydrosphere, the jerks seem to represent related flow changes in the outermost part of the outer core.

1. INTRODUCTION

The annual and intra-annual frequencies of changes in the Earth's length of the day (LOD) seem perfectly balanced by (or rather driven by) circulation changes in the atmosphere, primarily the westerlies (Barnes et al., 1983; Eubanks et al., 1983, 1984; Rosen & Salstein, 1983). For lower frequency changes, i.e. signals of time periods ranging from some years to a century or more, it has been generally assumed that they originate from the interchange of angular momentum between the mantle (+ lithosphere) and the core (e.g. Rochester, 1984; Stephenson & Morrison, 1984. The role of the hydrosphere has been discussed by Mörner (1984a, 1984b, 1987a, 1988a, 1988b, 1988c) who claimed that these slower LOD changes were primarily balanced by oceanic circulation changes. I will here expand on this question as it is of fundamental interest and importance for the analysis and understanding of short-term geomagnetic changes.

2. EL NIÑO SIGNALS

The El Niño/Southern Oscillation (ENSO) events represent interesting and important anomalies in the Earth's climatic, oceanographic and

45

F. J. Lowes et al. (eds.), Geomagnetism and Palaeomagnetism, 45–53.

biological systems. They have therefore attracted a significant
interest during the last decade. In 1982/83, the Earth experienced a
very strong ENSO event, studied in detail and given an impressive
description within the World Climatic Data Programme of WMO (WCDP,
1985). The ENSO events seem to provide a mechanism (Mörner, 1988c)
that also operates on longer time scales causing signals/ffects in the
time range of up to about 50–150 years (Mörner, 1984a, 1984b, 1987a,
1987b, 1988a, 1988b, 1988c).

The Pacific is an enormous ocean covering about 150 degrees in
longitude. It is bounded by the Asian–Australian continental system
in the west and the American continental barrier in the east. During
ENSO years the sea level rises, i.e. water is piled up, all along the
American coast, and falls during non–ENSO years. This is well
documented by Enfield & Allen (1980) and was used by Mörner (1987b),
1988b, 1988c) to demonstrate the interchange of angular momentum
between the "solid" Earth and the hydrosphere, the American coast
being the point of application of an effective oceanic–continental
torque. However within the Southern Oscillation region between Darwin
and Tahiti, much of the sea level changes is controlled by air
pressure changes.

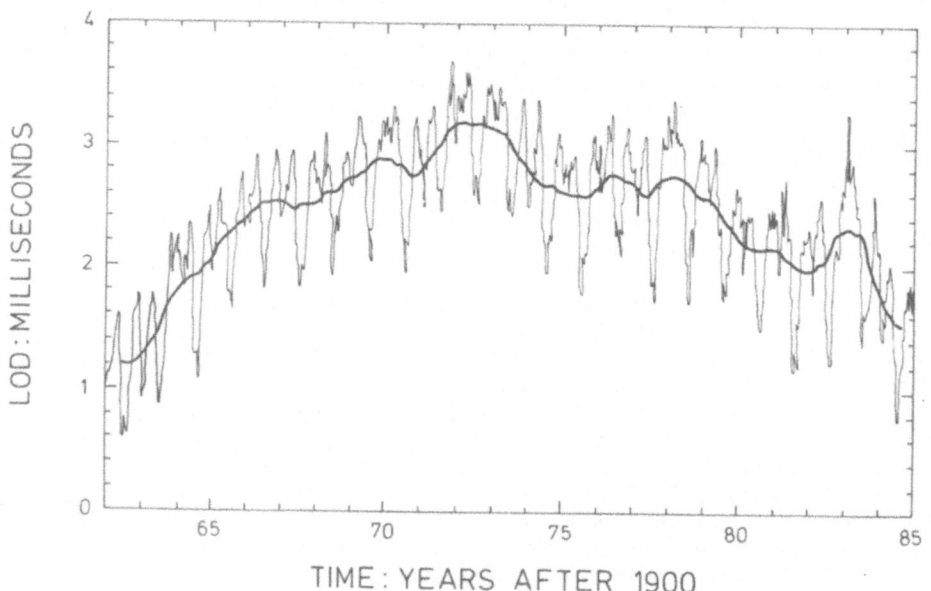

Fig. 1. LOD record 1962–1985 (redrawn from Eubanks et al., 1986)
showing the annual changes (thin line) and mean long-term non–tidal
changes (thick line) recording partly the main LOD change at around
1972 and partly a series of small amplitude oscillations of a duration
of 1–2 years which correlate with ENSO events (the 1982/83 event being
especially distinct).

In detailed LOD records (Fig. 1), the main non-tidal variations exhibit 1-2 year durations of increased LOD, i.e. losses of angular momentum from the "solid" Earth. These increases coincide with ENSO events (Mörner, 1988b). For example the 1982/83 signal amounted to +0.4 ms (Fig. 2).

From detailed analyses of the 1982/83 ENSO year (WCDP, 1985), we know that hot surface water started to pile up in the western Pacific in 1981 and reached a maximum of 3.5×10^{14} m^5 in mid 1982. During the later part of 1982 the LOD increased by 0.4 ms. This hot surface water was then displaced eastwards at a rate of about 100° longitude per year (Fig. 2). In early 1983, it reached the eastern Pacific and by mid 1983 it had reached the coastal area where sea level rose by about 30 cm. At the same time, angular momentum was transferred back to the "solid" Earth resulting in a LOD decrease by 0.5 ms. The hydrosphere appears to have taken up angular momentum in 1982 and then lost it again when the water mass hit the American coast in 1983 (Fig. 2). This example indicates that angular momentum is indeed interchanged between the "solid" Earth and hydrosphere (Mörner, 1984a, 1984b, 1987a, 1987b, 1988a, 1988b, 1988c).

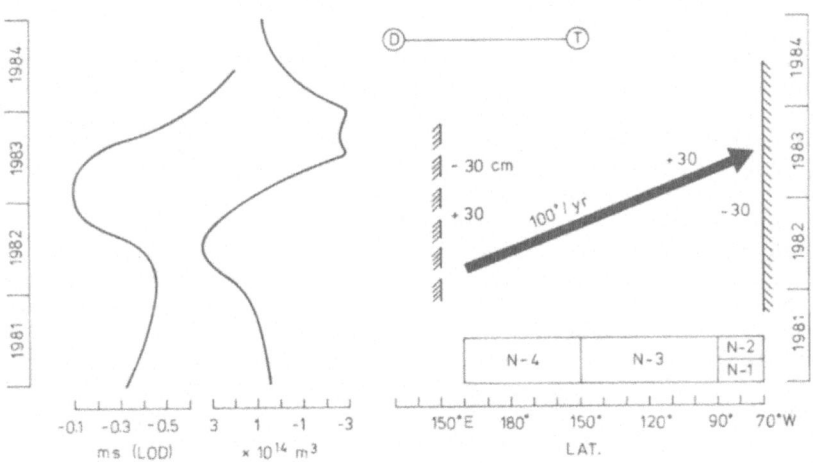

Fig. 2 Records indicating the transfer of angular momentum between the "solid" Earth and the hydrosphere during the 1982/83 ENSO event. The LOD curve gives a loss followed by a gain of its angular momentum. The middle curve gives the water volume changes between Lat. 10°N and 10°S, an increase followed by a decrease. The last curve gives the displacement of the hot surface water from the west to the east (with figures of corresponding sea level changes in cm). When the water masses reached (hit) the American coasts (sea level rose by at least 30 cm), angular momentum was immediately transferred to the "solid" Earth.

3. HOLOCENE 50-150 YEAR SIGNALS

Water masses that are displaced between high and low latitudes are far more effective in the interchange of angular momentum between the Earth's different layers and sublayers than east-west displacements along the same latitudes. The Gulf Stream in the Atlantic and the Kuroshio current in the Pacific both carry large quantities of water from low latitudes to high latitudes. Time variations in their strength and northeastward penetration should, therefore, be of great significance for the analyses of Earth's rotational changes (Mörner, 1984a).

The primary energy source for the short-term palaeoclimatic changes in northwestern Europe is the delicate balance between cold Arctic water penetrating southwards and warm Atlantic water (the Gulf Stream) penetrating northeastwards up to high latitudes (Mörner, 1984a, 1987a, 1988a, 1988b). During the Holocene, 16 pulses of the Gulf Stream have been recorded in deep-sea cores, in northwest European eustatic sea level, in south Scandinavian palaeoclimate (recorded by ^{18}O changes) and even in palaeomagnetic intensity (Mörner, 1984, Fig. 6). The circulation in the North Atlantic is in a feed-back loop with rotational changes; an intensified Gulf Steam pulse bringing more mass (seen as a sea level change) and energy (seen in palaeoclimate both in the sea and on land) would lead to an increased rate of rotation (via the interchange of angular momentum between the hydrosphere and the "solid" Earth), which in its turn would tend to reverse the process (Mörner, 1984a, Fig. 12). It seems significant that the 16 North Atlantic pulses recorded form frequency changing "cycles" (Mörner, 1973a, 1973b) just as one would expect from such a terrestrial feed-back system (and contrary to solar cycles and other astronomical cyclic changes).

The short-term palaeoclimatic changes and shifts during the last 20,000 years or so (Mörner & Karlén, 1984), ranging from a fraction of a degree Celsius up to 5 or even 10°C, were in no case found to be global but instead to represent local and regional events that were of a compensational type on a global scale; implying the redistribution of heat over the globe (and not global rises and falls in temperature). All these events had a duration of the order of 50-150 years (Mörner, 1984b). Redistributions of heat over the globe with such durations can only take place in the hydrosphere (Mörner, 1984b). This gives evidence of frequent short-term changes in the oceanic circulation systems must now be linked to corresponding rotational changes. The sudden warming at 13,000 BP, cooling at 11,000 BP, warming at 10,000 BP, cooling at 2500 BP and coolings at the two Little Ice Ages (in Europe), are all linked to clear changes in the Gulf Stream activity (Mörner, 1984a, 1987a, 1988a).

This model would lead to a general parallelism between the beating of the Gulf Stream and the Kuroshio Current, just as indicated by some observational data (e.g. Taira, 1981), and to a general dissimilarity between cimatic-eustatic records from the two hemispheres, just as indicated by observational data (Mörner, 1984a, 1984b, 1987a). It should be noted, however, that a strong signal

marking the onset of the Younger Dryas Stadial at about 11,000 BP
would affect the whole circulation system, giving rise to secondary
effects – coolings as well as warmings – in other parts of the globe.

4. THE LAST 350 YEARS' INSTRUMENTAL RECORDS

The Earth's rate of rotation has been measured since the beginning of
the 17th century (Stephenson & Morrison, 1984). Periods of
significant acceleration and deceleration are recorded. These LOD
variations cannot be understood only in terms of atmospheric changes
(Lambeck & Casenave, 1976). Core/mantle changes are usually
advocated, this seems hardly justifiable, except as a part solution in
connection with the so-called geomagnetic "jerks" at 1840, 1910 and
1970 (see e.g. Courtillot et al., 1978; Courtillot & Le Mouel, 1988).
I have previously shown (Mörner, 1988a, 1988b, 1988c) that these more
recent (instrumentally recorded) LOD variations also seem to be
primarily balanced by interchange of angular momentum between the
"solid" Earth and the hydrosphere. Changes in the coastal water
temperature, continental air temperature and precipitation, and sea
level changes in northwestern Europe all reflect a pulsation of the
Gulf Stream intensity (Mörner, 1988b). As with the Holocene and
major short-term climatic changes these changes possess a very good
correlation with the LOD variations (Fig. 3), indicating a causal
feed-back relation between a pulsation of the Gulf Stream and the
Earth's rate of rotation.

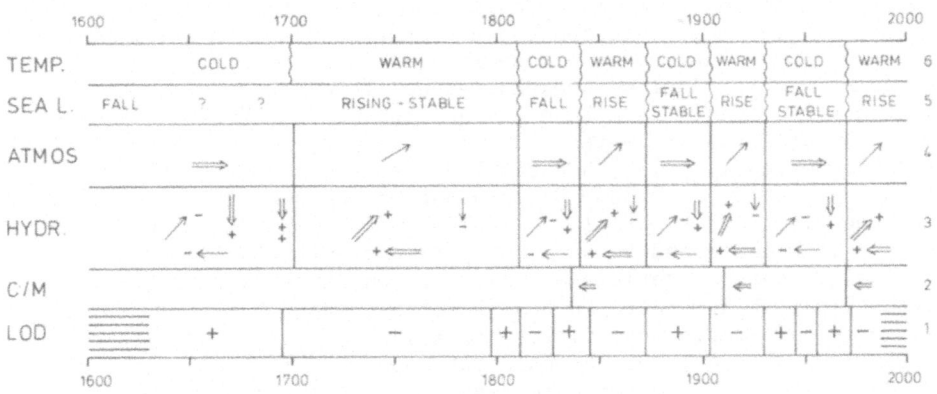

Fig. 3. Non-tidal fluctuations during the last 350 years (1),
geomagnetic jerks (2), hydrospheric ocean circulation changes in the
North-Atlantic (3), atmospheric changes in the westerly jet-stream
over the North Atlantic-European region (4), and causative eustatic
sea level changes (5) and general temperature changes (6) in
northwestern Europe (Mörner, 1988a, 1988b).

The amplitudes of the recorded LOD variations range from 6 ms to about 2 ms (excluding a few smaller ones) and the coresponding rates of change range between 0.4 and 0.2 ms/year. The corresponding losses and gains of angular momentum of the "solid" Earth may well be fully compensated by interchange of angular momentum with the hydrosphere (Fig. 3). The occurrences of geomagnetic "jerks" at about 1970, 1910 and maybe about 1840 suggest that there were at least some interchanges of angular momentum between the mantle and the core at this stage. Le Mouel & Jault (1988) proposed that the core effects were restricted to the flow in the outermost part of the outer core. This seems even more probable in light of the present results where most of the angular momentum transfer (at these three times of change from deceleration to acceleration) came from the hydrosphere. One might even argue that it was the hydrospheric changes that triggered the jerks, and not vice versa. However whilst the 1910 jerk lags the LOD signal, the 1970 jerk seems to lead the LOD change.

5. THE 950 AD LENGTH OF DAY CHANGE

Analyzing old astronomical data, Stephenson & Morrison (1984) were able to give a main LOD curve all the way back to 2650 BP (Fig. 4). From 2650 BP to 950 AD (1000 BP), a deceleration of 2.4 ms/century is recorded, which is exactly the value predicted from the tidal friction theory. It also implies that there are no signs of a residual effect from the last deglaciation. However, from about 950 AD up to about 1800 AD, the total deceleration only amounted to 1.4 ms/century. This implies that an acceleration effect of about 1.0 ms/century commenced at about 950 AD and probably ended at about 1800 AD. A transfer of angular momentum must have taken place corresponding to about a 10 ms LOD change and occurring at a mean rate of about 0.01 ms per year.

This feature remained unexplained until we got a detailed palaeomagnetic record from Lake Kassjön in Sweden (Mörner, & Sylwan, 1988). In an annually varved 6300 years record, it was shown that the virtual geomagnetic pole (VGP) circled around in the Kara Sea region (around Lat. 78°N) from 6300 BP to about 950 AD when it was rapidly displaced along the transpolar meridian 70°E-110°W to Arctic Canada where it stabilized and began a new period of circulation (Fig. 4).

This transpolar meridional shift is consistent with a triple-top motion (Flodmark & Mörner, to be submitted), where the inner core, outer core and mantle move differentially with respect to each other. The 950-1800 AD acceleration could correspond to a simultaneous inner core decleration, the outer core also being affected, but the three-layer motion giving the characteristics of this special event. As such, this would be a quite different mechanism than the one previously discussed.

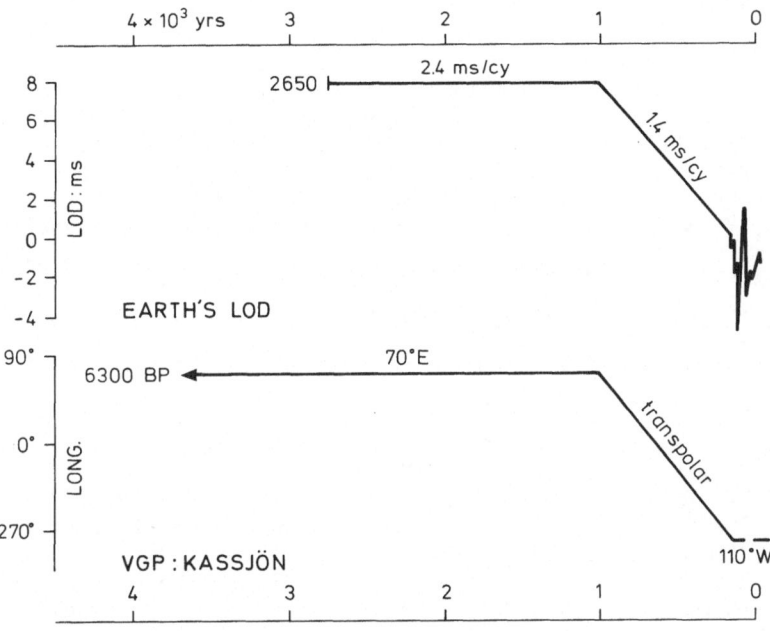

Fig. 4. Long-term LOD changes during the last 2650 years according to Stephenson & Morrison (1984), and the VGP displacement according to the 6300 years varved lake record from Kassjön (Mörner & Sylwan, 1988), indicating a major change at about 1000 BP that lasted between about 950 AD and 1800 AD. Time scale in 10^3 years BP.

6. CONCLUSIONS

The hydrosphere plays a very active role in the solid Earth's rotational changes, loosing and gaining angular momentum in a feed-back mechanism. The ocean circulation system exhibits a pulsation leading to the redistribution of mass (seen in sea level changes) and energy (seen in palaeoclimate and palaeoproductivity), and an interchange of angular momentum with the "solid" Earth. Multi-annual, decadal and century changes in the Earth's rate of rotation seem to be only secondarily, if at all, balanced by differential core/mantle motions as generally assumed. This has a clear bearing on the interpretation of short-term geomagnetic changes.

52

7. REFERENCES

Barnes, R.T.H., Hide, R., White, A.A. & Wilson, C.A., 1983.
    Atmospheric angular momentum correlated with length of the day
    changes and polar motion. Proc. Roy Soc. London, A 387, 31–73.
Courtillot, V., Ducruix, J. & Le Mouel, J-L., 1978. Sur une
    acceleration recente de la variation seculaire du champ
    magnetique terrestre. C.R. Acad. Sc. Paris 287:D, 1095–1098.
Courtillot, V. & Le Mouel, J.-L., 1988. A reversal of geomagnetic
    secular variations with an emphasis on impulses (or jerks).
    Abstracts, the NATO Symposium on Geomagnetism and
    Palaeomagnetism, Newcastle upon Tyne 1988.
Enfield, D.B. & Allen, J.S., 1980. On the structures and dynamics of
    monthly mean sea level anomalies along the Pacific coast of North
    America. J. Phys. Oceanogr., 10, 577–578.
Eubanks, T.M., Steppe, J.A., Dickey, J.O. & Callahan, P.S., 1983. A
    spectral analysis of the Earth's angular momentum budget. JPL
    Geod. Geophys. Preprint., no.102.
Eubanks, T.M., Dickey, J.O. & Steppe, J.A., 1984. The 1982.83 El
    Nino, The Southern Oscillation and changes in the length of the
    day. JPL Geod. Geophys. Preprint, no. 111.
Eubanks, T.M., Steppe, J.A. & Dickey, J.O., 1986. The El Nino, the
    Southern Oscillation and the Earth's Rotation. JPL Geod.
    Geophys. Preprint, No. 143.
Lambeck, K. & Casenave, A., 1976. Long term variations in the length
    of the day and climatic change. Geophys. J.R. Astr. Soc., 46,
    555–573.
Mörner, N.-A., 1973a. Climatic changes during the last 35,000 years
    as indicated by land, sea and air data. Boreas, 2, 33–53.
Mörner, N.-A., 1973b. Climatic cycles during the last 35,000 years.
    J. interdiscipl. Cycle Res. 4, 189–192.
Mörner, N.-A., 1984a. Planetary, solar, atmospheric, hydrospheric
    and endogene processes as origin of climatic changes on the
    Earth. In: Climatic Changes on a Yearly to Millenial Basis
    (N.-A. Mörner & W. Karlen, Eds.). Reidel Publ. Co., Dordrecht
    p. 483–507.
Mörner, N.-A., 1984b. Climatic changes on a yearly to millenial
    basis. Concluding remarks. Ibid., p. 637–651.
Mörner, N.-A., 1987a. Short-term palaeoclimatic changes.
    Observational data and a novel causation model. In: Climate,
    history, periodicity and predictability (M.R. Rampino, J.E.
    Sanders, W.S. Newman & L.K. Königsson, Eds.), van Nostrand
    Reinhold, New York, p.256–269.
Mörner, N.-A., 1987b. Dynamic Sea Surface Changes in the Past and
    Redistribution of Mass and Energy. In Late Quaternary Sea-Level
    Changes (Y. Qin & S. Zhao, Eds.), China Ocean Press, p.26–39.
Mörner, N.-A., 1988a. Ocean circulation changes and redistribution
    of energy and mass on a yearly to century time-scale. In: Long
    Term Changes in Marine Fish Populations (T. Wyatt, Ed.), Vigo,
    Spain, (in press).

Mörner, N.-A., 1988b. Terrestrial variations within given energy,
    mass and momentum budgets; palaeoclimate, sea level,
    palaeomagnetism, differential rotation and geodynamics. In:
    Secular Solar and Geomagnetic Variations in the last 10,000 years
    (F.R. Stephenson & A.W. Wolfendale, Eds.), Reidel Publ. Co.,
    Dordrecht, (in press).
Mörner, N.-A., 1988c. The Earth's differential rotation;
    hydrospheric changes. AGU, Geophys. Monograph. Series, (in
    press).
Mörner, N.-A., & Karlén, W. (Eds.), 1984. Climatic Changes on a
    Yearly to Millenial Basis. Reidel Publ. Co., Dordrecht, 667 pp.
Mörner, N.-A & Sylwan, C.A., 1988. Detailed palaeomagnetic record
    for the last 6300 years from varved lake deposits in northern
    Sweden. This volume.
Rochester, M.G., 1984. Causes of fluctuations in the rotation of the
    Earth. Phil. Trans. R. Soc. Lond., A 313, 95-105.
Rosen, R.D. & Salstein, D.A., 1983. Variations in atmospheric angular
    momentum on global and regional scales and the length of the day.
    J. Geophys. Res., 36, 5451-5470.
Stephenson, F.R. & Morrison, L.V., 1984. Long-term changes in the
    rotation of the Earth: 700 B.C. to A.D. 1980. Phil. Trans. Soc.
    Lond., A 313, 47-70.
Taira, K., 1981. Holocene tectonism in eastern Asia and geoid
    changes. Palaeogeogr. Palaeoclim. Palaeocol., 36, 75-85.
WCDP, 1985. The global climate system. Climatic System Monitoring
    (CMS) of the World Climate Data Program (WCDP), WMO, Geneva,
    52 pp.

GEOMAGNETIC SECULAR VARIATION IN BRITAIN DURING THE LAST 2000 YEARS

D.H. Tarling
Department of Geological Sciences
Plymouth Polytechnic
Plymouth PL4 8AA
England

ABSTRACT. The archaeomagnetic record of the change in declination and inclination in Britain during the last 2000 years is now adequate to consider the geomagnetic implications of these observations. Older periods, although becoming defined, are not yet adequate for such analyses. Some smoothing of the younger data is still required, for which a 25 year window has been used. The data have also been corrected for spatial variation, at any given time, using a geocentric dipole model. (This is shown to introduce little or no significant error when applied to the 1985.0 IGRF.) The British archaeomagnetic master curve yields annual rates of change of 0.08 ± 0.06°/yr in declination, 0.04 ± 0.03°/yr in inclination, and 0.06 ± 0.03°/yr for the total angle (all based on 25 year differences of the moduli). Although a short record, there are apparent regular fluctuations in these parameters with periods of 266 ± 27 years (total angle and inclination) and 400 ± 66 years (declination and distance from the axial geometric dipole field). For 25 year values the distance of the observed field direction from that of a central inclined dipole varies from 0.6 to 13.0°, with a mean of 6.5 ± 2.9°, and a mode (based on 1° bins) of 5°.

1. INTRODUCTION

The geomagnetic field varies on all time scales, from fractions of a second to billions of years, but the longer term (secular) variations are only poorly understood, as observatory records of directional change only commenced in 1600 in London and Paris, and later elsewhere, while intensity determinations were not made until the mid 19th century. The ability of rocks and archaeological materials to retain a remanent magnetization associated with the time that they were formed, particularly baked clays within known archaeological contexts, enables records of geomagnetic field parameters to be constructed for much earlier times. As the methods, techniques and principles of archaeomagnetic studies are well documented (Aitken, 1958, 1974; Tarling, 1975, 1983, 1988), only a very brief summary is

55

F. J. Lowes et al. (eds.), Geomagnetism and Palaeomagnetism, 55–62.
© *1989 by Kluwer Academic Publishers.*

provided here of those parts relevant to the reliability of the British secular variation record.

The present situation for determination of the ancient field intensity is considerably less certain than for directional studies, with specimen repeatabilities commonly being only of the order of ±20-30% (Walton, 1988a,b), although more recent determinations are suggesting that the repeatability may soon become nearer to ±2% (Aitken et al., 1988a,b). Such poor repeatability, certainly for observations made more than a few years ago, combined with the paucity of data, make it dangerous to infer geomagnetic field behaviour. Because of this, only the directional data are evaluated here.

Directions of remanences in samples of archaeological baked clays can usually be measured with a repeatability of better than ±2°, and each specimen can normally be oriented in the field with an accuracy of some ±1-3°, particularly if a sun compass is used. If both types of measurement error are random, then the mean direction based on >5 samples per site should be definable within ±1°. In practice, the observed scatter is much greater than expected, so that the mean direction is commonly only defined within about 3-5° at a 95% probability level (α_{95}). This unexpected scatter has commonly been attributed to magnetic "refraction" (Aitken & Hawley, 1971; Dunlop & Zinn, 1980; Abrahamsen, 1986), although there is increasing evidence that the major factor is differences in cooling rates between different parts of a fired structure, i.e. it is due to the interaction between fast and slow-cooled parts of the same structure, and this effect is considerably enhanced if the materials involved are inhomogenous (Tarling et al., 1986). As the reasons for such scatter are still subject to investigation, an uncertainty of some 2-3° should normally be assumed for each archaeological site, even if the precision is apparently higher. Until the cause of scatter is sufficiently known that it can be corrected by different sampling strategies or modes of analysis, it is not realistic to attempt analyses of geomagnetic field behaviour on a greater precision than this, other than where considerable numbers of sites are involved and such effects appear to be adequately averaged out - as in the present analysis.

As archaeologists also require a well defined geomagnetic secular variation curve for dating purposes, all available directional data for Britain have recently been examined by Clark et al. (1988) in order to construct such a curve for Britain during the last 3000 years, of which the data for the last 2000 years are considered adequate for preliminary evaluation in terms of generalised geomagnetic field behaviour.

2. THE BRITISH DATA BASE

The time variation of the geomagnetic field in Britain was originally established by Aitken & Weaver (1962), and in France by Thellier (1966, 1981). In both cases, the criteria for the selection of data for inclusion were primarily based on the magnetic behaviour of the

individual samples. These data still meet most modern requirements,
although the total data base and methods of mathematical analysis have
improved considerably. After the initial cessation of archaeomagnetic
studies at Oxford, the British data have been vastly expanded by
Tarling and colleagues, particularly A.J. Clark (then at the Ancient
Monuments' Laboratory of the British Department of the Environment).
Many of these data were derived from "rescue" sites, but priority was
given to well dated archaeological sites for periods when the original
British curve appeared most poorly constrained. In selecting the
data, similar emphasis was given to the magnetic reliability, the
number of samples involved, and their degree of scatter, etc., as in
previous summaries, but particular attention was also paid to the
reliability of the archaeological date. If there is a sufficiently
large data base, then such curves tend to "draw themselves", i.e. the
mean corrected directions all fall on the same curve, but calibrating
the curve requires precisely dated tie-points and so the temporal
reliability was given a high priority in constructing the curve,
although less constrained data often enabled alternative curve
patterns to be excluded. The pre-Roman part of the archaeomagnetic
curve is currently in the process of "drawing itself", but
insufficient well-authenticated site dates are available to calibrate
it adequately for meaningful geomagnetic analysis. Hence this section
of the curve is not considered here.

Of particular relevance to geomagnetic interpretations is the use
of an apparently more effective geomagnetic model for correcting the
determinations from geographically dispersed sites to some central
location. Previously, site directions were corrected to either Paris
or London on the assumption of an axial geocentric dipole field, i.e.
the inclinations were adjusted for the difference in latitude between
the sampling site and the central location, (using the standard
formula tan(Inclination) = 2 tan(Latitude)), but no correction was
made to the declination values. As most early British results were
primarily from southeastern England, relatively close to London, the
inclination corrections were small and the spatial variations in
declination are also likely to have been small, but the French data
sources extended from northern France to the Mediterranean, with
resultant greater uncertainty in the validity of the method of
inclination correction and almost certainly introducing errors into
the "Paris" declination values.

The correction of site observations to an inferred value at a
central site within England and Wales (few data were then available
from Scotland) was developed by M. Noel and described in Tarling
(1983). The central location was chosen as Meriden (52.43°N, 1.62°W)
and so the correction was termed the "Meriden" correction. It assumes
that the geomagnetic field, over an area of some 10^6 km^2 and at any
given time, can be adequately described by that of an inclined
geocentric dipole. Examination of the 1985.0 International
Geomagnetic Reference Field for values centred around Meriden shows
that the solid angular error introduced by such a correction for sites
600 km from Meriden (an area of 10^6 km^2) is, on average, 0.88° with
the maximum observed difference being 1.35°; for nearer sites the

error will be less. The "errors" introduced by this correction are
thus comparable with the theoretical measurement accuracy and
considerably lower than the observed accuracy when applied on this or
smaller regional scales.

As this correction has reasonable validity anywhere on the
Earth's surface for regions of similar size, it is proposed that it is
termed the "regional inclined geocentric dipole correction" (RIGDC).

3.   GENERALISED GEOMAGNETIC FIELD BEHAVIOUR

Clearly, considerable smoothing has been incorporated, in both space
and time, in defining this secular variation curve for Britain.
Furthermore, a specific geomagnetic model has been introduced in
correcting the observed data to a central location. These factors
urge caution in an geomagnetic interpretation and only the major
features of the secular variation trend are thought to be adequately
defined, such as the average annual rate of directional change
(determined from the absolute difference between the 25 year average
values) and the relationship between the observed field and that of
particular models, such as the axial geocentric dipole field.
Nonetheless, while the extant data base (Table I) still needs further
extension in both time and space, particularly for earlier times,
particular features of the field behaviour in Britain during the last
2000 years do seem to be clear, and likely to be only refined as more
data become available.

3.1 Average Rates of Directional Change

The rate of change in declination is not constant, but its modulus
averages $0.08 \pm 0.06°$/year, with the maximum rate of change of
$0.32°$/year between 1700 and 1725 A.D., and no change on several
occasions. Inclination also shows a variable rate, its modulus
averaging $0.04 \pm 0.03°$/year, with a maximum rate of change of
$0.20°$/year between 1550 and 1575 A.D., and several zero changes. (In
this region the rate of declination change is approximately twice that
of inclination because $1°$ of declination only corresponds to $0.38°$
solid angle.) A more physically meaningful statistic is that the
average solid angular change is $0.06 \pm 0.03°$/year, with a maximum of
$0.1°$/year between 1550 and 1575 A.D., and minima of $0.01°$ year between
600 and 675 A.D.

As the fluctuating rates of declination, inclination and solid
angular change show some visual indications of possible periodicities,
a series of power spectrum analyses, using fast Fourier transforms and
Parzen windows, have been undertaken. (The error given here for the
observed periods is the half width when the peak value has fallen to
66%.) All three records contain strong indications of the presence of
much longer periodicities or trends than the total record length
currently available. Within the available time-span, the change of
declination shows a weak periodicity around 106 ± 10 years, but a very
strong periodicity around 400 ± 66 years. In contrast, inclination

TABLE I   ARCHAEOMAGNETIC FIELD DIRECTION CORRECTED TO MERIDEN

Declination and inclination are the 25 year averages corrected to
Meriden (52.43°N, 358.38°E) and ADF is the angular distance of the
observed field direction from that of a central axial dipole (0.00°,
68.16°).   ΔF is the total angular change over the preceeding 25 years.

Year	Decl.	Incl.	ΔF	ADF	Year	Decl.	Incl.	ΔF	ADF
1975	-8.5	67.2	0.8	3.6	950	23.5	70.0	1.3	8.3
1950	-10.2	67.7	1.4	4.0	925	20.0	70.5	0.8	7.1
1925	-14.0	67.6	1.2	5.3	900	18.0	71.0	1.3	6.5
1900	-17.0	68.0	1.1	6.2	875	14.0	71.0	1.1	5.2
1875	-20.0	68.0	2.0	7.4	850	10.5	71.0	1.3	4.1
1850	-23.5	69.5	1.6	8.3	825	8.0	72.0	1.0	4.0
1825	-25.5	71.0	1.0	8.9	800	7.0	73.0	1.0	4.6
1800	-25.5	72.0	1.3	9.0	775	7.0	74.0	0.7	5.5
1775	-22.5	73.0	1.7	8.3	750	9.0	74.5	0.8	6.3
1750	-17.5	74.0	1.5	7.5	725	12.0	74.5	1.0	1.0
1725	-13.5	75.0	2.0	7.3	700	12.0	73.5	0.3	5.9
1700	-5.5	75.0	1.0	6.3	675	12.0	73.2	0.2	5.7
1675	-2.0	74.5	1.0	5.6	650	11.5	73.0	0.3	5.5
1650	1.0	74.0	1.2	5.0	625	11.5	72.7	0.2	5.3
1625	5.0	73.5	1.3	4.8	600	11.0	72.5	0.5	5.7
1600	9.0	73.0	2.3	5.0	575	10.5	72.0	1.0	4.6
1575	12.5	71.0	5.0	5.0	550	9.5	71.0	1.0	3.8
1550	12.0	66.0	2.6	5.5	525	8.5	70.0	0.7	3.2
1525	10.0	63.5	1.6	6.8	500	7.0	69.5	1.1	2.5
1500	9.0	62.0	1.7	7.9	475	5.5	68.5	1.5	2.0
1475	7.5	60.5	2.0	9.0	450	4.5	67.0	1.2	2.6
1450	5.0	59.0	1.6	10.2	425	3.0	66.0	0.6	3.6
1425	2.5	58.0	1.5	11.0	400	2.0	65.5	0.7	3.5
1400	0.5	57.0	2.2	12.0	375	1.0	65.0	0.5	4.0
1375	-3.0	56.0	2.3	13.0	350	1.0	64.5	0.5	4.5
1350	-7.0	56.5	3.5	12.8	325	1.0	64.0	0.5	5.0
1325	-2.5	59.0	2.5	10.0	300	1.0	63.5	2.5	5.5
1300	1.5	60.5	2.5	8.5	275	4.3	61.5	4.5	7.7
1275	6.0	59.5	2.1	9.8	250	4.0	57.0	2.4	12.1
1250	10.0	59.0	1.5	10.9	225	1.5	59.0	2.6	10.0
1225	13.0	59.0	2.0	11.4	200	0.0	61.5	1.5	7.5
1200	13.0	61.0	1.8	9.6	175	-0.5	63.0	1.5	6.0
1175	15.0	62.5	1.4	8.9	150	-1.0	64.5	1.0	4.5
1150	17.0	63.5	1.3	8.7	125	-1.0	65.5	1.0	3.5
1125	19.0	64.5	1.0	8.7	100	-0.5	66.5	1.0	2.5
1100	21.0	65.0	1.0	9.1	75	-0.5	67.5	1.0	1.5
1075	23.0	65.5	1.3	9.5	50	-1.0	68.5	0.9	0.6
1050	25.0	66.5	0.8	9.7	25	-3.0	69.0	0.7	1.1
1025	26.5	67.0	0.5	10.0	0	-5.0	69.0	1.2	1.8
1000	27.0	67.5	1.5	10.0	-25	-8.0	68.5	1.6	2.9
975	26.0	69.0	1.3	9.3	-50	-9.5	67.0		4.1

GEOMAGNETIC FIELD DIRECTION AT MERIDEN

Meriden at 52.43°N 358.38°E.   F = solid angular change,
ADF = angular distance from axial geocentric dipole field,
(0.00° 68.96°) - all angular changes over the preceding 25 yrs.

shows a very distinct periodicity around 266 ± 27 years, and it is
this periodicity that is also dominant for the solid angular rate of
change of the field.

3.2 Relationship to the Axial Geocentric Dipole

One of the basic assumptions about the (palaeomagnetic) long-term
average nature of the geomagnetic field is that it can be simulated by
an axial geocentric dipole field. Apart from the geomagnetic
implications of this hypothesis, it is also fundamental to virtually
all geological interpretation of palaeomagnetic studies. However, the
present geomagnetic field in Britain is at present shallower than the
axial geocentric dipole. The average observed declination has
similarly not been that of an axial field, but is westerly since 1660
A.D. Both discrepancies may merely reflect short-term, local
characteristics during the last 300 years or so, rather than
reflecting genuine geomagnetic properties, but some palaeomagnetic
observations have been interpreted as indicating that the average
geomagnetic field can best be simulated by a dipole field displaced
northwards, along the Earth's axis of rotation, from the Earth's
centre (Wilson, 1970, 1971) during the last 5 to 25 million. The mean
direction of the geomagnetic field at Meriden, giving equal weight to
each 25 year determination between 1975 A.D. and 50 B.C., is 4.1°,
67.5° (α_{95} = 1.3°). As the axial geocentric dipole field direction at
Meriden is 0.00°, 68.96°, the average field direction does not
correspond with that of the axial geocentric dipole field, but is some
2.1° away from it. The degree of shallowing is, however, small, 1.5°,
and only just outside the 95% confidence limit of 1.3°, and the
easterly offset of 4.1° is similarly small, as this corresponds to a
solid angular difference of only 1.6°. (If less well defined data for
50 B.C. to 350 B.C. are added, then the mean direction becomes
identical to that of the axial geocentric dipole field at the 95%
confidence level.)

However, while the mean direction appear to come increasingly
close to the axial geocentric dipole field as the time base is
extended, the field did not become aligned with the axial dipole field
at any time during the last 2000 years. The nearest approach was
within 0.6° around 50 A.D. During most of the last 2000 years, the
observed field has been some 5° (modal value based on 1° bins;
arithmetic mean 6.5° ± 2.9) away from the axial dipole field
direction. In view of the apparently periodic changes in declination
and inclination, it is not surprising that a periodicity is also
apparent in the changing distance from the axial geocentric dipole
field, with marked periodicity on a 266 ± 27 year cycle, although
there are also longer term trends that are not resolvable on the
current time base.

4. CONCLUSIONS

Archaeomagnetic data are now becoming increasingly well documented for particular areas of the world and are beginning to provide information that is directly relevant to considerations of the secular variations of the geomagnetic field. At the moment, the data are rather sparse or the archaeological dating too imprecise for many regions, but the British data are now adequate for preliminary analyses. However, there is a fundamental need to supplement such directional studies with more precise evaluations of the strength of the geomagnetic field and to examine the spatial variations of the geomagnetic field for particularly well documented periods, e.g. for Greco-Roman times.

The present review clearly demonstrates that the directional behaviour of the geomagnetic field is a multivariate problem and that some periodicities of change have much longer time scales than the 2000 years considered here. Nonetheless, the existence of distinct periodicities at about 266 and 400 years is considered to be reasonably well established. It is also clear that the average geomagnetic field direction, while becoming closer to that of an axial geocentric dipole, still does not correspond with it, even over this extended period. However, the movement of the field direction, relative to that of the axial geocentric dipole, suggests that the field direction has a preferred location some 5° away from the axial geocentric dipole field, but that such preference is not related to any particular location, i.e. it is apparently azimuthally random relative to the axial geocentric dipole field direction.

As the precision of the archaeomagnetic data inhibits resolution on time scales much less than 25 years, there is still a need to establish a more precise record of secular variation for Britain, but this largely depends on the "solution" of the more magnetic refraction (or differential cooling) effect and also more precise dating of the archaeological sites involved. The extension of the curve to earlier times is mainly inhibited by the uncertainty in the dating methods currently employed, but the ability of archaeomagnetism to incorporate all observations from the British Isles enables the evaluation of such dating and should allow an extension of the record of geomagnetic secular variation back for a further 2000 or 3000 years.

ACKNOWLEDGEMENTS

The Meriden correction was originally developed by Dr. Mark Noel and is a fundamental factor in the construction of the British secular variation curve. I am grateful to him and Dr. A.J. Clark for their assistance in constructing this curve and to numerous students whose work contributed to it. R. Eddies undertook the power spectrum analyses.

REFERENCES

Abrahamsen, N., 1986. 'On shape Anisotropy', Geoskrifter, 24, 11–21.

Aitken, M.J., 1958. 'Magnetic Dating', Archaeometry, 1, 16–20.

Aitken, M.J., 1974. Physics and Archaeology, Clarendon Press, Oxford, p. 291.

Aitken, M.J. and Weaver, G.H., 1962. 'Magnetic dating: Some archaeomagnetic measurements in Britain', Archaeometry, 5, 4–22.

Aitken, M.J. and Hawley, H.N., 1971. 'Archaeomagnetism; evidence for magnetic refraction in kin structure', Archaeometry, 13, 83–85.

Aitken, M.J., Allsop, A.L., Bussell, G.D., Winter, M.B., 1988a. Determination of the intensity of the Earth's magnetic field during archaeological times: Reliability of the Thellier technique'. Rev. Geophys., 26, 3–12.

Aitken, M.J., Allsop, A.L., Bussell, G.D., Winter, M.B., 1988b. Comment on "The lack of reproducibility in experimentally determined intensities of the Earth's magnetic field" by D. Walton, Rev. Geophys., 26, 23–25.

Clark, A.J., Tarling, D.H. and Noel, M.J., 1988. 'Developments in Archaeomagnatic Dating in Britain'. J. Archaeo. Sci. (in press).

Dunlop, D.J. and Zinn, M.B., 1980. 'Archaeomagnetism of a 19th century pottery kiln near Jordan, Ontario'. Canad. J. Earth Sci., 17, 1275–1285.

Tarling, D.H., 1975. 'The dating of archaeological materials by their magnetic properties'. World Archaeology, 7, 185–197.

Tarling, D.H., 1983. Palaeomagnetism, Chapman & Hall, London, p. 379.

Tarling, D.H., 1988. 'Secular variations of the geomagnetic field: The archaeomagnetic record'. In Secular solar and geomagnetic variations in the last 10,000 years (Ed. F.R. Stephenson and A.W. Wolfendale), pp. 349–365, Kluwer, Dordrecht.

Tarling, D.H., Hammo, N.B. and Downey, W.S., 1986. 'The scatter of magnetic directions in archaeomagnetic studies'. Geophysics, 51, 634–639.

Thellier, E., 1966. 'Magnétisme Interne'. In Encyclopedie de la Pleiade: Geophysique, 235–376.

Thellier, E., 1981. Sur la direction du champ magnétique terrestre, en France, durant les deux derniers millenaires'. Phys. Earth Planet. Inter., 24, 89–132.

Walton, D., 1988a. Comments on "Determination of the intensity of the Earth's magnetic field during archaeological time: Reliability of the Thellier technique. Rev. Geophys., 26, 13–14.

Walton, D., 1988b. The lack of reproducibility in experimentally determined intensities of the Earth's magnetic field. Rev. Geophys., 26, 15–22.

Wilson, R.L., 1970. 'Permanent aspects of the Earth's non-dipole magnetic field over Upper Tertiary times'. Geophys. J. Roy. astr. Soc., 19, 417–437.

Wilson, R.L., 1971. 'Dipole offset – the time-average palaeomagnetic field over the past 25 million years'. Geophys. J. Roy. astr. Soc., 22, 491–504.

DETAILED PALAEOMAGNETIC RECORD FOR THE LAST 6300 YEARS FROM VARVED
LAKE DEPOSITS IN NORTHERN SWEDEN

N.-A. Mörner and C.A. Sylwan
Stockholm Palaeomagnetic Laboratory
University of Stockholm
S-106 91 Stockholm
Sweden

ABSTRACT. A detailed palaeomagnetic record of the secular geomagnetic
variations during the last 6300 years was obtained from annually
varved lake sediments in Lake Kassjön, north-eastern Sweden. A firm
chronology was established by varve counting. Several swings in
declination and inclination were recorded and dated. Spectral
analyses reveal a dominant periodicity of about 2150 years both in
declination and in inclination (but not of the same phase). The VGP
path records an irregular circulation in the Kara Sea region up to
about 1000 BP, a rapid transpolar displacement and a new period of
irregular circulation in the Canadian Arctic region during the last
centuries.

1. INTRODUCTION

Lake Kassjön (63°55'N, 20°01'E) is located in the Umeå region of
north-eastern Sweden. It has an elevation of 83.8 m and was, due to
uplift, isolated from the Baltic at a period which is recorded by the
change from unvarved clayey sediments to varved lake deposits. In
total, 6321±70 varves have been counted (Renberg & Segerström, 1981).
    The sedimentological characteristics of the lake deposits have
been studied in great detail by Renberg and Segerström (Renberg &
Segerström, 1981; Renberg, 1982; Segerström, in prep.). They
classified the varves in Lake Kassjön as the best and most easy to
count within the region. We therefore started our palaeomagnetic work
on annually varved lake sediments in this lake.
    Northern Sweden has a fortunate position with little or no
influence from any major non-dipole field structure, at least at
present. This may be of great significance for the interpretation of
recorded changes.

F. J. Lowes et al. (eds.), Geomagnetism and Palaeomagnetism, 63–70.
© 1989 by Kluwer Academic Publishers.

2. SAMPLING

A 1 m long and 10 cm wide Russian corer was used for the sampling
(Fig. 1a). This coring device had been used for all the previous
sedimentological studies of varved lake sediments (e.g. Renberg, 1982)
and had been shown to give excellent cores.

The coring was undertaken from the winter ice. The water depth
is 12 m. Two holes were used alternatively so that fully undisturbed
cores with 20 cm overlap could be obtained (Fig. 1b). Seven cores
were taken through the entire varved sequence down to the upper part
of the unvarved clay deposits.

The corer was carefully oriented and kept in the same position
during the coring operation. The cores were wrapped in plastic film,
placed on planks, wrapped in metallic foil and placed in special boxes
for transport to the laboratory.

At the laboratory, cubic plastic boxes (2.1 cm) were carefully
pressed into the sediment in the N–S direction. From the total core
length of 5.5 m, 261 samples were obtained for palaeomagnetic
analysis.

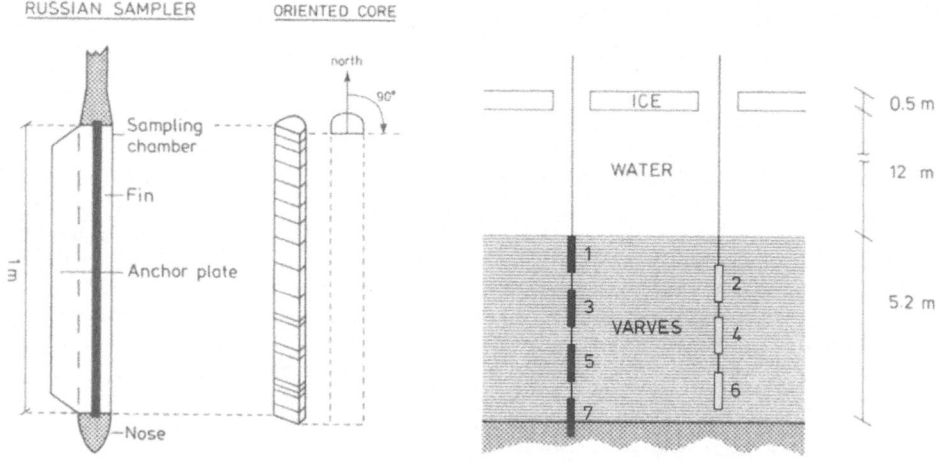

Figure 1. The investigation of Lake Kassjön; (a) coring device
(left), (b) coring operation in two alternate holes with overlapping
core segments (right), and (c) declination and inclination records of
the 7 overlapping core segments plotted versus depth in metres
(opposite page).

3. SEDIMENTOLOGY AND AGE

The lake sequence is annually thin-varved; 6321±70 varves (= years) had been counted from the year 1978 downwards. The lake sequence also exhibits an irregular thick banding between darker and lighter bands representing slower climatic-environmental changes (probably century-signals). These have not yet been studied in detail.

The core also included two "marker" levels; one being the isolation level and corresponding change from Baltic clay to lacustrine varves, and the other a distinct grey clay layer in the year 1901-1902 representing the digging of a ditch in the adjacent farmland.

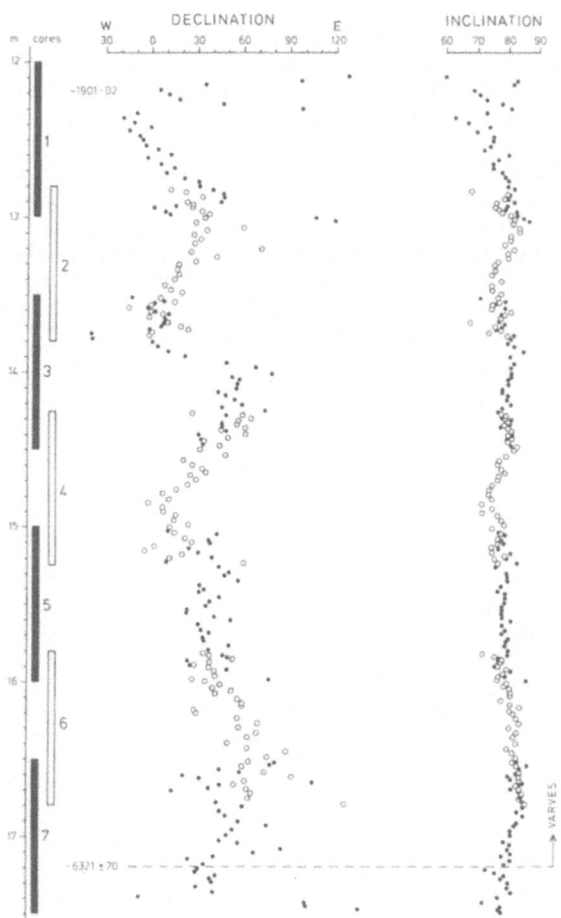

Figure 1c.

The grey clay layer of 1901–1902 is an excellent marker horizon from which studies can begin, avoiding the uppermost part of high water content and little stability.

Segerström (Renberg & Segerström, 1981; Segerström, personal communication) counted 6321 varves over a core length of 5.43 m. The corresponding section in our core is 5.17 m. The difference has been compensated for by linear adjustment.

The sediment quality is excellent and no trace of disturbance has been recorded. The horizontality of the individual varves and their mutual parallelism provide a good tool for checking the sedimentological stability.

4. DEMAGNETIZATION

Out of our 261 samples, 27 were subjected to complete alternating field (AF) demagnetization. The process was done stepwise with measurements at 5, 10, 20, 30, 40, 50, 60, 80, and 100 mT in order to choose the best standard peak field to be applied to the entire set of samples.

In general, demagnetization had a very small effect on the NRM values. All the 23 demagnetization curves from the varved lake sequence fall within a narrow band (Fig. 2) indicating very stable conditions. The 4 samples from the Baltic clay below the isolation level give a wide scatter, however, indicating unstable and variable conditions. We therefore only used the data from the varved lake sequence in our final palaeomagnetic plot (Fig. 3).

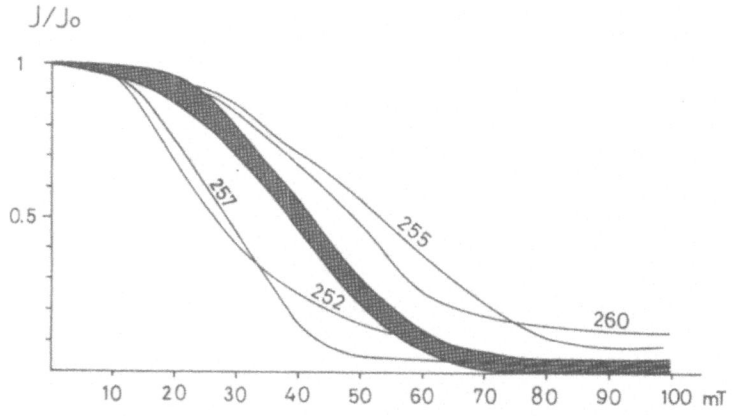

Figure 2. AF demagnetization curves with black zone representing all 23 samples from the varved sequence and the other 4 curves representing the Baltic clay at the base.

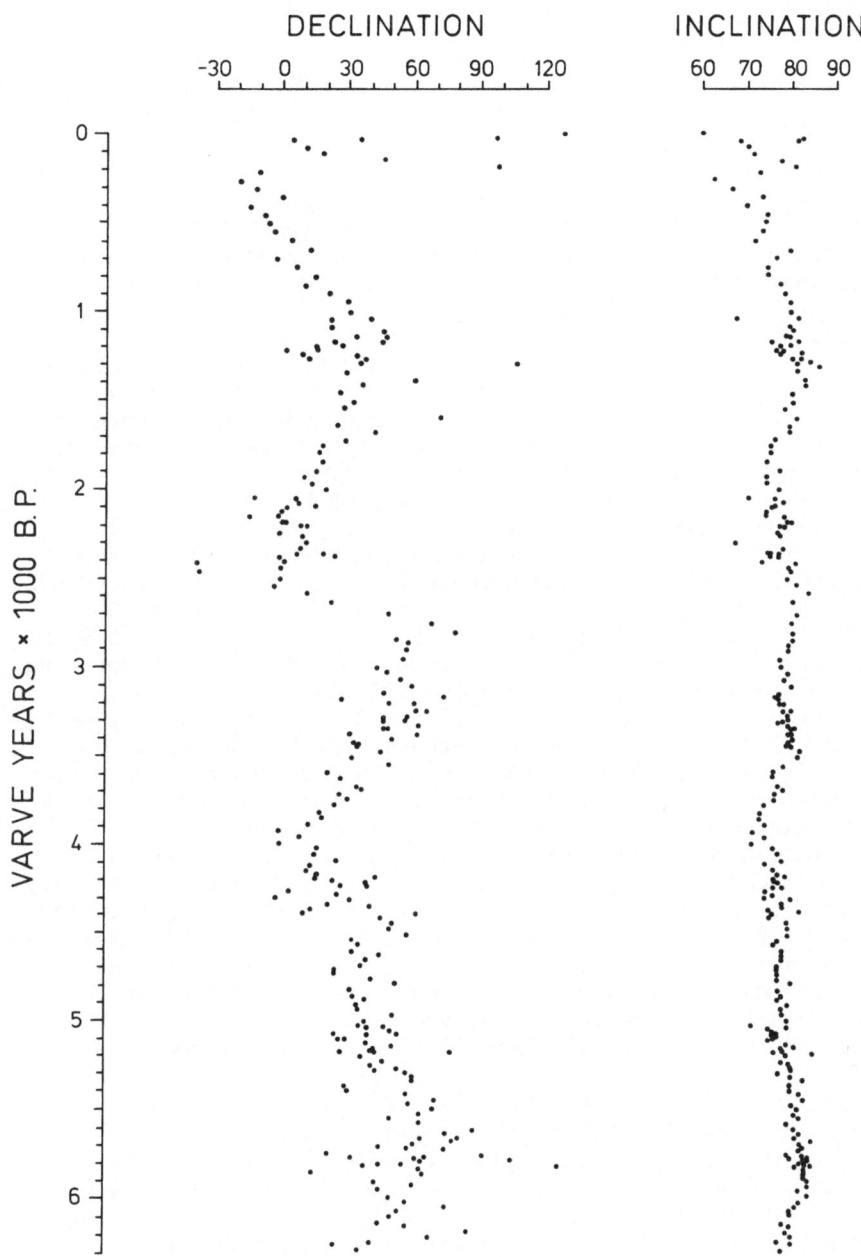

Figure 3.   Declination and inclination versus time in varve years BP.

5. PALAEOMAGNETIC RESULTS

The intensity measurements have given relatively high values varying between 25 and 90 x 10^{-6} emu cm^{-3} (1 emu cm^{-3} = 10^{3} Am^{-1}) (the sediments of high water content in the top and the Baltic clay at the base giving lower values, however).

The variation in the declination and inclination after 20 mT AF demagnetization is shown in Fig. 1c. The consistency between the results obtained from overlapping sections is in general very good indicating that the individual core segments were taken with a consistent orientation.

In Fig. 3, declination and inclination are plotted versus time. Well-defined swings can be seen both in declination and inclination.

The uppermost part (last 200 years) is clearly affected by disturbances due to the high water content and semi-liquid state. Only the 1901-1902 clay layer provides reliable values. For the rest we can give the following characteristics (Fig. 3).

Declination fluctuates between 15°W and 65°E with a clear dominance for the eastern side. The following peaks are recorded: about 5-10°E in 1901/2, about 15°W around 300 BP, about 30°E around 1150 BP (1000-1500 BP), around 0 between 2100 and 2500 BP, about 60°E around 3000 BP (2900-3400), about 10°E around 4000 BP, about 45°E around 4450 BP, about 35°E around 4900 BP, about 65°E around 5600 BP (5500-5850 BP) and around 35°E at the base (6300 BP). Power spectrum analysis gives a dominant peak at about 2150 years.

Inclination fluctuates between 72° and 85°. The following peaks are recorded: 72° in 1901/02, 82-84° around 3100 BP, 80° around 3400 BP, 73-74° around 3900 BP, 85° around 5850 BP and around 80° at the base (6300 BP). Power spectrum analysis gives a dominant peak at about 2150 years with a weaker peak at around 1000 years.

The time control even allows the calculation of corresponding rates of the recorded swings in declination and inclination. We may, for example note the very rapid 50-60° swing in declination between 2800 and 2550 BP, and the slow and steady 11-12° shift in inclination between 5800 and 3900 BP. Before expanding further on rates and amplitudes of changes, we want to duplicate our record in other nearby lakes with annually varved sediments, especially one covering the last 9300 years and one covering the last 2000 years with higher resolution.

Our palaeomagnetic record (Fig. 3) may be compared with other records obtained from Europe; e.g. Lake Windermere (Mackereth, 1971; Turner & Thompson, 1981), Loch Lomond (Turner & Thompson, 1979) and Lac du Bouchet (Creer et al., 1986). In general, all major swings in Fig. 3 can be identified in the other records, too. The chronological positions often differ, however. We believe that is because it is only our core that really has firm dating control.

6. VGP DISPLACEMENT

Our declination/inclination record (Fig. 3) gives a corresponding VGP displacement that is characterized by (1) irregular movements within the Kara Sea region around a center at about Lat. 78°N and Long. 70°E from 6300 BP up to about 1000 BP, (2) a rapid transpolar meridional displacement along Long. 70°E–110°W from about 950 AD to about 1700 AD, and (3) movements around a new center in Arctic Canada during the last centuries.

Because northern Sweden seems to have been fairly unaffected by the influence of local non-dipolar structures, we believe that our record from Lake Kassjön provides a good picture of geomagnetic pole behaviour during this period of time.

7. CONCLUSIONS

We have obtained a palaeomagnetic record of an annually varved lake sequence covering the last 6300 years. Strong and stable magnetization has provided reliable curves of declination and inclination swings from 6300 BP to about 300 BP (and the 1901/02 marker bed). Annual varve counting provides a very close dating control. Our record may therefore be regarded as the best standard curve for northern Europe. Duplication (in progress) of the palaeomagnetic records in other annual varved sequences is needed, however.

8. ACKNOWLEDGEMENT

We acknowledge with thanks the good collaboration with I. Renberg and U. Segerstrom, from the Department of Ecological Botany, Umeå University.

9. REFERENCES

Creer, K.M., Smith, G., Tucholka, P., Bonifay, E., Thouveny, N. & Truze, E., 1986. A preliminary palaeomagnetic study of the Holocene and Late Würmian sediments of Lac du Bouchet (Haute Loire, France)'. Geophys. J.R. Astr. Soc., 86, 943–964.
Mackereth, F.J.H., 1971. 'On the variation in direction of the horizontal component of remanent magnetisation in lake sediments'. Earth Planet. Sci. Lett., 12, pp.332–338.
Renberg, I., 1982. 'Varved lake sediments – geochronological records of the Holocene'. Geol. For. Stockholm Förh., 104, 275–279.
Renberg, I. & Segerström, U., 1981. 'Applications of varved lake sediments in palaeoenvironmental studies'. Wahlenbergia, 7, 125–133.

Turner, G.M. & Thompson, R., 1979. 'Behaviour of the Earth's magnetic field as recorded in the sediment of Loch Lomond'. <u>Earth Planet. Sci. Lett.</u>, **42**, 412-426.

Turner, G.M. & Thompson, R., 1981. 'Lake sediment record of the geomagnetic secular variation in Britain during Holocene times'. <u>Geophys. J.R. Astr. Soc.</u>, **65**, 703-725.

THE LAC DU BOUCHET PALAEOMAGNETIC RECORD: ITS RELIABILITY AND SOME
INFERENCES ABOUT THE CHARACTER OF GEOMAGNETIC SECULAR VARIATIONS
THROUGH THE LAST 50000 YEARS

K.M. Creer
Department of Geophysics
University of Edinburgh
Edinburgh   EH9 3JZ

ABSTRACT. Palaeomagnetic data assembled from fifteen cores collected
during three field campaigns are stacked into three data sets at
average time horizons of 100 years. The first set, derived from six
9 m Mackereth cores covers the interval back to 38000 years BP. The
second and third sets extend back to about 50000 years BP. They are
derived respectively from six 12 m Mackereth cores and from five
Livingston cores. Holocene data have been removed from the records.
There is good agreement between the three stacked logs of declination
and inclination. The stacked directions have oscillated through
amplitudes of up to about 50° (declination) and 30° (inclination).
The time series is clearly not stationary: the general character of
the variations, in particular the dominant periodicities, amplitudes
and the phase relationship between declination and inclination
variations, changes on a time scale which is longer than the longest
periods of the individual oscillations. The Laschamp 'event' is not
recorded, though there is some evidence that the inclinations measured
in individual cores attained negative values at some of the more
pronounced minima of the secular variation oscillations. Some
smoothing of the recorded signal is to be expected as a normal
consequence of the sediment magnetization process for 'short' periods
(a few hundred years) in the geomagnetic input signal. However,
longer periods of the order of 1000 years should not be appreciably
attenuated. Overall, there is a very strong bias to clockwise
rotation of the geomagnetic vector, particularly between about 35000
and 12000 years BP. One interpretation of this is that westward
drifting nondipole field sources were an important feature of the
geomagnetic secular variations.

1.   INTRODUCTION

1.1. Geomagnetic secular variations through pre-historic times

Historically, geomagnetic secular variations have been recorded by
purposely designed scientific instruments installed in magnetic

F. J. Lowes et al. (eds.), Geomagnetism and Palaeomagnetism, 71–89.
© 1989 by Kluwer Academic Publishers.

observatories. If we are to extend our knowledge of the behaviour of
the field back beyond historic times, we must rely on natural
processes which happen to have recorded past variations of the
intensity and direction of the field. Thus we must look to
palaeomagnetism. Sequences of sediments deposited in lakes provide
the best source of records of secular variations (SV) on the time
scale of the order of 10^3 years (Creer, 1983), while deep sea
sediments have allowed the pattern of polarity reversals through the
last 200 million years to be well established. Transitional between
instrumental and natural records are those inadvertently left behind
by ancient civilizations, but the time span of these records is
limited to the latter half of the Holocene.

On the whole, lake sediments have gained a rather doubtful
reputation as carriers of past geomagnetic SV. This is essentially
because palaeomagnetic records may be incomplete or overprinted due to
erosion of bottom sediments by under-water currents and they may be
discontinuous due to slumping episodes etc. Two important criteria
must be satisfied for the recovery of reliable palaeomagnetic secular
variation records: the _first_ concerns the choice of lake; the _second_,
the method of coring.

1.2 Essential criteria

First, small lakes, isolated from the local water table, have been
found to yield the best quality records because rates of sediment
deposition are much less variable than in large lakes which are
usually connected to the regional drainage system (inflowing or
outflowing rivers), and because the water level has not undergone
major long term changes over times of the order of 10^3 to 10^4 years.

Second, it is essential that sediment cores to be used for
palaeomagnetic secular variation studies should be extracted with
special care, preferably using custom-designed coring equipment. The
cores from which the results discussed in this paper were derived were
collected either with pneumatically powered corers (Mackereth, 1958)
in single lengths of 6, 9 or 12 m, or with modified Livingston type
coring equipment in sections of 1 or 2 m length. Cores for
palaeomagnetic SV study should be of at least 5 cm diameter to ensure
that the fabric of the recovered sediment should not be appreciably
affected by 'arching', due to friction at the inner wall of the coring
tube. This rules out most cores taken for purely palynological study
since these are normally of only about 2.5 cm diameter.

1.3 Pitfalls

Attempts to recover palaeomagnetic secular variations from cores taken
with gravity piston corers from oceanographic vessels have not yielded
good results, because of damage caused to the sediment fabric during
the coring operation and because of careless handling of the cores
subsequent to collection (shock, partial drying out etc). Damage to
the fabric of the sediment that may have serious consequences for
palaeomagnetism may not be apparent to the naked eye, especially when

the sediments are homogeneous. Invisible damage to the magnetic
fabric can be caused if excessive vibration is used to aid penetration
of the corer. Thus it is advisable to carry out magnetic anisotropy
experiments as a check of suitability of the cored material.
Palaeomagnetic directions strongly oblique to the range of values
expected for normal secular variation superimposed on the axial dipole
field direction occurring within records which otherwise exhibit
little or no secular variation signal are surely artefacts of bad
coring procedures.

Problems may arise for other reasons: even with good coring
techniques palaeomagnetic records which are not repeatable from core
to core within a given lake have sometimes been obtained due to high
energy sedimentation dynamics – a quiet depositional environment is
essential to allow the sediment to record the geomagnetic variations
faithfully.

1.4 Further reading

An abundance of poor quality results in the published data has led
many geomagnetists and Quaternary stratigraphers to become
disillusioned with the palaeomagnetic method, to the extent even that
they do not believe any secular variation results obtained in this
way. Thus one of the objectives of this paper is to consider one
particular case history which demonstrates clearly that repeatable
secular variation can be extracted from lake sediments. For broad
reviews of contributions to the subject, especially of inter-lake
correlation and inter-continental comparisons, see Creer (1985) and
Creer, Tucholka and Barton (1983).

2. LAC DU BOUCHET

2.1 The lake

Lac du Bouchet is a small maar lake of 800 m diameter, situated at a
height of 1200 m near Cayres (44.9°N, 3.8°E) in the Province of Haute
Loire, France. The water depth in the central basin is ca. 27 m and
the overall diameter of the crater is 1.2 km. Two samples from the
basaltic flows cut by the explosion crater have been dated by the K/Ar
method at 800 ± 40 kyr and 870 ± 60 kyr, placing an upper bound on the
age of the crater so that it is likely that sediment deposition
started early in the Brunhes epoch. It is estimated that the lake
contains up to about 100 m of sediment, which potentially constitute a
unique source of a long palaeomagnetic SV record spanning up to about
700 kyr. Throughout its early history, water level, as identified by
ancient beach deposits, was some 12 m higher than the present level,
and remained so for most of the Pleistocene. It has remained
remarkably constant, to within ± 2 m of the present level, through the
Late Glacial and Holocene. Figure 1 illustrates the geographical
environment of Lac du Bouchet.

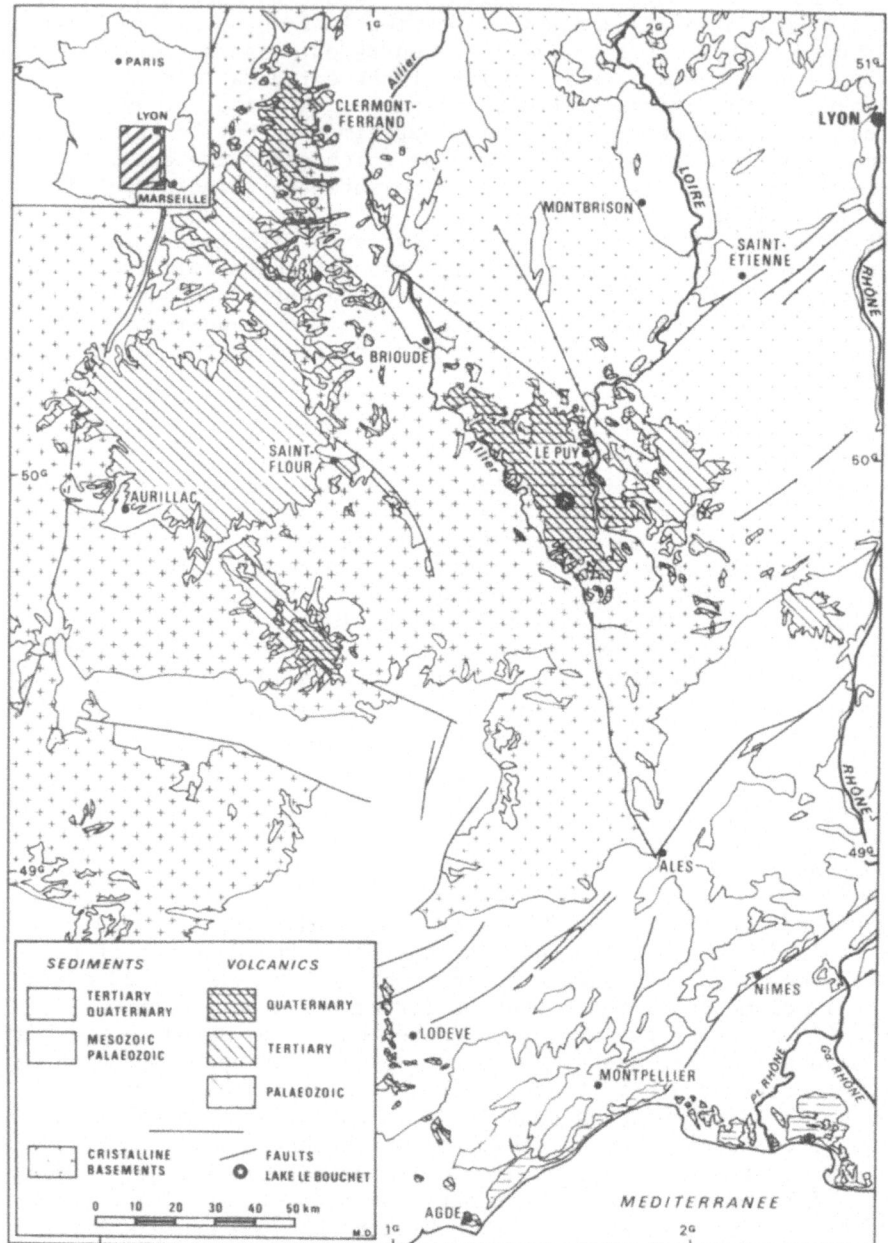

Figure 1. Sketch map of the area around Lac du Bouchet.

2.2 The first three phases of the study

A multi-disciplinary research programme into the palaeoenvironment of
Lac du Bouchet, involving palynological, sedimentological, diatom,
ostracod and other studies, was started in 1982. The three initial
phases were carried out on cores taken with pneumatically operated
Mackereth type corers. In the first phase, cores of length 6 m were
shown to span the last 22000 years (Creer et al., 1986; Bonifay et
al., 1987). The second phase of the study was carried out on a series
of 9 m cores (Smith and Creer, 1986), extending the records back to
about 35000 years before the present. Results from a set of 12 m
cores collected in 1984 and 1985 are described by Creer et al. (1988).
All the above-mentioned research was carried out jointly by the
Department of Geophysics, University of Edinburgh and the Laboratoire
de Geologie du Quaternaire de CNRS at Luminy, Marseille. The work was
financed principally by the NERC (UK) and by CNRS (France), with
support from the local community.

2.3 The fourth phase

In 1986, a fourth phase of the study, called Geomaarsl, was financed
by a grant from the European Community under the 'Stimulation Action'
Programme. This allowed the collaboration to be extended to include
the Department of Geology, University of Trier and the Department of
Botany, University of Kiel, Federal Republic of Germany. A set of
four cores (A, B, C, D and F) of length 14 m to 20 m were collected in
1 and 2 m sections using a modified Livingston corer operated from a
raft. This work is not yet complete, though some palynological data
have been reported by Reille and de Beaulieu (1986, 1988), de Beaulieu
and Reille (1987) and Pons et al (1987). An initial report on the
palaeomagnetic results is in course of preparation (Thouveny et al.,
1988).

3. RESULTS

3.1 Construction of stacked data sets

The data from six cores from the 9 m set (Smith and Creer, 1986), from
six different cores from the 12 m set (Creer et al., 1988), and from
five cores from the Geomaarsl study (Thouveny et al., 1988) have been
separately stacked on a common depth scale based on core D, which is
the longest core (20 m) from the Geomaarsl study. Transformation to a
common depth scale within each of these three sets of cores was
affected first by direct visual and photographic observation of the
opened cores and of photographs of them. Since the sediments are
rather homogeneous in grain size and colour, only a rather coarse
correlation can be made in this way, and to proceed further it is
necessary to use down-core patterns of magnetic susceptibility (k) and
intensity (J). In the final depth transform functions, k and J marker
horizons were defined every 20-30 cm.

3.2 Transformation to a time scale

For the results to be of use geomagnetically rather than of purely
local statigraphic use, it is necessary to define a depth/time
transform. For this purpose 5 radiocarbon dates were determined
(ranging from 280 to 19400 years BP) at the Centre de Faibles
Radioactivités at Gir-sur-Yvette using the conventional method (see
Creer et al., 1988), 8 dates were determined (range 9700 to 30600
years BP) with the High Energy Mass Spectrometer facility at Oxford
(see Smith and Creer, 1986) and 6 dates were determined with the
Accelerator Mass Spectrometer facility at Tucson (range 31080 –
41500 yr BP). In addition there was some direct input from the
palynology which provided a tight age control between the present and
15000 yr BP, and indirectly by linkage with the oxygen isotope stages
identified from ^{18}O studies of deep sea cores and of the Vostok ice
core (Lorius et al., 1988), giving the following dates: 62000 yr for
the stage 3/4 boundary, 73000 yr for stage 4/5 (start of Wurm),
107000 yr for stage 5c/d and 124000 yr for stage 5d/e (top of Eemian.)
The last three points are not used in the context of the present
paper.

  Thus the three stacked data sets shown in Figure 2 were
transformed to this time-scale following a procedure described in
Creer et al. (1988). The 9 m core series extends back to 35000 yr BP
and the 12 m series back to about 50000 yr BP. The Geomaars1 series
extends back to about 124000 yr BP, but the data discussed here have
been truncated at 50000 yr BP.

  The bars in Figure 2 indicate standard errors for the data points
stacked at each horizon. The stacking levels were evenly spaced at
25 mm intervals on the depth scale, back to 12 m (50000 years BP).
Since the logs were stacked at evenly-spaced intervals in the depth
domain, the levels on the time-plots of Figure 2 are not evenly
spaced.

3.4 Attenuation and phase of the recorded signal

The declination and inclination variations consist essentially of
irregular oscillations with characteristic times ('periodicities')
typically of the order of a few thousand years with amplitudes up to
about 50° in declination and 30° in inclination. The pattern does not
repeat itself and its general appearance changes on a time scale of
several 'periodicities'.

  The time constants associated with the magnetization process,
'magnetization lock-in times' are of the order of a few hundreds of
years (Tucker, 1979; 1980a,b), so that we should not expect the signal
from the part of the geomagnetic spectrum with 'periodicities' of the
order of a thousand years to have been appreciably attenuated by the
recording process. On the other hand, the higher frequency part of
the geomagnetic input signal, characterized by 'periodicities' of the
order of a few hundreds of years, should have been appreciably
attenuated.

3.5 Discussion

The most prominant feature exhibited by the inclination and
declination logs occurs in the age range 13000 to 16000 yr BP. It
consists of two large inclination oscillations of amplitude 40° (on
the stacked logs) with rounded maxima (π, σ) and pointed minima
(ξ, ρ, τ). These are accompanied by declination oscillations
(I, K, Λ, M, N) which are out of phase by 1/4 period such that they
can be modelled by a westward drifting source (e.g. a radial dipole or

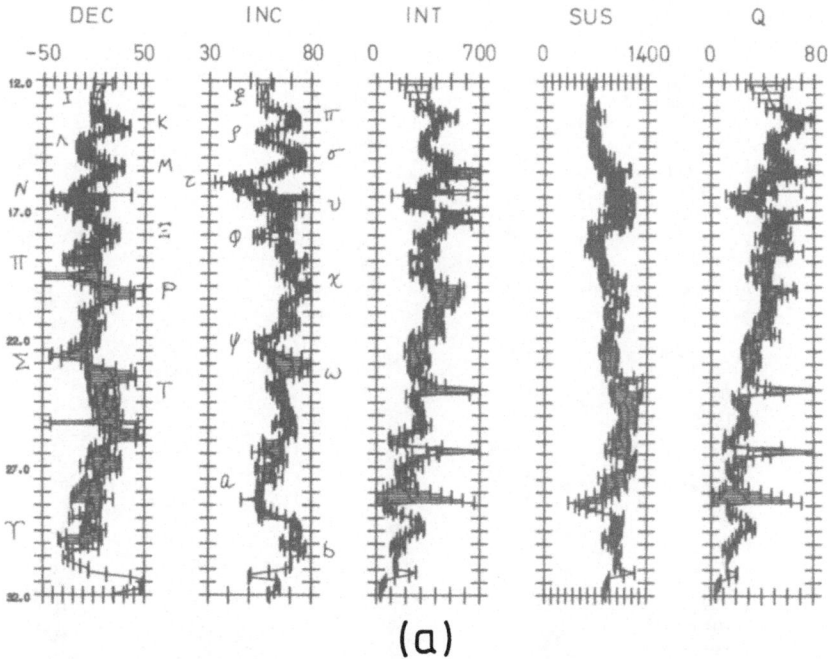

(a)

Figure 2a. Stacked plots of: declination (DEC); inclination (INC);
intensity after AF demagnetization in 10 mT (INT) – units 10^{-3} A m^{-1};
susceptibility (SUS) – units 10^{-5} SI; Q-ratio (Q = J/k) – units A m^{-1}.
The three sets of plots are compiled from three different sets of
cores taken at different times with different corers. The set of
plots on Fig. 2a is from six 9 m Mackereth cores (Smith and Creer,
1986); the set on Fig. 2b (over the page) is from six 12 m Mackereth
cores plus one of the cores from the 9m set (Creer et al., 1988); the
set on Fig. 2c (over the page) is from five Livingston cores
(Thouveny et al., 1988). The depth scales of all 15 cores were
transformed on to a common depth scale before applying a further
transform to the time scale. The depth/time transform function is
described in Creer et al. (1988).

78

a current loop in the outer core – see Creer, 1983). This combination
of features can also be recognized in the palaeomagnetic SV record
from Meerfelder Maar (Eifel region of the Federal Republic of
Germany), and also in the N. American record for Anderson's Pond (Lund
and Banerjee, 1979) – see Creer (1985).

Let us now consider the older part of the records. The
relationship between the inclination and declination oscillations
exhibited between about 16000 and 22000 years BP is not so clear cut

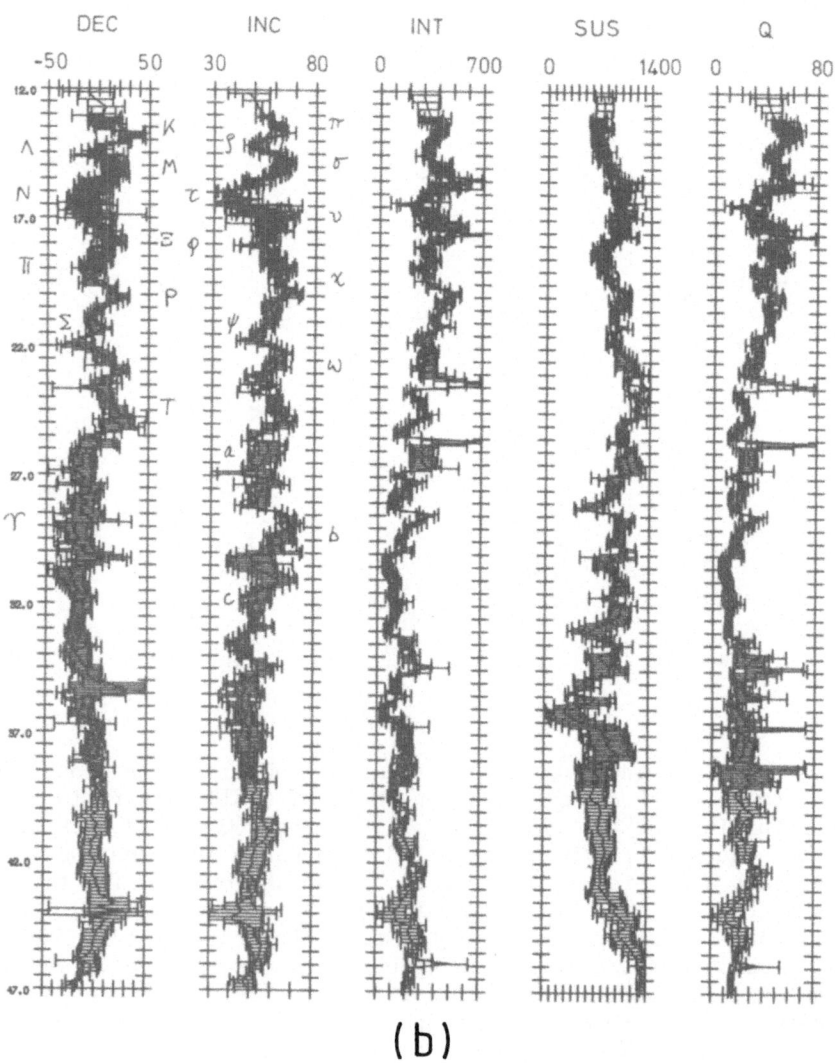

(b)

Figure 2b.

as for the preceding interval discussed in the last paragraph; there are two large amplitude declination swings (N, Ξ, Π, P, Σ) which are accompanied by a number of small amplitude, short period swings grouped together in Figure 2 under the labels υ, φ, χ, and ψ. Between 22000 and 32000 years the large amplitude declination oscillations continue (T, Υ, Φ). They are accompanied by a flat topped inclination maximum ω, a minimum a and another flat topped maximum b. Thus the patterns of directional variations are certainly not repetitive.

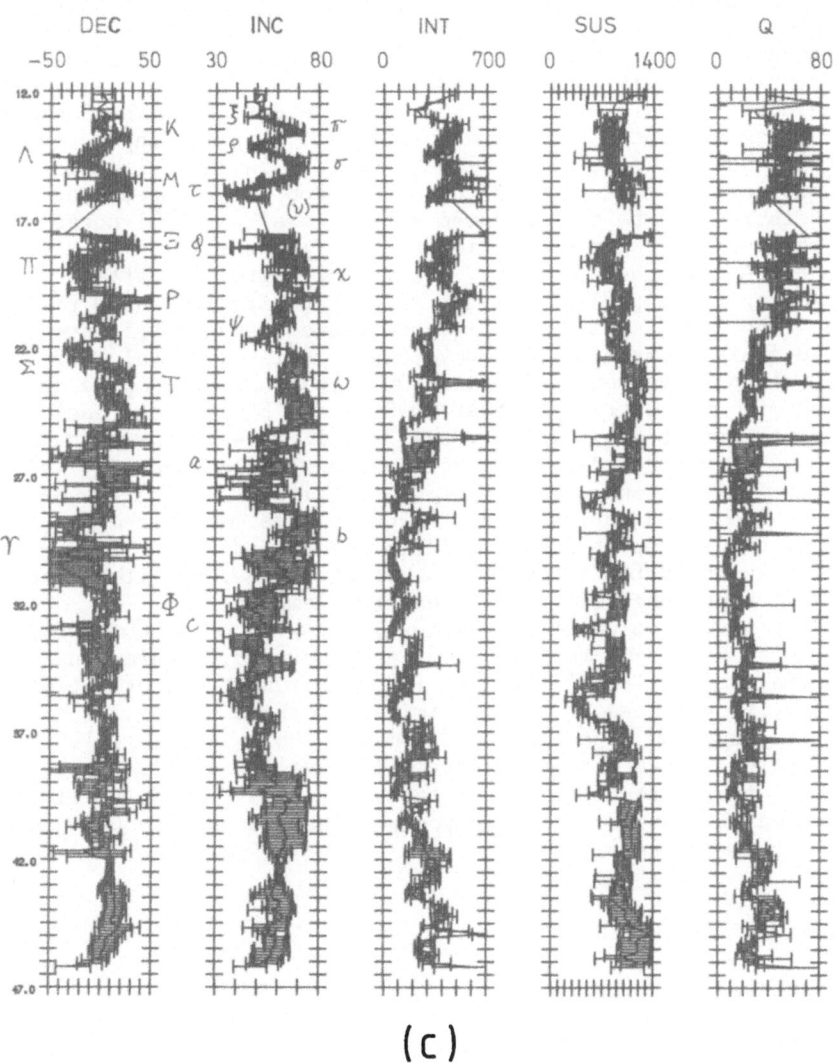

(c)

Figure 2c.

Since the three sets of stacked logs were constructed from data obtained from different cores taken at different times they are quite independent. The overall agreement is good noting that data from 15 different cores are represented. However some minor differences are apparent from a closer inspection of the three sets of logs. These originate in part from the presence of long wavelength trends in the raw Mackereth core data, originating from a modest but unavoidable azimuthal twisting and longitudinal bending of the core tube during each coring operation. This effect can be significant in 9 m and 12 m coring, but not in 6 m coring. When data from a number of cores are available, as in the present study, these effects can be partly corrected for, though not entirely eliminated, by detrending the individual core data before stacking, and then retrending on to the average trend. This procedure assumes that successive misalignments of the core tube are small and random. The data from the 9 m and 12 m core series have been processed separately, so providing a test of this procedure. There can be some ambiguity in the azimuthal alignment of adjacent sections of Livingston type cores, though such uncertainties can be much reduced when several parallel cores have been taken with staggered section boundary depths.

Let us now consider variations in intensity (J) (after alternating field (AF) cleaning), susceptibility (k) and Q ratio (J/k). The three stacks exhibit good overall agreement: intensity and Q ratio exhibit parallel long wavelength variations, values being relatively high between 13000 and 20000 years BP, then passing through a broad minimum centred at about 30000 years BP, and increasing again to high values at about 45000 years BP. It is well known that intensity magnitudes depend on both ancient field intensity and magnetic mineral content. Q ratios are calculated with the object of removing the contribution of variations in the magnetic mineral content, but susceptibility is widely recognized not to be the best magnetic quantity for intensity normalization because it is influenced by a different range of grain sizes than remanence.

With a view to gaining a better insight into palaeointensity variations, down-core variations of anhysteretic, saturation isothermal and 'stirred' (Tucker, 1980b) remanent magnetization are currently in progress. Some preliminary work on deconvolving the intensity data has been completed (Smith, 1985; Thouveny, 1987). The long wavelength pattern of susceptibility variations is clearly quite different from intensity, and probably reflects primarily climatic and environmental changes. These aspects of the Lac du Bouchet study will be reported elsewhere.

4. THE LASCHAMP EXCURSION

4.1 Excursions in general

Sharp deviations in palaeomagnetic directions recorded in recent sediments and in recent lava flows have been reported on occasion in the published literature. The questions is whether these deviations

really represent geomagnetic field behaviour or whether they represent
aberrations of the magnetization processes which operate in sediments
and lavas. More often than not they can be shown not to be of
geomagnetic origin (e.g. see Verosub and Banerjee, 1977 for review),
though when such suspect reports are discarded, a few intriguing cases
still remain. One of these is the 'Laschamp event'

4.2 Its discovery

Some twenty years ago a 'polarity event' or 'excursion' was reported
by Bonhommet and Babkine (1967) and by Bonhommet and Zahringer (1969)
for two sites, Laschamp and Olby, in the Chaine des Puys, France.
These sites are of interest in the context of the present study
because they are situated less than 150 km from Lac du Bouchet.
Repeat measurements of the Laschamp and Olby lavas confirmed the
existence of these almost reversed palaeomagnetic directions.
However, palaeomagnetic studies at other sites in Europe, and in other
continents, have failed to provide confirmatory evidence, but other
apparently reliable 'excursions' have been reported in lavas, baked
clays and sedimentary sequences elsewhere, of which the best known are
at Lake Mungo, Australia (Barbetti and McElhinny, 1976) and Mono Lake,
California (Denham, 1974). Crucial to the resolution of the problem
of correlation of the Laschamp and Olby events with other reported
'excursions' and with the Lac du Bouchet record is their dating.

4.3 Its age and duration

Over the decade which followed its initial discovery, intense efforts
were made to date the Laschamp and Olby flows. A variety of methods
were employed, ranging from ^{14}C measurements on wood burnt by the
spreading lava flow, through thermoluminescence on clays burnt by the
lavas and on plagioclases extracted from the lavas themselves, to K/Ar
and ^{40}Ar/^{39}Ar radiometric determinations (Hall and York, 1978;
Huxtable et al., 1978; Hall and York, 1979; Gillot et al., 1979;
Guerin and Valladas, 1980). These developments are reviewed by Jacobs
(1984).
    For the Laschamp flow, thermoluminescence measurements now
suggest an age in the range 32500 to 35000 years BP and K/Ar
determinations an age in the range 35000 to 43000 years BP, with
associated measurement errors of the order of 10%. Corresponding
estimates of the age of the Olby flow are 37300 to 44000 years BP (TL)
and 42000 to 50000 years BP (K/Ar). Thus, the Olby flow seems to be
significantly older than the Laschamp flow. It follows that three
possibilities should be considered: first, if the excursion was a
single event, it must have been of rather long duration, of the order
of 5000 years at least; second, it may have been a double event, each
one having been of short duration; and third, it is possible that no
geomagnetic event occurred at all and that the 'reversed'
palaeomagnetic directions are due to a self-reversal process, as
argued by Heller (1980) and by Heller and Petersen (1982) from results
of carefully planned thermal experiments on samples of these lavas.

4.4 Excursions in the Lac du Bouchet record?

Thus it is pertinent to examine the Lac du Bouchet records carefully to see whether they provide any substantive evidence of aberrent palaeomagnetic directions.

A single geomagnetic event, lasting about 5000 years, would be represented on the palaeomagnetic logs (Figure 2) by 50 or so consecutive data points, so that a single Laschamp/Olby event would be revealed very clearly indeed. It is quite apparent that no 'excursion' of such long duration is recorded, at least back to 47000 years ago, which is older than the oldest of the determinations of the age of the Laschamp flow. (However it is about 3000 years younger than the oldest determination of the age of the Olby flow, but even so it is stressed that the vast majority of the many age determinations for both flows are younger than 47000 years.)

Regarding the possibility of two separate short events, the key question is 'how short?'. We have discussed some essential properties associated with the sediment magnetization process in section 3.4, where the point is made that there should be no appreciable attenuation of that part of the palaeomagnetically recorded signal having 'periodicities' more than about a few thousand years. On the other hand, that part of the spectrum represented by shorter 'periodicities' should be subject to progressively more attenuation which should be significant for periodicities of a few hundreds of years. Thus, if the geomagnetic inclination actually did reverse briefly during the time spanned by the Lac du Bouchet record, it could only have been during the short intervals around the observed secular variation inclination minima. Progressive AF demagnetization experiments on samples taken from those levels corresponding to the more pronounced inclination minima suggest that these are somewhat smoothed in our stacked plots, and that negative inclinations were attained briefly within the plotted minima (Smith, 1985). This implies that the 'excursions' recorded by the Laschamp and Olby lavas should be regarded as large amplitude secular variations of the geomagnetic field. This type of 'excursion' should be rare: they should be of regional geographic extent only. This contrasts with the other suggestion that they are aborted polarity reversals: these would be expected to be rather rare events of global significance. It would, however, be a rather remarkable coincidence if the times of cooling of both Laschamp and Olby flows occurred at the precise times of two successive minima in the secular variation pattern, since these minima cover only about 10% of the secular variation cycle.

We are left with the third possibility, that the Laschamp and Olby excursions are due to self-reversal, in which case the (almost) reversed palaeomagnetic directions have no connexion with the past behaviour of the geomagnetic field.

5. WESTWARD AND/OR EASTWARD DRIFTING NONDIPOLE FIELDS?

5.1 Drifting and standing sources

The observed westward drift of the nondipole part of the geomagnetic field through historic time was firmly established by Vestine et al. (1947) and by Bullard et al (1950), although the concept originated several centuries ago. An abundant literature has grown up over the last few decades. To summarize, the drift rate, averaged over latitudinal bands and over time through the first half of the 20th century, is estimated at 0.18° per year, so that a complete revolution of the outer core (the source of the field) relative to the Earth's mantle (on which the observer sits) would take 2000 years. This is within the range of the 'periodicities' which characterize the palaeomagnetic secular variation curves (Figure 2).

The size and intensity of nondipole foci of the field through historic time have varied appreciably: changes in the general topography of the nondipole field from epoch to epoch are immediately obvious from casual observation of geomagnetic charts. This immediately suggests that a simple model of drifting sources is not capable of explaining the spatial and temporal historic observations. This led Yukutake and Tachinaka (1969) to suggest that the historically observed nondipole field can usefully be separated into drifting and standing parts.

Palaeomagnetically, the longer term co-existence of drifting and standing sources is evidenced by the observation that the phase relationship between declination and inclination changes with time down the logs for any given geographical site. In particular, a study of the correlation between the patterns of variations recorded through Holocene time in western Europe and in North America has demonstrated clearly that a distribution of standing nondipole sources which fluctuated in intensity together with drifting sources are required to model the relationships between the SV patterns on either side of the North Atlantic (Creer and Tucholka, 1983).

5.2 Palaeomagnetic detection of sense of drift

Nondipole sources, drifting in the Earth's outer core westwards relative to an observer on the surface, will perturb the main dipole field such that the instantaneous geomagnetic vector precesses in a clockwise sense around the main field vector; eastward drifting sources will cause counter-clockwise rotation (Bauer, 1895; Skiles, 1970). This rule is generally valid, though some rare exceptions may occur (Dodson, 1979). The converse of this rule does not hold because rotation of the geomagnetic vector can also be caused by interference between two spatially separated oscillating or fluctuating standing sources (Creer, 1983).

5.3   Dominance of clockwise rotation (35 - 12 kyr BP)

We now examine the Lac du Bouchet records with a view to assessing the
relative importance of clockwise as opposed to counter-clockwise
rotation of the palaeomagnetic vector. The sense of rotation has been
calculated from virtual geomagnetic pole (VGP) paths computed for
successive points along the stacked curves of Figure 2. (VGP paths
are equivalent to paths of time sequences of the geomagnetic vector
plotted in declination, inclination space.) The sign of the curvature
(+ or -) reflects the sense of rotation (clockwise or
counter-clockwise) of the geomagnetic vector about its time averaged
direction. Curvatures calculated for successive declination and
inclination levels along the smoothed stacked logs for the 12 m series
are shown in Figure 3 for the interval 37 to 12 kyr BP. It should be
pointed out that the calculation of curvature is very sensitive to
small errors in the palaeomagnetic data and to the methods of
smoothing (the method used was to fit cubic splines at evenly spaced
intervals of 400 years). 'Curvature' plots such as Figure 3
accentuate short episodes of sharp curvature ('hairpins' along the VGP
plot) rather than longer intervals of persistent though gentle
curvature in one particular sense.

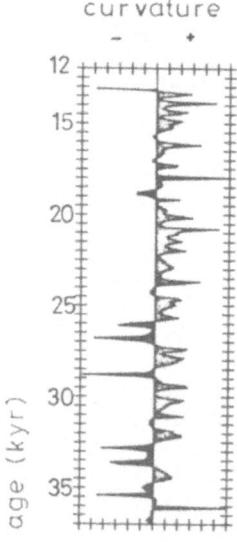

Figure 3.   This plot shows the sense of rotation of the palaeomagnetic
vector as defined by the stacked D and I curves from the 12 m core
series of Figure 2 after smoothing using cubic splines with knots at
equally spaced intervals of 400 years. The interval covered is 13000
to 37000 years.   Age is plotted downwards on the vertical scale;
clockwise rotation (positive curvature) is plotted to the right and
counter-clockwise rotation (negative curvature) to the left.

Thus it is better to concentrate on rather general characteristics of the sense of rotation. Calculations of the integrated durations of clockwise and counter–clockwise rotation through different time windows indicates that there is a very clear bias to clockwise rotation from 35 to 12 kyr BP as shown in Table I.

TABLE I. Percentage of clockwise rotation of the palaeomagnetic vector for 5 kyr time windows moved in 2.5 kyr steps between 13 and 38 kyr BP.

midpoint of time window	% clockwise rotation through 5 kyr window		
	9 m series	12 m series	Geomaars1 series
15.5 kyr	98.0	84.3	84.3
18.0	80.4	78.4	78.4
20.5	76.5	90.2	82.3
23.0	72.3	90.2	68.6
25.5	76.5	64.7	52.9
28.0	72.8	64.7	65.8
30.5	39.2	47.1	45.1
33.0	34.8	47.1	45.1
35.5	–	43.1	52.9

Possible variations in bias to clockwise curvature with spectral band have been investigated by sliding a band-pass filter in the frequency domain (0.2 cycles per kyr has been used in the example illustrated in Figure 4) across the calculated curvature data down to a frequency of 2 cycles per kyr (equivalent to a 'period' of 500 years). A strong bias to clockwise curvature is clearly apparent over all frequency bands, but particularly strong in the ranges around 0.2 and 1 cycles per kyr (equivalent 'periods' 8000 and 1000 years). It is necessary now to investigate whether these characteristics are invariant properties of the geomagnetic field, and therefore other time windows are being investigated with this approach.

In this paper, the word 'curvature' refers to the shape of the VGP path. This curvature is a consequence of the 'rotation' of the geomagnetic vector as it goes through its secular variations. 'Rotation' is sometimes referred to as a 'precession' of the instantaneous geomagnetic vector about at a time–averaged (e.g. axial dipole) direction.

6. ACKNOWLEDGEMENTS

The substantial palaeomagnetic data set assembled to date from measurements on cores extracted from Lac du Bouchet has involved the long term efforts of several palaeomagnetists, particularly

86

N. Thouveny, G. Smith and P. Tucholka. In addition several workers
have participated in the research for shorter periods, notably
G. Turner and I. Blunk. The reader is advised to study the published
papers referenced in this article describing the original research,
and also to note that several papers are about to be published. The
value of the palaeomagnetic data has been enhanced by the efforts of

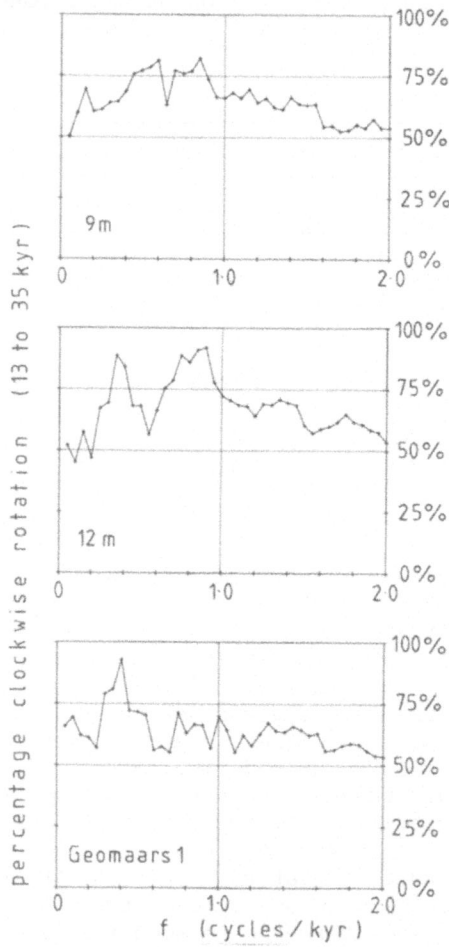

Figure 4. These plots show the percentage of <u>clockwise</u> rotation as a
function of frequency as seen through a band pass filter of 0.2 cycle
per kyr width. The data sets have been truncated in the frequency
domain at 2 cycles per kyr removing all periodicities shorter than
500 yr. All these plots show a bias to rotation in the clockwise
sense. Note that the percentage for the higher frequencies tends to
50% as would be expected for random noise. It is stressed that the
bias to clockwise rotation (westward drift?) may not hold for other
time windows: this question is currently being investigated.

specialists in other disciplines, notably the palynologists, J-L. de Beaulieu, M. Reille and A. Pons, the Marseilles group of Quaternary geologists led by E. Bonifay, and by a range of radiocarbon age determinations made at Gif-sur-Yvette, Oxford and Tucson. The work has been financed principally by grants from the NERC (UK), CNRS (France) and from the Stimulation Action Programme of the European Community (Geomaarsl Project), with additional support from local and regional organizations.

7. REFERENCES

Barbetti, M. and McElhinny, M., 1976. 'The Lake Mungo geomagnetic excursion'. Phil. Trans. Roy. Soc., **1281**, 515-542.

Bauer, L.A., 1895. 'On the distribution and the secular variation of terrestrial magnetism'. Amer. J. Sci., 50, 314-325 .

Bonhommet, N. and Babkine, J., 1967. 'Sur la presence d'aimantations inversees dans les Chaines des Puys'. C.R. Acad. Sci., Paris, **264**, 92-94.

Bonhommet, N. and Zahringer, J., 1969. 'Palaeomagnetism and potassium - argon age determinations of the Laschamp geomagnetic polarity event'. Earth Planet. Sci. Lett., 6, 43-46.

Bonifay, E., Creer, K.M., de Beaulieu, J.L., Casta, L., Delibrias, G., Perinet, G., Pons, A., Reille, M., Servant, S., Smith, G., Thouveny, N., Truze, E. and Tucholka, P., 1987. 'Study of the Holocene and Late Wurmian sediments of Lac du Bouchet, (Haute Loire, France): First Results'. Climate - History, Periodicity and Predictability, eds. Rampino, M.R., Sanders, J.E., Newman, W.S., and Konigsson, L.K., van Nostrand Reinhold Co., New York, pp.90-116.

Bullard, E.C., Freedman, C., Gellman, H. and Nixon, J., 1950. 'The westward drift of the Earth's magnetic field'. Phil. Trans. Roy. Soc., **A243**, 67-92.

Creer, K.M., (1983). 'Computer synthesis of geomagnetic palaeo-secular variations'. Nature, **304**, 695-699.

Creer, K.M., 1985. 'Review of lake sediment palaeomagnetic data (Part I)'. Geophys. Surveys, 7, 125-160.

Creer, K.M., Tucholka, P. and Barton, C.E., 1983. Geomagnetism of baked clays and recent sediments, Elsevier, Amsterdam, 324pp.

Creer, K.M., Smith, G., Tucholka, P., Bonifay, E., Thouveny, N., and Truze, E., 1986. 'A preliminary palaeomagnetic study of the Holocene and Late Glacial sediments of Lac du Bouchet (Haute Loire), France'. Geophys. J. Roy. astr. Soc., **86**, 943-964.

Creer, K.M., Thouveny, N., Blunk, I. and Turner, G., 1988. 'Palaeomagnetic results back to 50000 years BP from 12m cores from Lac du Bouchet'. In preparation.

de Beaulieu, J.L. and Reille, M., 1987. 'Histoire de la vegetation Wurmienne et Holocene du Velay occidental (Massif Central, France): analyse pollinique comparee de trois sondages du Lac du Bouchet'. Documents du C.E.R.L.A.T., Memoire no.1, 113-132.

Denham, C.R., 1974. 'Counter clockwise motion of paleomagnetic directions 24000 years ago at Mono Lake, California'. J. Geomagn. Geoelect., 26, 487-498.

Dodson, R.E., 1979. 'Counter-clockwise precession of the geomagnetic field vector and westward drift of the geomagnetic non-dipole field'. J. Geophys. Res., 84, 637-644.

Gillot, J., Labeyrie, C., Laj, C., Valladas, G., Guerin, G., Poupeau, G. and Delibrias, G., 1979. 'Age of the Laschamp palaeomagnetic excursion revisited'. Earth Planet. Sci. Lett., 42, 444-450.

Guerin, G. and Valladas, G., 1980. 'Thermoluminescence dating of volcanic plagioclases'. Nature, 274, 462 -.

Hall, C.M. and York, D., 1978. '^{40}Ar/^{39}Ar ages of the Laschamp geomagnetic polarity reversal'. Nature, 274, 462 -.

Heller, F., 1980. 'Self-reversal of natural remanent magnetization in the Olby - Laschamp lavas. Nature, 284, 334 - .

Heller, F. and Petersen, N., 1982. 'The Laschamp excursion'. Phil. Trans. Roy. Soc., A306, 169 - .

Huxtable, J., Aitken, M.J. and Bonhommet, N., 1978. 'Thermoluminescence dating of sediment baked by the lava flows of the Chaine des Puys'. Nature, 275, 207-209.

Jacobs, J.A., 1984. 'Reversals of the Earth's Magnetic Field'. Adam Hilger, Ltd., Bristol, (see pp.87-91).

Lorius, C., Barkov, N.I., Jouzel, J., Korotkevich, Y.S., Kotlyakov, V.M. and Raynaud, D., 1988. 'Antarctic ice core: CO_2 and climatic change over the last climatic cycle'. EOS, 69, 681-684.

Lund, S.P. and Banerjee, S.K., 1979. Paleosecular variation from lake sediments'. Rev. Geophys. Space Phys., 17, 244-249.

Mackereth, F.J.H., 1958. 'A portable core sampler for lake deposits'. Limnol. Oceanogr., 3, 181-191.

Pons, A., de Beaulieu, J.L., Guenet, P. and Reille, M., 1987. 'Les enseignements de l'analyse pollinique des anciens lacs du Massif Central'. Documents du C.E.R.L.A.T., Memoire no.1, 97-111.

Reille, M. and de Beaulieu, J.L., 1986. 'The Velay Maars, key sites for Pleistocene chronology'. Zurich in press.

Reille, M. and de Beaulieu, J-L., 1988. 'La fin de L'Eemian et les interstades du Pre-Wurm mis pour la première fois en evidence dans le Massif Central francais par l'analyse pollinique'. C.R. Acad. Sci., in press.

Skiles, D.D., 1970. 'A method of inferring the direction of drift of the geomagnetic field from paleomagnetic data'. J. Geomagn. Geoelect., 22, 441-462.

Smith, G., 1985. 'Late glacial palaeomagnetic secular variations from France'. Ph.d. Thesis, University of Edinburgh, 115pp.

Smith, G. and Creer, K.M., 1986. 'Analysis of geomagnetic secular variations, 1000 - 3000 years b.p., Lac du Bouchet, France'. Phys. Earth Planet. Int., 44, 1-14.

Thouveny, N., 1987. 'Variations of the relative intensity of the geomagnetic field in western Europe in the interval 25-10 kyr BP as deduced from analyses of lake sediments'. Geophys. J. Roy. astr.Soc., **91**, 123-142.

Thouveny, N., Creer, K.M., and Blunk, I., 1988. 'Initial palaeomagnetic results for cores A-D from Lac due Bouchet, 0-125 Myr BP'. C.E.R.L.A.T., Memoire no.2, in preparation.

Tucker, P., 1979. 'Selective post-depositional alignment in a synthetic sediment'. Phys. Earth Planet. Int., **20**, 11-14.

Tucker, P., 1980a. 'A grain mobility model of post-depositional realignment'. Geophys. J. Roy. astr. Soc., **63**, 149-163.

Tucker, P., 1980b. 'Stirred remanent magnetization: a laboratory analogue of post-depositional alignment'. J. Geophys., **48**, 153-157.

Verosub, K.L. and Banerjee, S.K., 1977. 'Geomagnetic excursions and their paleomagnetic record'. Rev. Geophys. Space Phys., **15**, 145-155.

Vestine, E.H., Laporte, L., Lange, I., Cooper, C., and Hendrix, W.C., 1947. 'Description of the Earth's main magnetic field, 1905-1945'. Carnegie Institution of Washing Pub. **578**.

Yukutake, T. and Tachinaka, H., 1969. 'Separation of the Earth's magnetic field into drifting and standing parts'. Bull. Earthquake Res. Inst. Tokyo, **47**, 65-97.

DEVELOPMENTS IN CAVE SEDIMENT PALAEOMAGNETISM

Mark Noel and Laurence Thistlewood
Department of Geology
The University of Sheffield
Sheffield
S3 7HF

ABSTRACT. In the deep interior of caves, sediments and spelaeothems
are preserved in an environment of high humidity, slow weathering,
near-constant temperature and negligible bioturbation. Consequently,
the remanent magnetisation of these materials can yield high fidelity
records of the geomagnetic field which spans a far longer timescale
than is accessible by lake sediment coring. This paper describes
sampling techniques developed for cave sediment palaeomagnetism and
presents results recently obtained from three sites in Britain
together with a complete summary of world spelaeomagnetic data.

1. INTRODUCTION

Continuous palaeomagnetic records from lake sediment cores have
enabled detailed studies of Holocene geomagnetism and provide a
comparative check on historical and archaeomagnetic observations.
Since the pioneer work by Mackereth (1971), on a core from Lake
Windermere, research in this field has progressed rapidly to include
material, for example, from Europe, Australasia and the Americas. The
large dataset now enables analysis of spatial and temporal
characteristics of the palaeomagnetic field (Thompson 1984) and also
constrains theoretical models for the geodynamo.
    High-resolution geomagnetic timescales are becoming an important
component in comparative studies of Quaternary deposits and
environments. However, extending the lake sediment record much beyond
~13 ka has proved difficult for a number of reasons. There are
technical problems in designing equipment capable of recovering
longer, oriented core samples with minimal disturbance. Moreover, in
North European lake basins there is a strong possibility that older
sediments will be disturbed, discontinuous or completely absent as a
result of Late Glacial erosion.
    In contrast, cave sediments and spelaeothems are preserved under
conditions of high humidity and stable temperature with minimal
bioturbation and weathering, particularly in the deep interior zone.
Here, clastic sediments are composed of solution residues, externally

91

F. J. Lowes et al. (eds.), Geomagnetism and Palaeomagnetism, 91–106.
© *1989 by Kluwer Academic Publishers.*

derived fluvial material and aeolian grains transported from the entrance zone by phreatic (conduit) or vadose (stream) flow. Spelaeothems, on the other hand, are solid deposits formed by chemical precipitation from percolation or pore water with calcium carbonate being the predominant mineral (Ford and Cullingford 1978).

U^{238}/Th^{230} disequilibrium dating provides the main evidence for the age of spelaeothems and associated cave sediments. Spelaeothem dates in Britain cluster into groups whose mean ages coincide with the timing of warm intervals identified in surface terrestrial deposits and the marine oxygen isotope record (Atkinson et al. 1978). Some material is found to be older than the 350 ka limit of the Uranium-series method. A similar trend is seen in an analysis of global spelaeothem age data (Gordon et al. in press). The presence of reversed polarity remanence in some cave sediments and spelaeothems (Table I) points to ages beyond 720 ka.

These lines of evidence show that spelaeomagnetic data could extend existing secular variation chronologies into the Middle or Lower Quaternary and provide independent verification of reported geomagnetic 'events'. Furthermore, cave sediments provide a good opportunity to study the mechanics of detrital magnetisation in natural material since deposition often occurred in 'flume' environments where flow orientations and relative velocities are known (Noel 1986).

This paper will describe the techniques we have developed for use in cave sediment palaeomagnetism and summarise the results of some recent studies of British cave material.

2. FIELD METHODS

Sampling cave sediments and spelaeothems, particularly in the deep interior zone, presents an unusual set of problems. The equipment must be light, rugged, portable and waterproof and capable of obtaining accurately oriented specimens, sometimes in a confined space, as quickly as possible.

Unconsolidated sediments are sampled by pressing open-ended, 25 mm, chamfered cylinders into a clean surface which has been prepared with a taut-wire planing tool. Orientation is recorded using a spirit level and fluxgate compass to an accuracy of 1° and specimen depths logged with an electronic tape measure. Samples are labelled by scribing through a contrasting paint film, removed and then sealed end to end in storage tubes, so avoiding the need to fix endcaps.

Spelaeothem hand samples, when dry, can be oriented using a tripod table or by adapting the button method developed for archaeomagnetism (Clark et al. in press). However, when spelaeothems are wet or interbedded with sediment these methods become impractical and an alternative technique is to obtain an oriented mould of part of the specimen with fast-setting epoxy putty. The mould can later be keyed with the sample for reorientation and subsampling in the laboratory, usually by drilling. A more detailed description of these field methods and of a device for obtaining oriented cave sediment cores for palaeomagnetism are given in Noel (in press).

3. RECENT RESULTS FROM BRITAIN

Since 1978, palaeomagnetic investigations have been carried out at a
number of cave sites in Britain (Fig. 1). Most of this work is of a
preliminary nature and includes studies of the depositional remanent
magnetisation process, magnetic stratigraphy and correlation. The
following sections review briefly some of the results and conclusions.

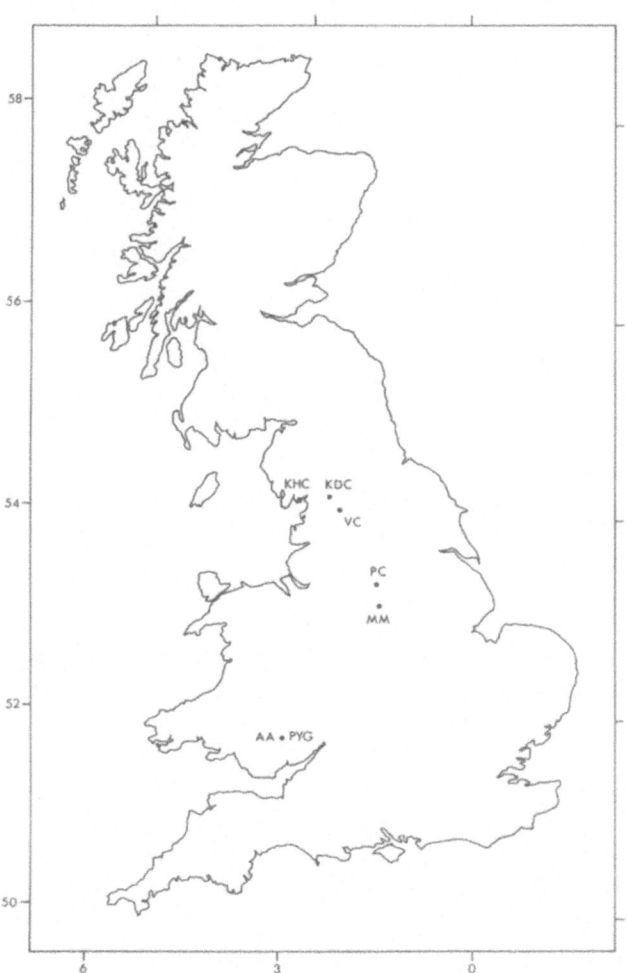

Figure 1. Cave sites in Britain where samples have been collected for
palaeomagnetic investigation. PC, Peak Cavern; KHC, Kirkhead Cave;
KDC, Kingsdale Cave; AA, Agan Allwedd; PYG, Pwyll y Gwynt; MM, Masson
Mine; VC, Victoria Cave. For references see Table I.

TABLE I. Research in cave sediment palaeomagnetism

Cave	Location	Material	Polarity	Reference
Victoria Cave, England	54°04'N, 2°16'W	Seds	N	Homonko (1978); Stober (1970)
Kingsdale Cave, England	54°10'N, 2°27'W	Stal	N	Latham et al. (1979)
Kirkhead Cave, England	54°13'N, 2°54'W	Seds	N	Gale et al. (1984, 1985)
Peak Cavern, England	53°21'N, 1°46'W	Seds	N	Noel (1986)
Masson Mine, England	53°08'N, 1°32'W	Seds	N & R	Noel (1987); Noel et al. (1984)
Pwyll y Gwynt, Wales	51°53'N, 3°09'W	Seds	N & R	Noel (1986)
Agen Allwedd, Wales	51°53'N, 3°09'W	Seds	N	Noel (1983)
Skjonghelleren Caves, Norway	62°17'N, 6°7'E	Seds	N & R	Lovlie & Sandnes (in press)
Arago Cave, France	Tautavel	Stal	N	de Lumley et al. (1984)
Tito Bustillo, Spain	Asturias	Seds	N	Creer & Kopper (1974); Kopper & Creer (1976)
Arbreda Cave, Spain	42°10'N, 2°42'E	Seds	N	Creer & Kopper (1976); Kopper & Creer (1976)
Hermit's Cave, Spain	42°16'N, 2°34'E	Seds	N	Creer & Kopper (1976)
Cova dets Alexandres, Spain	Majorca	Seds	N	Kopper & Creer (1973, 1976)
Mystery Cave, Canada	50°30'N, 126°45'W	Stal	N	Latham et al. (1987)
Bear Jaw Cave, Canada	49°43'N, 114°39'W	Stal	N	Latham et al. (1979, 1982)
Eagle Cave, Canada	49°39'N, 114°35'W	Stal	N	Latham et al. (1979, 1982); Latham (1981)

Location	Coordinates	Material	Polarity	References
Cascade Cave, Canada	49°31'N, 124°45'W	Stal	N	Latham et al. (1982); Latham (1981)
Mammoth Cave, U.S.A.	37°10'N, 86°08'W	Seds	N & R	Schmidt (1982)
Lilburn Cave, U.S.A.	California	Seds	N	Ulfeldt (1978)
Sotano del Arroyo, Mexico	22°05'N, 99°00'W	Stal	N	Latham et al. (1986)
Chiapas Cave, Mexico		Stal	N	Latham (1981)
Tunnel Hill Caves, China	25°21'N, 110°11'E	Stal & Seds	N & R	Williams (1987); Williams et al. (1986)
Coincidence Cavern, New Zealand	30°17'S, 175°8'E	Seds	N	Turner & Lyons (1986)
Punchbowl System, Australia	35°01'S, 148°33'E	Seds	N & R	Schmidt et al. (1984)
Komori-ana Cave, Japan	34°13'N, 131°19'E	Stal	N	Morinaga et al. (1985a,b)
Ryuga-do Cave, Japan	Kochi	Stal	N	Morinaga et al. (1985a)
Komori-ana Cave, Japan	34°13'N, 131°19'E	Stal	N	Morinaga et al. (1986)
Gujo-Hachiman Cave, Japan	Gifu	Stal	N	Inokuchi et al. (1981)
Jeita Cave, Lebanon	33°55'N, 35°38'E	Seds	N	Creer & Kopper (1976); Kopper & Creer (1976)
Makapansgat Site, S. Africa		Sed & Stal	N & R	Brock et al. (1977); McFadden et al. (1979)
Clearwater Cave, Sarawak	4°10'N, 115°10'E	Seds	N	Noel & Bull (1982); Smart et al. (1985)

ABBREVIATIONS:

Seds — Sediments
Stal — Stalagmite
N — Normal polarity remanence
R — Reversed polarity remanence

3.1 Agen Allwedd

Agen Allwedd ('Keyhole Cave') is a 26 km cave system which forms part of a more extensive complex of passages within Mynydd Llangattwg, Powys. Much of the cave is developed on a single level as a network of large, fossil phreatic tubes which contain an abundance of clastic sediments in the form of bedded clays, silts and sands, up to 5 m thick in places. These have been studied in detail by Bull (1981) who considers these deposits to have been derived from periglacial weathering of overlying sandstones followed by transport into the cave through joints and fissures. This premise is supported by the absence of any proximal-distal sediment assemblages.

The final phase of sedimentation is marked by the deposition of a 5-15 cm thick silt band which forms a conspicuous 'cap mud' unit throughout much of the cave. This layer contains 196 laminations which are possibly annual.

40 oriented samples of cap mud were obtained from the five localities shown in Fig. 2. At sites 2 and 3 specimens were collected on the sloping flanks of a mud cone which marks a point where sediment entered Main Passage through an open joint. The sediments contained a natural remanent magnetisation intensity in the range 4.8 to 10.2 x 10^{-3} Am^{-2}, composed of a single component as shown by alternating field demagnetisation. A Curie temperature determination indicated that the remanence was largely due to magnetite grains (Noel 1983).

The directions of remanence after partial demagnetisation in alternating field of 20-35 mT are shown in Fig. 2. The directions at sites 1, 4 and 5 were indistinguishable within their circular standard errors. The anomalous directions at sites 2 and 3 are due to 'bedding errors' caused by the remanence grains rolling on slopes of 8° and 29° respectively. When correction is made for this effect (King 1955) the mean vectors are in agreement with the remaining directions (Fig. 1). The palaeomagnetic results thus support the hypothesis based on lamination matching that cap mud deposition was simultaneous throughout Agen Allwedd (Bull 1981). The data also demonstrate that the simple grain rolling model of bedding error is valid for silts deposited on slopes up to 29°.

3.2 Masson Hill

West of the Derwent Gorge in Derbyshire, a complex system of disused lead and fluorspar mines worked ore deposits in the form of fissure, pipe and flat veins in the Carboniferous limestone (Warriner et al. 1981). Post-mineralization cavernisation has occurred by solutional enlargement of voids left by the hydrothermal fluids and these then acted as pathways for groundwater movement. This phase was probably initiated during the late Tertiary or early Pleistocene when incision of the Derwent Gorge commenced and the necessary hydraulic gradients became established (Ford and Worley 1977). Most of the phreatic network was subsequently filled with clastic deposits and the cave passages then abandoned by active streams. The narrow Clay Shaft in Old Jant Mine (Fig. 3) penetrates about 4 m of indurated sands, silts

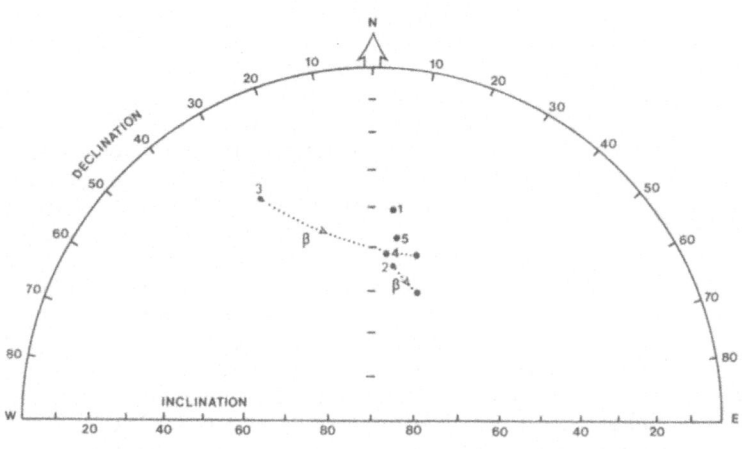

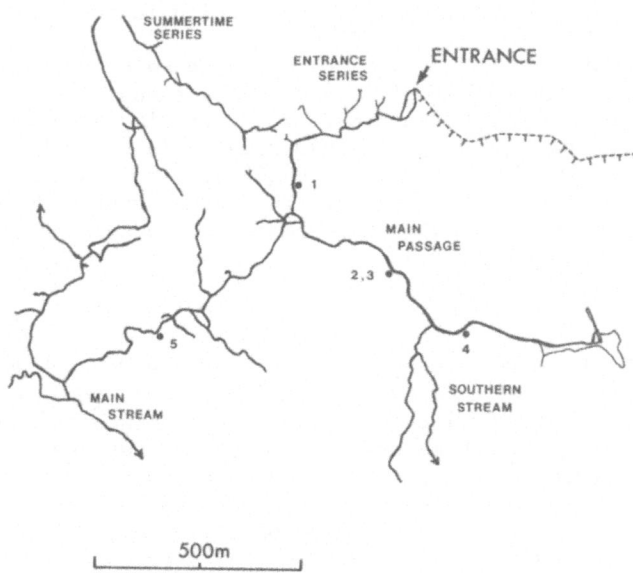

Figure 2. Location of five sediment sampling sites in Agen Allwedd, Powys. Sites 2 and 3 were on the flanks of a mud cone. Mean site vectors are shown after partial demagnetisation in an alternating field of 20 or 35 mT. The dotted lines show how a correction for 'bedding error' (B=1.4 x slope) has been applied to the remanence directions at sites 2 and 3. The mean vectors from the five sites were then indistinguishable at the 95% confidence level.

and clays within a pipe vein cavity situated 40 m above the modern
River Derwent.

In a pilot study, a vertical profile of 124 specimens were
collected down the section using a hydraulic jack to help insertion.
The directions of remanent magnetisation after partial demagnetisation
in maximum fields of 15 mT are shown in Fig. 3.

At depths below 250 cm the sediments record a period of reversed
geomagnetic polarity as shown by the negative inclinations and
oscillating, southerly declinations. Between 250 cm and 140 cm there
is a steady northward trend in declination while the inclination
remains negative. Finally, above 140 cm there is a rapid swing to
positive inclination accompanied by an eastward drift in declination
signifying a transition to the normal polarity field recorded by the
remainder of the section. The calculated virtual geomagnetic pole
path during the polarity transition was confined to the site longitude
(Noel 1987). This agrees with the observation that, in general, R   N
polarity transitions have virtual poles which are near-sided (Fuller
et al. 1979). More detail of the field behaviour during the
geomagnetic reversal is provided by a second set of 27 specimens
inserted in the depth range 91-174 cm (Fig. 3). Reversed polarity
magnetisation was also found in a 2 m clay section in the neighboring
Clay Cavern (Noel 1987).

The low concentration of pollen and the absence of dateable
spelaeothem has made the dating of this polarity transition difficult.
However, quartz grain surface features imply a proglacial origin,
although the timing must predate the Wolstonian cold stage because
Devensian tills are locally absent (Burek 1977). The Clay Shaft
sediments may thus be the first U.K. record of the Brunhes/Matuyama
polarity transition, although this hypothesis must await corroborative
evidence.

3.3 Peak Cavern

Peak Cavern is the longest cave system (5 km) in the Derbyshire karst
region. Development, within Carboniferous limestones, spans a
vertical range of 140 m and clastic sediments are common throughout
the cave. Although spelaeothems are not abundant, they are
occasionally interbedded with sediments and preliminary Uranium-series
determinations have provided ages of 1.1 to 59.0 ka for this material
(Ford 1986). The abundance of clastic deposits in a large and complex
cave system makes Peak Cavern an ideal pilot site for an intensive
palaeomagnetic and sedimentological study whose main aims are:
1. To use palaeomagnetic secular variation records and other
   indices to determine the relative timing and duration of
   sedimentation events in the cave and their geomorphic and
   climatic associations.
2. Validate and extend the existing U.K. lake sediment
   magnetostratigraphy from remanence measurements of sediments
   and spelaeothems.
3. To study the mechanics of the detrital remanent magnetisation
   process in passages where the orientation and relative

velocity of the flow vector are known. Hence derive an improved model to correct the remanence for rotations arising from flow and grain rolling and which provides palaeohydrological information.

In cave sedimentology, modern analogues provide comparative material to assist interpretation although there is much scope for further research. The initial strategy has therefore been to study in detail deposits along a single passage where there are likely to be only small variations in sedimentation history and lithology. The passage chosen, Maypole Inlet, is a fossil phreatic tube, about 15 m above the present day stream and almost completely filled with 1-2 m of fine grained sediments with intercalated spelaeothems. These deposits are now exposed in continuous section along an 80 m excavation (Fig. 4). A total of 826 oriented samples have been obtained at six locations spaced at 5 to 10 m intervals along the passage and the results from two of these are shown in Fig. 4.

The samples all contain a normal polarity natural remanent magnetisation of depositional origin, as shown by a 'primary style' magnetic fabrics and remanence 'bedding errors' on sloping surfaces (Hamilton and Rees 1970). Partial demagnetization in alternating magnetic fields removes a small viscous component leaving a very stable primary magnetization. Hence the remanence directions shown in Fig. 4 are thought to record the geomagnetic secular variation. Oscillations in susceptibility and remanence intensity probably reflect changes in magnetic mineralogy and grainsize.

Similar features can be identified in the magnetic parameters at each sampling location in Maypole Inlet and it is clear that sedimentation was broadly contemporaneous along this passage. However, conduit depositional history is revealed in more detail by carefully comparing the relative depths at which these features appear. In Fig. 4 it is apparent that a depositional break occurred in the PEAK3 section (relative to PEAK6) at a depth of 72 cm. This event corresponds to a marked change in grain size and a horizon of intraclasts, indicative of erosion and reworking.

Extending this simple approach further, it becomes possible to adjust the length of each time series by 'stretching', until the correlation of magnetic indices is maximised relative to a fixed 'master' timescale (e.g. PEAK3). Stacking of records will then establish an averaged time series which is both more complete and of higher fidelity. This work continues as part of the first comprehensive palaeomagnetic study of a British cave system.

4. ACKNOWLEDGEMENTS

The authors are grateful to T.D. Ford, R.P. Shaw, P.A. Bull and A. Kendal for helpful discussions and assistance under the field. The research is supported by the NERC and the Royal Society.

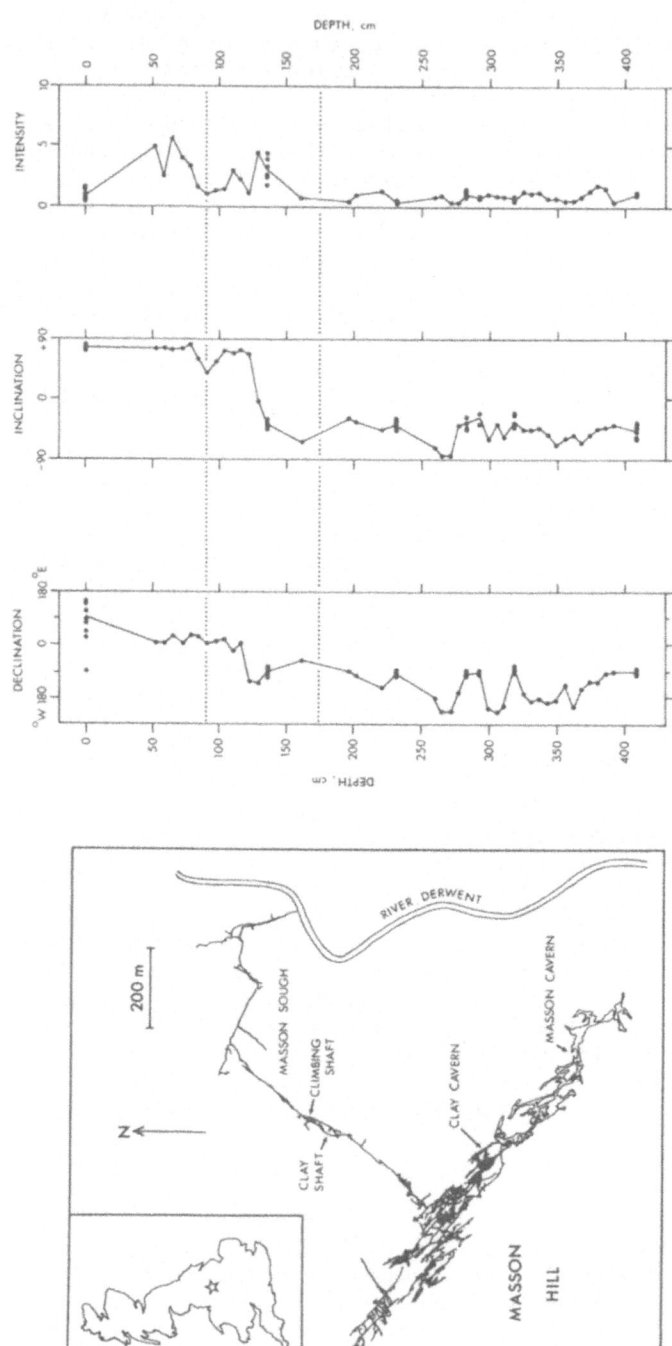

Figure 3. Masson Mine at Matlock, Derbyshire. Palaeomagnetic samples were obtained from a 4.5 m section of layered sands, silts and clay in the Clay Shaft which intersects a natural karst cavity. The directions and intensity of the remanence (after partial demagnetisation) record a geomagnetic polarity transition. This is revealed in more detail (opposite page) by results from a further 27 samples in the depth range dotted. Intensity in units of 10^{-5} Am^2kg^{-1}.

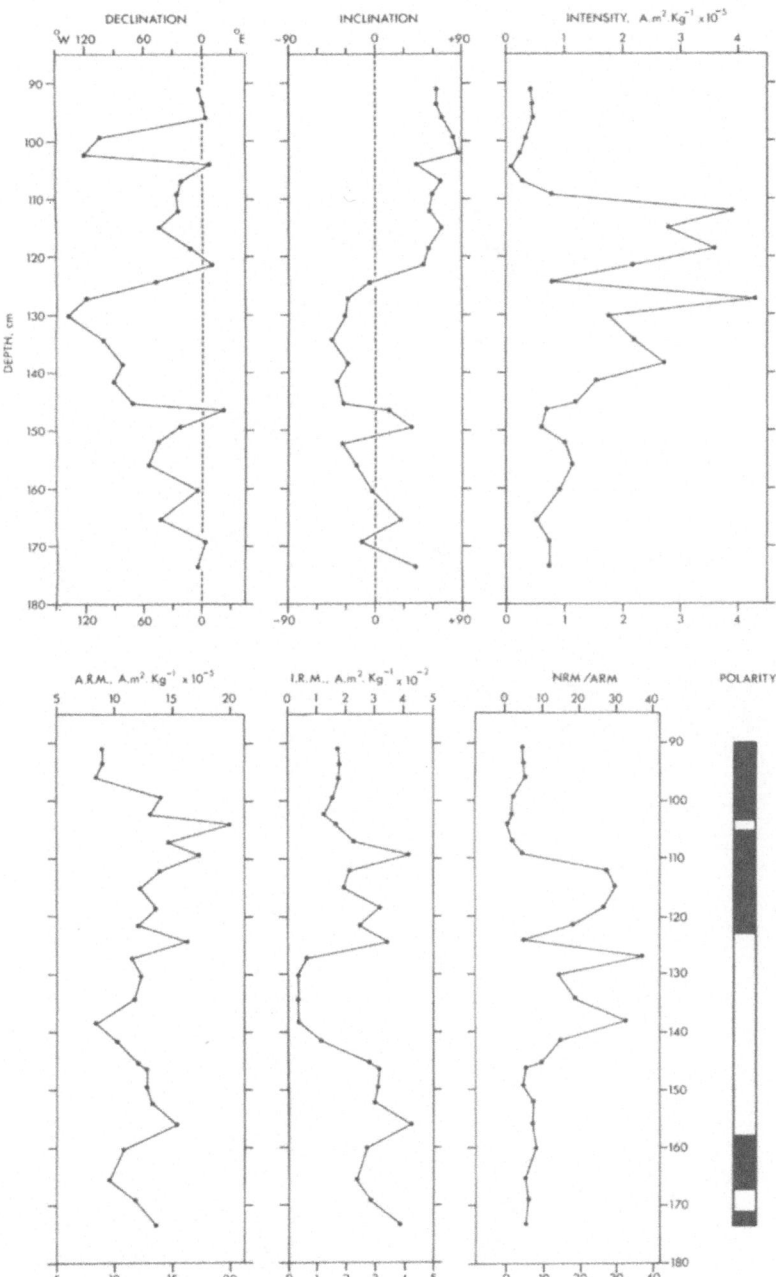

Figure 3 continued.

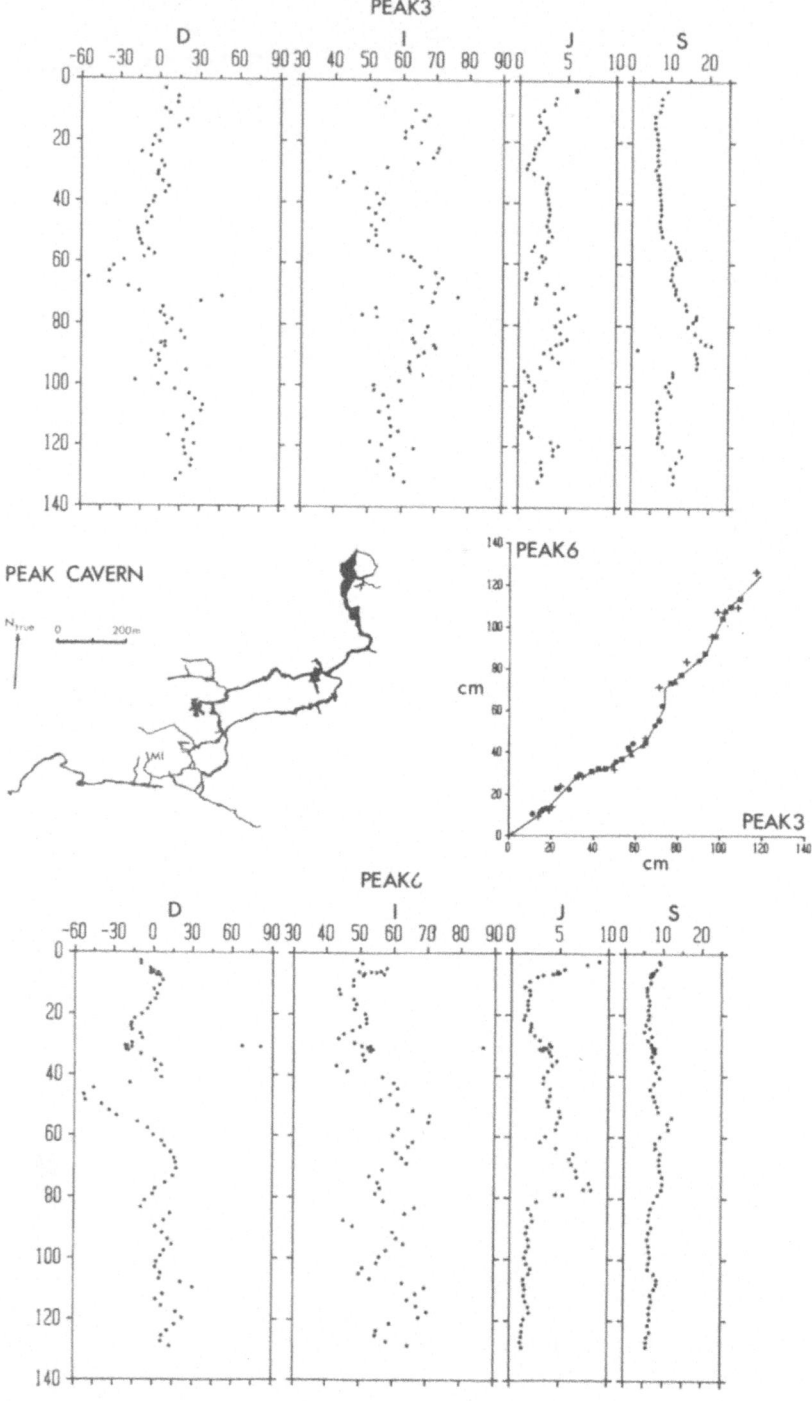

Figure 4. (See opposite for caption.)

Figure 4 (opposite). Palaeomagnetic results from two sediment sections in Maypole Inlet (MI), Peak Cavern. D=declination, I=inclination, J=intensity of natural remanent magnetisation in units of 10^{-6} Am^{2}kg^{-1}, S=susceptibility in units of 10^{-8} m^{3}kg^{-1}. Depth in cm. Directional data are smoothed with a two point running vector mean. Intensity and susceptibility data are 'smoothed' with a two point running arithmetic mean. Changes of sedimentation through section PEAK6 are shown relative to PEAK3 by plotting the corresponding depths of identifiable features in the magnetic parameters: declination (crosses), inclination (squares), intensity (circles) and susceptibility (triangles). The solid line shows the change in relative sedimentation rate.

5. REFERENCES

Atkinson, T.C., Harmon, R.S., Smart, P.L. and Waltham, A.C., 1978. 'Palaeoclimatic and geomorphic implications of Th^{230}/U^{234} dates on speleothems from Britain', Nature, 272, 24-28.

Brock, A., McFadden, P.L. and Partridge, T.C., 1977. 'Preliminary palaeomagnetic results from Makapansgat and Swartkrans', Nature,, 266, 249-250.

Bull, P.A., 1981. 'Some fine-grained sedimentation phenomena in caves', Earth Surf. Proc. Landforms, 6, 11-22.

Burek, C.V., 1977. 'The Pleistocene Ice Age and after', in: Ford, T.D., (ed.), Limestones and caves of the Peak District, GeoAbstracts, Norwich, pp87-128.

Clark, A.J., Tarling, D.H. and Noel, M., in press. 'Developments in archaeomagnetic directional dating in Great Britain', Jour. Arch. Sci.

Creer, K.M. and Kopper, J.S., 1974. 'Paleomagnetic dating of cave paintings in Tito Bustillo Cave, Asturias, Spain', Science, 186, 348-350.

Creer, K.M. and Kopper, J.S., 1976. 'Secular oscillations of the geomagnetic field recorded by sediments deposited in caves in the Mediterranean region', Geophys. J. R. astr. Soc., 45, 35-58.

De Lumley, H., Fournier, A., Park, Y.C., Yokoyama, Y. and Demouy, A., 1984. 'Stratigraphie du remplissage Pleistocene Moyen de la Caune de l'Arago a Tautavel etude de huit carottages effectues de 1981 a 1983', L'Anthropologie, 88, 5-18.

Ford, T.D., 1986. 'The evolution of the Castleton cave systems and related features, Derbyshire', Mercian Geologist, 10, 91-114.

Ford, T.D. and Cullingford, C.H.D., 1978. 'The science of speleology', Academic Press, London.

Ford, T.D. and Worley, N.E., 1977. 'Phreatic caves and sediments at Matlock, Derbyshire', Proc. 7th Int. Speleol. Congr., Sheffield, 194-196.

Fuller, M., Williams, I. and Hoffman, K.A., 1979. 'Palaeomagnetic records of geomagnetic field reversals and the morphology of the transitional fields', Rev. Geophys. Space Phys., 17, 179–203.

Gale, S.J., Hunt, C.O. and Southgate, G.A., 1984. 'Kirkhead Cave: Biostratigraphy and magnetostratigraphy', Archaeometry, 26, 192–198.

Gale, S.J., Hunt, C.O. and Southgate, G.A., 1985. 'The stratigraphy of Kirkhead Cave, an Upper Palaeolithic site in Northern England', Proc. Prehistoric Soc., 51, 283–304.

Gordon, D., Smart, P.L., Ford, D.C., Andrews, J.N., Atkinson, T.C., Rowe, P.J. and Christopher, N.S.J., in press. 'The dating of Late Pleistocene interglacial periods in the United Kingdom from speleothem growth frequency', Quat. Res.

Hamilton, N. & Rees, A.I., 1970. 'The use of magnetic fabric in palaeocurrent estimation', pp.445–464 in Runcorn, S.K. (ed.) Palaeogeophysics, Academic Press, London.

Homonko, P., 1978. 'A palaeomagnetic study of lake and cave sediments in Britain', unpublished M.Sc. Thesis, University of Newcastle upon Tyne.

Inokuchi, H., Morinaga, H. and Yaskawa, K., 1981. 'Preliminary report on palaeomagnetism of cave deposit', J. Geomag. Geoelectr., 33, 325–327.

King, R.F., 1955. 'The remanent magnetism of artificially deposited sediments', Mon. Not. R. astr. Soc. geophys. Suppl., 7, 115–134.

Kopper, J.S. and Creer, K.M., 1973. 'Cova dets Alexandres, Majorca: Palaeomagnetic dating and archaeological interpretation of its sediments', Caves and Karst, 15, 13–18.

Kopper, J.S. and Creer, K.M., 1976. 'Palaeomagnetic dating and stratigraphic interpretation in archaeology', MASCA Newsl., 12, 1–3, University of Pennsylvania.

Latham, A.G., 1981. In: Beck, B.F. (ed.) Proceedings of the 8th International Congress of Speleology, Kentucky, USA.

Latham, A.G., Schwarcz, H.P. Ford, D.C. and Pearce, G.W., 1979. 'Palaeomagnetism of stalagmite deposits', Nature, 280, 383–385.

Latham, A.G., Schwarcz, H.P., Ford, D.C. and Pearce, G.W., 1982. 'The palaeomagnetism and U–Th dating of three Canadian speleothems: evidence for the westward drift, 5.4–2.1ka BP', J. Earth Sci., 19, 1985–1995.

Latham, A.G., Schwarcz, H.P. and Ford, D.C., 1986. 'The paleomagnetism and U–Th dating of Mexican stalagmite, DAS2', Earth Planet. Sci. Lett., 79, 195–207.

Latham, A.G., Schwarcz, H.P. and Ford, D.C., 1987. 'Secular variation of the Earth's magnetic field from 18.5 to 15.0ka BP, as recorded in a Vancouver Island stalagmite, Can.', J. Earth Sci., 24, 1235–1241.

Løvlie, R. and Sandnes, A., in press. 'Palaeomagnetic excursions recorded in Mid–Weichselian cave sediments from Skjonghelleren, Valderoy, N. Norway', Geophys. J. R. astr. Soc.

Mackereth, F.J.H., 1971. 'On the variation in direction of the horizontal component of remanent magnetisation in lake sediments', Earth planet. Sci. Lett., 12, 332–338.

McFadden, P.L., Brock, A. and Partridge, T.C., 1979. 'Palaeomagnetism and the age of the Makapansgat hominid site', Earth Planet Sci. Lett., 44, 373–382.

Morinaga, H., Inokuchi, H., Yaskawa, K., Ikeya, M., Miki, T. and Kusakabe, M., 1985a. 'Paleomagnetism, paleoclimatology and ESR dating of stalagmite deposits', In: Ikeya, M. and Miki, T. (eds.) ESR dating and dosimetry, IONICS, Tokyo.

Morinaga, H., Inokuchi, H. and Yaskawa, K., 1985b. 'Paleomagnetism and paleotemperature of a stalagmite', J. Geomag. Geoelectr., 37, 823–828.

Morinaga, H., Inokuchi, H. and Yaskawa, K., 1986. 'Magnetization of a stalagmite in Akiyoshi Plateau as a record of the geomagnetic secular variation in West Japan', J. Geomag. Geoelectr., 38, 27–44.

Noel, M., 1983. 'The magnetic remanence and susceptibility anisotropy of cave sediments from Agen Allwedd, South Wales', Geophys. J. R. astro. Soc., 72, 557–570.

Noel, M., 1986. 'The palaeomagnetism and magnetic fabric of sediments from Peak Cavern, Derbyshire', Geophys. J. R. astr. Soc., 84, 445–454.

Noel, M., 1986. 'The palaeomagnetism and magnetic fabric of cave sediments from Pwyll y Gwynt, South Wales', Phys. Earth planet. Inter., 44, 62–71.

Noel, M., 1987. 'The magnetostratigraphy of cave sediments in Masson Hill, Derbyshire', Proc. Yorks. geol. Soc., 46, 193–201.

Noel, M., in press. 'Field equipment and techniques for cave sediment palaeomagnetism', J. Geol.

Noel, M. and Bull, P.A., 1982. 'The palaeomagnetism of sediments from Clearwater Cave, Mulu, Sarawak', Trans. Br. Cave Res. Ass., 9, 134–141.

Noel, M., Shaw, R.P. and Ford, T.D., 1984. 'A palaeomagnetic reversal in Early Quaternary sediments in Masson Hill, Matlock, Derbyshire', Mercian Geologist, 9, 235–242.

Schmidt, V.A., Jennings, J.N. and Bao, H., 1984. 'Dating of cave sediments at Wee Jasper, New South Wales, by magnetostratigraphy', Australian J. earth Sci., 31, 361–370.

Smart, P.L., Bull, P.A., Rose, J., Laverty, M., Friederich, H. and Noel, M., 1985. 'Surface and underground fluvial activity in the Gunung Mulu National Park, Sarawak', In: Douglas, I. and Spencer, T. (eds.), Environmental change and tropical geomorphology, Allen and Unwin.

Stober, J.C., 1978. 'Palaeomagnetic secular variation studies of Holocene lake sediments', Unpublished PhD Thesis, University of Edinburgh.

Thompson, R., 1984. 'Geomagnetic evolution: 400 years of change on planet Earth', Phys. Earth Planet. Inter., 36, 61–77.

Turner, G.M. and Lyons, R.G., 1986. 'A palaeomagnetic secular variation record from c.120000 yr-old New Zealand cave sediments', Geophys. J. R. astr. Soc., 87, 1181-1192.
Ulfeldt, S.R., 1978. 'Regional geomagnetioc variations as a dating and correlative tool in cave sedimentology: preliminary results from Lilburn Cave, California', Bull. Nat. Speleol. Soc., 40, 83.
Warriner, D., Willies, L. and Flindal, R., 1981. 'Ringing Rake and Masson Soughs and the mines on the east side of Masson Hill, Matlock', Bull. Peak Dist. Mines hist. Soc., 8, 65-102.
Williams, P.W., 1987. 'Geomorphic inheritance and the development of tower karst', Earth surface proc. landforms, 12, 453-465.
Williams, P.W., Lyons, R.G., Wang, X., Fang, L. and Bao, H., 1986. 'Interpretation of the palaeomagnetism of cave sediments from a karst tower at Guilin', Carsologica Sinica, 6, 119-125.

EVIDENCE FOR WAVE PROPAGATION IN THE HOLOCENE PALAEOMAGNETIC FIELD

V. Lee Hagee and Peter Olson
Department of Earth and Planetary Sciences
The Johns Hopkins University
Baltimore, MD 21218
U.S.A.

ABSTRACT. Analysis of Holocene lake sediment palaeomagnetic records
to 13,000 years BP from nine sites on four continents reveals that
prominent 1000-4000 year oscillations in declination and inclination
can ·be modelled as propagating wave-like disturbances superimposed on
an axial dipole field. Single site modelling and cross-correlation
analyses among intracontinental sites yield apparent propagation
directions with poleward components in both hemispheres, with phase
speeds in the range 0.03-$0.06°$/yr. We present results of numerical
calculations of non-linear $\alpha^2\omega$-dynamos, which exhibit the principal
features inferred from the palaeomagnetic data, including poleward
propagating quadrupolar dynamo waves coexisting with a dc dipolar
field.

Time series records of Quaternary palaeomagnetic secular
variation (PSV) which have been derived from lake sediment cores at
many locations are dominated by 1000-4000 year long oscillations in
both inclination and declination. This oscillatory character, along
with the phase relationships between records from sites located near
one another, suggests that the source of these non-dipole field
fluctuations consists of propagating wave-like disturbances which
originate in the outer core. In other papers [Olson and Hagee, 1987;
Hagee and Olson, 1988] we have suggested that these disturbances may
represent dynamo waves. To examine the validity of the propagating
wave interpretation, in section 1 we analyze time series records of
Holocene palaeomagnetic secular variation to determine propagation
characteristics such as propagation direction and phase velocity. In
section 2 we present results of numerical calculations on dynamo wave
propagation in the outer core.

1. DATA ANALYSIS

We have analyzed palaeomagnetic secular variation records retrieved
from lake sediment cores at sites located on four continents – North
America, South America, Eurasia, and Australia. The goal of the
analysis has been to identify and characterize forms of systematic

F. J. Lowes et al. (eds.), Geomagnetism and Palaeomagnetism, 107–121.
© *1989 by Kluwer Academic Publishers.*

behaviour of secular variation using multi-site cross-correlation
analyses and single site modelling of the records.

Cross-correlation analyses allow us to determine the degree of
correlation between any two records, as well as their phase
relationship. In addition, we have devised a modelling procedure
which allows us to estimate PSV drift (or propagation) directions from
data at a single site, since the present distribution of high quality
records is not sufficient to provide a fully global picture of
palaeomagnetic secular variation. In this model, which has been
described elsewhere [Olson and Hagee, 1987; Hagee and Olson, 1988],
the surface field disturbances are represented as propagating plane
waves originating in the outer core. The disturbance field Fourier
components at the Earth's surface are given by

$$B_x = -A\cos(a)\sin(nt + \phi)$$

$$B_y = -A\sin(a)\sin(nt + \phi)$$

$$B_z = A\cos(nt + \phi) \tag{1.1}$$

where A is the disturbance amplitude, a is the local propagation
direction, n is the apparent frequency (the frequency as seen at the
Earth's surface), and ϕ is a phase angle. For this model, we have
chosen to represent the total magnetic field as consisting of two
parts, a fluctuating disturbance field as given by the Fourier
components in equation (1.1) and a steady axial dipole field with
northern and vertical components given by

$$X(\theta) = B_o \sin \theta$$

$$Z(\theta) = 2B_o \cos \theta \tag{1.2}$$

where   is the colatitude of the site and B_o is the amplitude of the
axial dipole at the equator, 3.07×10^{-5} T. The inclination of the
total field is then given by

$$I = \tan^{-1} (Z + B_z)/[(X + B_x)^2 + B_y^2]^{\frac{1}{2}} \tag{1.3}$$

and the total declination is given by

$$D = \tan^{-1}[B_y/(X + B_x)]. \tag{1.4}$$

These equations give the time variation of the field at a single site
due to a single Fourier component. More complex sources can be
simulated by replacing the individual Fourier components in equations
(1.3) and (1.4) with the sum of two or more Fourier components, each
with its own propagation azimuth, amplitude, frequency, duration and
phase angle.

Palaeomagnetic secular variation records from three sites in North
America have been analyzed (Figure 1A): Fish Lake (42.5°N, 241.5°E;

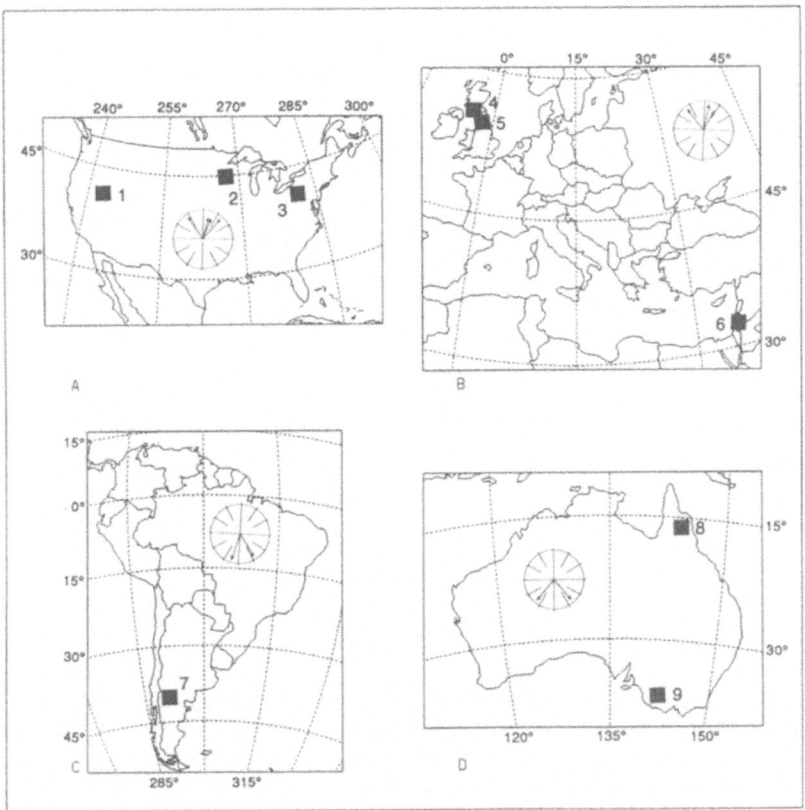

Figure 1. Location of sites from which palaeomagnetic secular
variation records have been analyzed. A. North America: 1-Fish Lake
(Oregon, USA), 2-Lake St. Croix (Minnesota, USA) and 3-Lake LeBoeuf
(Pennsylvania, USA); B. Eurasia: 4-Loch Lomond (Scotland), 5-Lake
Windermere (England) and 6-Lake Kinneret (Israel); C. South America:
7-Lake Moreno (Argentina); and D. Australia: 8-Lake Eacham
(Queensland) and 9-Lake Keilambete (Victoria). Insets show the
propagation directions derived from best-fitting wave models (long
arrows) and from phase delays (short arrows).

Verosub et al., 1986), Lake St. Croix (45.0°N, 267.2°E; Lund and
Banerjee, 1985), and Lake LeBoeuf (41.9°N, 280.0°E; King et al.,
1988). The North American records (Figure 2) exhibit a number of
features characteristic of the records from all four continents.
Records from sites located on the same continent tend to be similar
with respect to frequency, amplitude, and general wave shape; records
from widely separated sites are less similar. This tendency is
confirmed by cross-correlation analysis. In general, two dominant

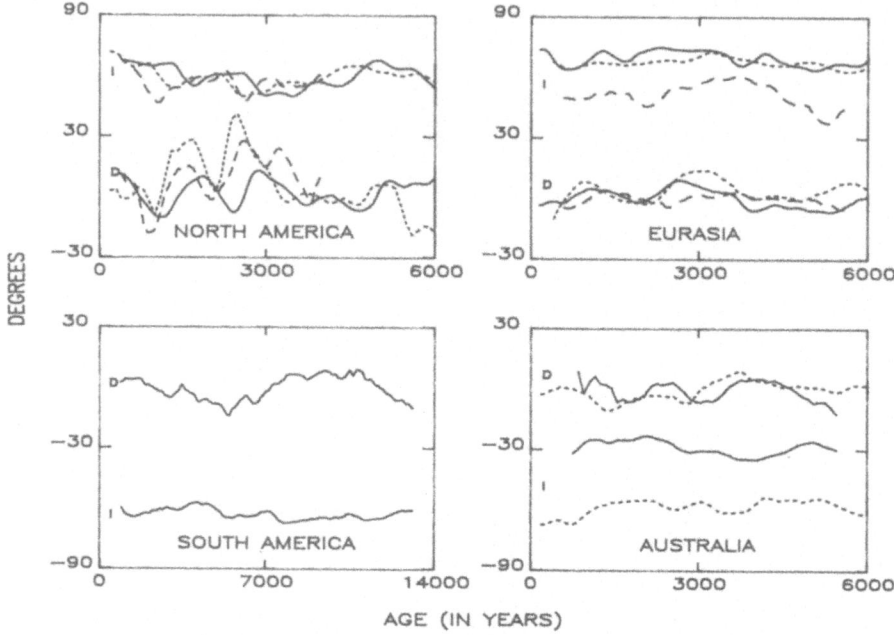

Figure 2. Inclination (I) and declination (D) palaeomagnetic secular variation records from the sites analyzed. North America: Fish Lake (solid lines), Lake St. Croix (dotted lines) and Lake LeBoeuf (dashed lines). Eurasia: Loch Lomond (solid lines), Lake Windermere (dotted lines) and Lake Kinneret (dashed lines). South America: Lake Moreno (solid lines). Australia: Lake Eacham (solid lines) and Lake Keilambete (dotted lines).

frequencies can be identified in any individual record; the higher frequency is about two times the lower frequency. Typically, a record contains only a few oscillations at any one frequency and a change in frequency content is common in records extending back more than 6 to 8 thousand years.

Comparing the North American records, we note a phase difference between the Fish Lake record and the others which suggests a time difference of about 350 years, with features in the Fish Lake record (from the western U.S.) preceding the related feature in records from sites further east. If the dating of these records is presumed to be accurate, the features of late Holocene palaeomagnetic secular variation in North America appear to be drifting with an eastward component over the last 3000 years. This is contrary to the generally westward drift that is thought to dominate secular variation.

Cross-correlation analysis generally confirms the overall similarities between the records. Averaging the phase delays at which the maximum cross-correlations of inclination and of declination occur for each comparison confirms the eastward drift of the features and suggests a phase delay of 350 and 400 years between records from Fish Lake and the other two sites. From these phase delays, an apparent propagation (or drift) direction of N21°E and a phase velocity of 0.034°/year can be inferred.

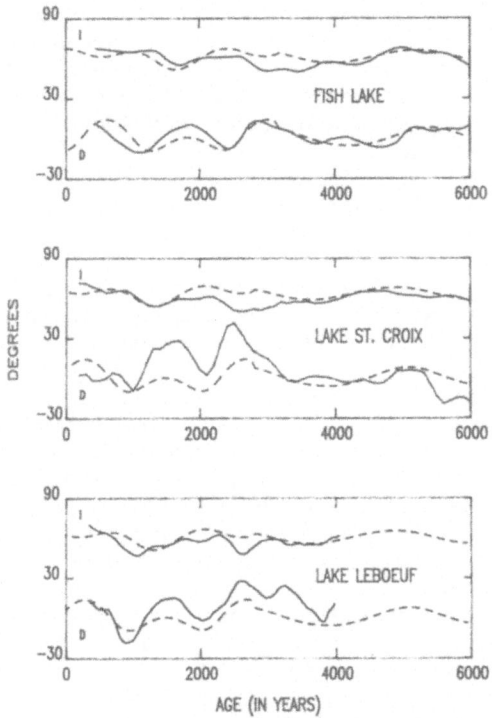

Figure 3. Best fitting plane wave models of the North America PSV records. Solid lines are data, dashed lines are plane wave models; inclination is indicated by I, declination by D.

The North American data can be modelled by a disturbance consisting of two propagating plane waves (Figure 3). One wave, which is evident throughout the duration of the record, has a period of 2400 years and a N35°W propagation direction. During the most recent 2800 years, the best fitting model is obtained by adding a second wave with a period of 1200 years propagating along an azimuth of N40°E. The phase difference between the best-fitting models of the Fish Lake record and those of the other two sites represents 330 years. From the phase delays an apparent propagation direction of N14°E and a phase velocity of 0.027°/year are inferred. The analysis suggests that the PSV record in North America over the past 6000 years may be represented by northward propagating disturbances (Figure 1a). The plane wave models contain both eastward and westward components, but these are smaller than the northward component. The propagation directions inferred from phase delays as determined both by

cross-correlation analysis and by modelling at individual sites
confirm the strong northward component.

Data from three sites in western Eurasia have been analyzed
(Figure 1B) - Loch Lomond (56.0°N, 355.0°E; Turner and Thompson,
1979), Lake Windermere (54.5°N, 356.2°E; Turner and Thompson, 1981),
and Lake Kinneret (32.4°N, 35.7°E; Thompson et al., 1985). Comparing
these records in Figure 2, it is apparent that although the phase
difference between the Loch Lomond and Lake Windermere records is
negligible, the phase delay between the Lake Kinneret record and the
others represents a time difference of about 400 years, with events
occurring first at Lake Kinneret.

The cross-correlation analysis indicates that the correlation
between the Eurasian records is quite good. The analysis suggests a
small phase delay between the U.K. sites, but it falls within the
range of statistical error of the dating procedure and so remains
uncertain. The analysis does, however, indicate a phase delay of 200
and 300 years between the Lake Kinneret records and the others. These
phase delays suggest an apparent propagation direction of N27°E and a
phase velocity of 0.062°/year.

Single site modelling of the Eurasian records indicates that
these data can be represented by a disturbance consisting of two
waves, both of which are present throughout the duration of the
record. One wave has a period of 3200 years and a N10°E propagation
direction. The second has a 1600 year period and is propagating along
an azimuth of N40°W. The phase delay between the Lake Kinneret record
and the other two is 400 years. The phase delay between the U.K.
sites is not, however, resolved.

Propagating plane waves representations of the Eurasian data and
cross-correlation analysis of these records indicate that, like the
North American data, the palaeomagnetic secular variation of western
Eurasia over the last 6000 years may be characterized by northward
propagating disturbances, which include both eastward and westward
components of motion.

We have analyzed southern hemisphere Holocene secular variation
data from two sites in Australia (Figure 1D) - Lake Eacham (17.3°S,
145.6°E; Constable and McElhinny, 1985) and Lake Keilambete (38.2°S,
142.9°E; Barton and McElhinny, 1981). The records from these lakes
are generally similar (Figure 2) except for a phase delay which
suggests a time difference of about 200 years, with events occurring
first at Lake Eacham. Cross-correlation analysis indicates that these
records do not correlate as well as the northern hemisphere records,
but it does confirm the 200 year time delay between the records.
Since only two records were compared, the propagation characteristics
could not be determined using phase delays. The phase delay resolved
between the records does suggest a southward propagation direction,
but this cannot be confirmed without data from additional sites.

The Australian data was modelled by a disturbance consisting of
two plane wave components. One wave, which is present throughout the
duration of the records, has a period of 2800 years and is propagating
along an azimuth of S40°E. The more recent part of the records can be
modelled by adding to this a second wave with a period of 1400 years

and a S40°W propagation direction. The phase difference between the
best fitting waves represents 230 years with events in the north
preceding those in the south. Unlike the generally northward
propagation directions inferred from northern hemisphere data, late
Holocene palaeomagnetic secular variation in Australia may consist of
southward propagating wave-like disturbances.

Records from one additional southern hemisphere site have been
analyzed - Lake Moreno (41.0°S, 288.5°E; Creer et al., 1983) in
Argentina (Figure 1C). This record extends back 13,000 years and is
dominated by a long period oscillation that is particularly prominent
in declination. The Lake Moreno record can be modelled by a
disturbance consisting of two waves, both of which persist through the
entire record. One wave has a period of 8000 years and is propagating
along an azimuth of S35°E; added to this is a second wave with a
period of 3100 years and a S15°W propagation direction. This suggests
that the Holocene PSV record of this region may consist of generally
southward propagating features, similar to what was found in
Australia.

Analysis and modelling of the data from four continents allows us
to make some observations about the nature of Holocene PSV.
Propagation characteristics that have been determined are summarized
in Table I and below.

TABLE I

Location	Propagation Directions from model wave components		from phase delays		Phase Velocity
North America	N35°W	N40°E	N14°E	N21°E	0.03°/year
Eurasia	N10°E	N40°W	---	N27°E	0.06°/year
Australia	S40°E	S40°W	---	---	---
South America	S35°E	S15°W	---	---	---

1. Holocene palaeomagnetic secular variation appears to consist
of poleward propagating wave-like disturbances. As shown in Figure 1,
the propagation directions derived from the best-fitting plane wave
models and from phase delays resolved by both plane wave modelling and
by cross-correlation analysis suggest that the motion of PSV features
has poleward components in both hemispheres. In both hemispheres all
propagation directions are within ±40° of the pole.

2. Both eastward and westward components of propagation are
evident during the Holocene.

3. The records suggest a transient nature for these
disturbances, with individual wave components persisting for only a
few cycles.

2. A MODEL OF DYNAMO WAVE SECULAR VARIATION

In another paper [Hagee and Olson, 1988] we point out that these characteristics are suggestive of dynamo wave propagation, as first proposed for the solar dynamo by Parker [1955]. There are some important differences in this case, however, which make it impossible to apply results from solar dynamo theory directly to secular variation of the terrestrial field. In the first place, the solar dynamo cycle involves migration of poloidal field structures from mid-latitudes towards the equator in both hemispheres [Parker, 1979], opposite to the propagation directions inferred in our study. Secondly, the entire solar field is ac (oscillatory), whereas in the terrestrial field the dominant dipole term is dc on thousand year timescales, and only the non-dipole components show ac behaviour. Thus it is important to establish from first principles that dynamo waves can coexist simultaneously with a strong dc dipolar field under conditions likely to be found in the Earth's core. It is far from obvious a priori that such coexistence is possible. Indeed, based on results from the linear kinematic theory of $\alpha\omega$-dynamo waves, this might appear unlikely. In the linear theory, the growth rate of each mode is independent of, and in general is different from, all others [Moffatt, 1978]. It would be very fortuitous (and in fact very artificial) if the growth rates for the dc dipole and ac nondipole fields were exactly equal, and yet, in the context of linear kinematic theory, any other arrangement would preclude their coexistence for times comparable to the diffusion timescale. However, linear kinematic theory is valid only for weak magnetic fields, for which the Lorentz force is negligible. This is certainly not the case in the Earth's core and consequently it is not sufficient to use linear theory to judge whether or not such an arrangement is possible.

Because non-linear effects associated with the Lorentz force are critical in determining the equilibrium core field configuration, we have modified the usual kinematic dynamo equations by introducing into the induction (α-effect) tensor an inverse power-law dependence on the local magnetic energy density. We have integrated the resulting non-linear equations numerically in an axisymmetric spherical domain representing the Earth's core. Based on previous experience with kinematic dynamos, it is known that an α-effect acting on both toroidal and poloidal field components can produce dc dipolar fields [Roberts, 1972; Busse, 1983] and that strong toroidal shear flows acting on a poloidal field, in conjunction with an α-effect, results in $\alpha\omega$ dynamo waves. It is therefore reasonable to presume that a non-linear dynamo with both α^2 and $\alpha\omega$ induction can sustain both dc and ac fields simultaneously.

Good arguments can be made that both effects are active in the Earth's core. It is generally agreed that free convection is occurring throughout much or all of the outer core, although it may be compositionally rather than thermally driven. Compositional buoyancy forces are produced near the inner core boundary as a product of inner core solidification [Loper, 1978; Gubbins et al., 1979; Verhoogen, 1980]. The inner core is enriched in iron relative to the outer core

liquid [Jephcoat and Olson, 1987] and as it crystallizes a residual
liquid is produced, enriched in light elements such as sulphur or
oxygen. This liquid is gravitationally unstable near the inner core
boundary, and in the process of rising into and mixing with the denser
outer core liquid, a large amount of kinetic energy is produced in the
form of turbulent motion, plus inertial and internal gravity waves.
This buoyancy-driven motion, acting on large scale poloidal and
toroidal magnetic fields, results in an α-effect induction,
particularly in the deeper parts of the outer core. Loper [1978] has
termed this the gravitational dynamo. Calculations indicate that it
may be the dominant form of convection in the core [Stevenson, 1981;
Loper and Roberts, 1983].

There are a variety of possible mechanisms for generating the
toroidal flows necessary for the ω-effect. One proposal is that they
represent thermal winds driven by lateral temperature variations just
beneath the core-mantle boundary, products of core-mantle thermal
interaction [Bloxham and Gubbins, 1987; Olson and Hagee, 1987]. This
is an attractive possibility because it implies the source of toroidal
flow is to a degree independent of the source of convection, and
consequently secular variation (which is strongly influenced by
toroidal flow) is partly independent of the generation mechanism for
the main dc field. It suggests a core dynamo model with secular
variation originating in a field of toroidal flow just beneath the
mantle boundary, and with the source of the main dc field,
compositionally-driven convection, concentrated closer to the inner
core boundary.

We have constructed a numerical, mean field, model of the core
dynamo which incorporates the basic physical processes just described.
In a system of spherical coordinates (r, θ, ϕ) the large scale or
macroscopic magnetic field can be written in dimensionless form as a
sum of axisymmetric toroidal and poloidal parts as

$$\vec{B} = \hat{\phi} B_T + \nabla \times (\hat{\phi} B_p) \tag{2.1}$$

in terms of which the macroscopic induction equations are

$$\left(\frac{\partial}{\partial t} - \nabla^2 + \frac{1}{r^2 \sin^2 \theta} \right) B_T = \hat{\phi} \cdot \nabla \times (\vec{U} \times \vec{B} + \vec{E}) \tag{2.2}$$

and

$$\left(\frac{\partial}{\partial t} - \nabla^2 + \frac{1}{r^2 \sin^2 \theta} \right) B_p = \hat{\phi} \cdot (\vec{U} \times \vec{B} + \vec{E}) \tag{2.3}$$

where $\vec{U}$ is the macroscopic toroidal velocity field, and $\vec{E} = \langle \vec{u} \times \vec{b} \rangle$ is
the large scale EMF produced by interaction of small scale velocity
field $\vec{u}$ and magnetic field $\vec{b}$. Equations (2.1-2.3) are
nondimensionalized using the core radius r_c and the Ohmic diffusion
timescale $\mu \sigma r_c^2$ where σ is electrical conductivity and μ is

permeability. We express the mean EMF as proportional to the mean
magnetic field through an induction tensor α with the following form

$$E_i = \alpha_{ij} B_j = R_\alpha \frac{(1-r)\cos\theta}{1-r_{ic}} \frac{(\delta_{ij} - r_j r_j / r^2)}{1+(B_k B_k)^n} B_j \tag{2.4}$$

In this expression, R_α is the magnetic Reynolds number for the
α-effect, the cosine angular dependence is due to the influence of
rotation, the radial anisotropy is intended to model effects of weakly
stable density stratification produced by chemical differentiation,
while the radial distribution reflects the origin of the driving force
at the inner core boundary, r_{ic}. The inverse power law dependence on
the local magnetic energy density models the dynamic reduction of
convective velocities by the Lorentz force, and serves as the
non-linear mechanism necessary for field equilibration. Values of n
between 1 and 2 have been proposed [Jepps, 1975; Moffatt, 1978]. In
the calculations presented here, a value of unity has been adopted,
although experimentation with other values indicates that the gross
behaviour of the solution is not sensitive to this parameter in the
range $1 \leq n \leq 2$. For purposes of computation, we have scaled the
magnetic energy density in Elsasser number units, $B^2 \sigma / 2\Omega\rho$, where is
outer core density, and Ω is angular velocity of rotation. In the
limit of zero Elsasser number (weak magnetic field) equation, (2.4)
reduces to the α-effect for a linear kinematic dynamo. The large
scale velocity field consists of a toroidal shear flow concentrated
beneath the core-mantle boundary, of the form

$$\vec{U} = \hat{\phi} R_\omega r^m \sin\theta \tag{2.5}$$

with R_ω denoting the magnetic Reynolds number of the macroscopic flow.
    Boundary conditions reflect the difference in solidity and
electrical properties of the inner core, outer core, and mantle. We
assume the same conductivity in the inner and outer core, so equations
(2.1-2.5) can be applied directly to the inner core, with the magnetic
Reynolds numbers set to zero. Near the Earth's centre the field
components must behave as

$$B_r, B_\theta \to 0 \quad \text{as } r \to 0 \tag{2.6}$$

We assume the lower mantle to be an insulator, giving

$$B_T = 0 \quad, \quad r = 1 \tag{2.7}$$

The poloidal field must be continuous with an external potential field
at the core-mantle boundary. This is enforced by decomposing the
field at r=1 into spherical harmonic components, and requiring
continuity with the corresponding spherical harmonic terms in the
exterior potential field. In terms of the expansion

$$B_p = \sum_\ell A_\ell(r,t) P_\ell^1(\cos) \tag{2.8}$$

in which P_ℓ^1 is a Legendre polynomial of degree ℓ and order 1, continuity can be expressed as

$$\left(\frac{\partial}{\partial r} + \frac{\ell + 1}{1}\right) A_\ell = 0 \quad , \; r = 1 \qquad (2.9)$$

Equations (2.1-2.9) were solved numerically on finite difference grids with 40 angular and 25 radial points, using second order central differences and a predictor-corrector explicit time step. The exterior field was resolved up to degree $\ell=8$. We have benchmarked this method against analytical solutions for freely decaying fields, plus linear α^2- and $\alpha\omega$-dynamos [Moffatt, 1978]. It yields decay and growth rates within 2% of the analytically-derived values.

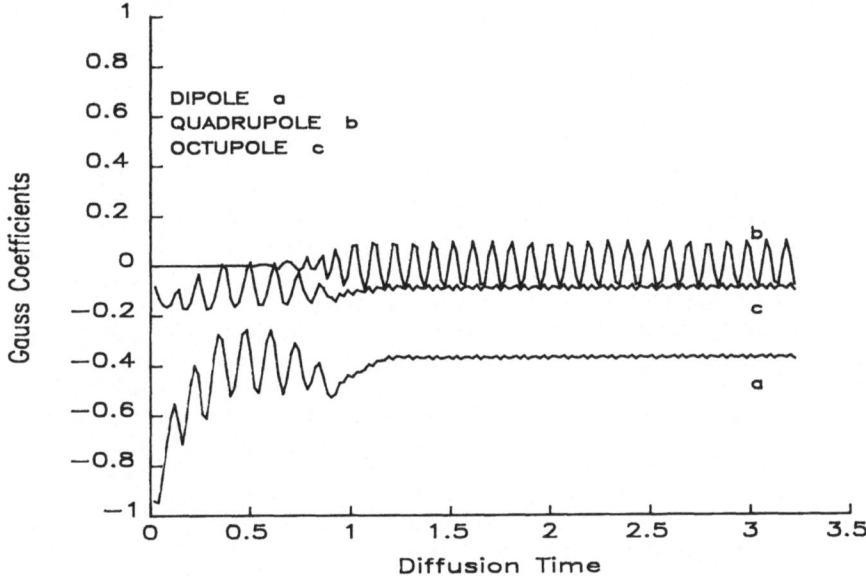

Figure 4. Time series of Gauss coefficients (unit 10^{-4} T) from the dynamo calculation described in the text. One time unit represents approximately 100,000 years.

Figure 4 shows time series of Gauss coefficients, the surface amplitude of the principal spherical harmonic terms, from a calculation with $R_\alpha = -300, R_\omega = -500$, and power law indices n=1 and m=6. (Negative R_ω corresponds to westward flow. Negative R_α is chosen so that the product $R_\alpha R_\omega$ is positive, the condition for poleward dynamo wave propagation.) The initial conditions consisted of equal strength, free decay dipole and toroidal fields. In this calculation, R_ω was chosen to give westward drift of the outermost core fluid relative to the mantle at a rate of 0.18°/year, similar in magnitude to the mean value inferred from the historical field

[Thompson, 1984]. The magnitude of R_α is chosen to give an equilibrium dipole moment close to the present day value. The time axis is dimensionless, with one unit corresponding to the Ohmic diffusion timescale. Based on an outer core conductivity of 6×10^5 mho/m [Stevenson, 1981] one unit corresponds to approximately 100,000 years.

The time series in Figure 4 clearly show a permanent secular variation in the quadrupole term, coexisting with a nearly stationary dipole/octupole family. After an initial transient adjustment period, both the dipole and the octupole equilibrate as dc fields, with only small fluctuations. The ratio of octupole to dipole amplitudes, g_3/g_1 is approximately 0.15, which is somewhat larger than the ratio inferred for the palaeomagnetic field over the past 5 Myr [Merrill and McElhinny, 1983; Lee and Lilly, 1986]. Numerical experiments with the α-effect concentrated near the outer core boundary, simulating thermal convection associated with heat loss to the mantle as the primary driving force, produced unacceptably large values of this ratio. In contrast to the dipole family, the quadrupole family is oscillatory (ac), with a periodicity of approximately 9500 years. This is longer than the periodicities observed in the Holocene palaeomagnetic record by a factor of 3–8, but the discrepancy can be reduced or eliminated by adopting different parameters, particularly the index m governing the radial distribution of toroidal flow, although we have not attempted to do this. The ratio of quadrupole to dipole terms in this calculation is in fact similar to the ratio of nondipole to dipole fields inferred from the Holocene records.

To demonstrate that the quadrupole oscillations in Figure 4 are indeed dynamo waves with the required propagation directions, we show in Figure 5 the evolution of the toroidal field intensity (left halves) and the lines of force of the poloidal field (right halves) through one complete oscillation, in steps of $\pi/4$. Disturbances in the poloidal field are generated at low latitudes, out of phase in each hemisphere. The disturbances propagate poleward and intensify, reaching a maximum intensity near $\pm 60°$ latitude, at which point they attenuate and are absorbed into the main dc field. The toroidal field is concentrated into propagating flux tubes. Its rms intensity, averaged over the core, varies over a range corresponding to magnetic energy density of between 1 and 4 Elsasser number units (1.5–3.0 mT approximately). In this calculation the dynamo waves are predominately quadrupolar. By varying the distribution of toroidal shear and the α-effect it is possible to generate waves which are predominantly hexadecapolar. We also note that in a fully three-dimensional simulation the waves would likely have an azimuthal dependence, and, because of transport by the toroidal flow, would have a component of westward motion, relative to the mantle, in addition to their poleward propagation.

Figure 5. Contours of toroidal field intensity B_T in intervals of $2(2\Omega\rho/\sigma)^{\frac{1}{2}}$ (left halves) and contours of poloidal function B_p in intervals of $0.1(2\Omega\rho r_c^2/\sigma)^{\frac{1}{2}}$ (right halves) for one complete oscillation in time series shown in Figure 4. Time advances from left to right, top to bottom, in increments of $\pi/4$. Circles indicate inner and outer core boundaries. Note the presence of poleward propagating dynamo waves.

References

Barton, C.E. and M.W. McElhinny, 1981. 'A 10,000 yr geomagnetic secular variation record from three Australian maars', Geophysics, J. Roy. Astron. Soc., **67**, 465-485.

Bloxham, J. and D. Gubbins, 1987. 'Thermal core-mantle interactions, Nature, **325**, 511-513.

Busse, F.H., 1983. 'Recent developments in the dynamo theory of planetary magnetism', Ann. Rev. Earth Planet. Sci., **11**, 214-268.

Constable, C.G. and M.W. McElhinny 1985. 'Holocene secular variation
    records from northeastern Australian lake sediments', Geophys.
    J. Roy. Astron. Soc., 81, 103–120.
Creer, K.M., D.A. Valencio, A.M. Sinito, P. Tucholka and J.F.A.
    Vilas, 1983. 'Geomagnetic secular variations 0–1400 yr B.P.
    as recorded by lake sediments from Argentina', Geophys. J. Roy.
    Astron. Soc., 74, 199–221.
Gubbins, D., T.G. Masters and J.A. Jacobs, 1979. 'Thermal evolution of
    the Earth's core', Geophys. J. Roy. Astron. Soc., 59, 57–99.
Hagee, V.L. and P. Olson, 1988. 'An Analysis of Paleomagnetic Secular
    Variation in the Holocene', submitted to Phys. Earth Planet.
    Int.
Jephcoat, A.P. and P. Olson, 1987. 'Is the inner core of the Earth
    pure iron?', Nature, 325, 332–335.
Jepps, S.A., 1975. 'Numerical models of hydromagnetic dynamos', J.
    Fluid Mech., 67, 625–646.
King, J.W., N. Holschuch, S.K. Banerjee, J. Marvin and A. Long, 1988.
    'Geomagnetic secular variation curves for northeastern North
    America for the last 9,000 years', Geophys. J. Roy. Astro.
    Soc., in press.
Lee, S. and F.M. Lilly, 1986. 'On paleomagnetic data and dynamo
    theory', Geomag. and Geoelectr., 38, 797–806.
Loper, D.E., 1978. 'The gravitationally powered dynamo', Geophys. J.
    Roy. Astron. Soc., 54, 389–404.
Loper, D.E. and P.H. Roberts, 1983. 'Compositional convection and the
    gravitationally powered dynamo', in Stellar and Planetary
    Magnetism, A.M. Seward, ed., Gordon and Breach, London.
Lund, S.P. and S.K. Banerjee, 1985. 'Late Quaternary paleomagnetic
    field secular variation from two Minnesota lakes', J. Geophys.
    Res., 90, 803–825.
Merrill, R.T. and M.W. McElhinny, 1983. The Earth's Magnetic Field,
    Academic Press, London.
Moffatt, H.K., 1978. Magnetic Field Generation in Electrically
    Conducting Fluids, Cambridge University Press.
Olson, P. and V.L. Hagee, 1987. 'Dynamo waves and paleomagnetic
    secular variation', Geophys. J. Roy. Astron. Soc., 88,
    139–159.
Parker, E.N., 1955. 'Hydromagnetic dynamo models', Astrophys. J., 122,
    293–314.
Parker, E.N., 1979. Cosmical Magnetic Fields, Oxford University
    Press.
Roberts, P.H., 1972. 'Kinematic dynamo models', Phil. Trans. Roy.
    Soc. Lond., A272, 663–698.
Stevenson, D.J., 1981. 'Models of the earth's core', Science, 208,
    611–619.
Thompson, R., 1984. 'Geomagnetic evolution: 400 years of change on
    Planet Earth', Phys. Earth Planet. Int., 36, 61–77.
Thompson, R., G.M. Turner, M. Stiller and A. Kaufman, 1985. 'Near East
    paleomagnetic secular variation recorded in sediments from the
    Sea of Galilee (Lake Kinneret)', Quaternary Research, 23,
    175–188.

Turner, G.M. and R. Thompson, 1979. 'Behaviour of the Earth's magnetic field as recorded in the sediments of Loch Lomond', Earth Planet. Sci. Lett., 42, 412–426.

Turner, G.M. and R. Thompson, 1981. 'Lake sediment record of the geomagnetic secular variation in Britain during Holocene time', Geophys. J. Roy. Astron. Soc., 65, 703,725.

Verhoogen, J., 1980. Energetics of the Earth, National Academy of Sciences, Washington, DC.

Verosub, K.L., P.J. Mehringer and P. Waterstraat, 1986. 'Holocene secular variation in western North America paleomagnetic record from Fish Lake, Harney County, Oregon', J. Geophys. Res., 91, 3609–3623.

BRUHNES CHRON GEOMAGNETIC EXCURSION RECORDED DURING THE LATE
PLEISTOCENE, ALBUQUERQUE VOLCANOES, NEW MEXICO, U.S.A.

John W. Geissman[1], Steve S. Harlan[1], Laurie Brown[2],
Brent Turrin[3], Leslie D. McFadden[1]

[1]Department of Geology
University of New Mexico
Albuquerque, NM 87131

[2]Department of Geology and Geography
University of Massachusetts
Amherst, MA 01003

[3]U.S. Geological Survey
Mail Stop 941
345 Middlefield Road
Menlo Park, CA 94305

ABSTRACT. An excursion of the geomagnetic field is recorded by all
basaltic lava flows of the Albuquerque Volcanoes, New Mexico. K–Ar
age data (weighted mean: 155 ± 47 ka) and evaluation of flow surface
soil profiles suggest the late Pleistocene (ca. between 100 and
150 ka) as the time of extrusion. Means from 63 sites (eight flows)
yield a grand mean of $D=101.1°$, $I=-36.1°$, $\alpha_{95}=1.2°$ ($\alpha^1_{95}=0.7$,
$\alpha^2_{95}=1.2$), $k=2219$ ($N=8$) and a corresponding VGP of 354.1E, 20.2S,
$dp=0.8°$, $dm=1.4°$ ($A^1_{95}=0.5$, $A^2_{95}=1.4$), VGP ASD=1.8°. TRM acquisition
experiments suggest that the basalts are high fidelity recorders of an
ambient field. Because all flows yield statistically
indistinguishable directions, an excursion morphology clearly cannot
be defined. The Albuquerque Excursion may correlate with other
excursions or short polarity episodes of late Pleistocene age (e.g.
Blake). Though limited in morphology, well-dated excursions and
polarity episodes recorded in volcanic rocks may provide information
on the frequency of significant dynamo instabilities.

F. J. Lowes et al. (eds.), Geomagnetism and Palaeomagnetism, 123–136.
© 1989 by Kluwer Academic Publishers.

1. INTRODUCTION

Geomagnetic excursions represent major deviations in field behaviour over relatively short periods of geologic time [Merrill and McElhinny, 1983]. Virtual geomagnetic pole positions (VGPs) differ by more than 45° from the time–averaged position for that chron [Watkins, 1976]. Excursions are not necessarily associated with transitions; the full inversion of magnetic field direction has been discussed as a polarity event or, where regarded as a stratigraphic feature, "episode" [Denham, 1976; Tuchulka et al., 1987]. On the basis of palaeomagnetic records of variable quality, excursions and polarity episodes have been suggested for several distinct times during the Brunhes chron [summaries by Jacobs, 1983; Tarling, 1983]. That most features have been reported from terrestrial and marine sediments has left the fidelity, or even the validity of each record open to question [Verosub and Banerjee, 1977; Hoffman, 1981]. The few records based on thermoremanent magnetization (TRM) data from lava flows, baked sediments, or artificially baked clays all suffer from the important drawback that only a few instantaneous parts of each feature are documented. Documentation of well–dated complete or even partial excursions in sequences of lava flows is nonetheless of considerable importance. The availability of records of identical excursions at different locations may provide considerable insight into the nature of excursion phenomena and the generation and frequency of dynamo instabilities [e.g. Liddicoat and Coe, 1979; Merrill and McElhinny, 1983; Champion et al., 1988].

Flows of the latest Pleistocene Albuquerque Volcanoes, west of Albuquerque (Figure 1), recorded an unusual characteristic remanent magnetization (ChRM). The low within–flow (VGP angular standard deviations (ASD) typically less than 7°) and between–flow (VGP ASD value of 1.8°) dispersions of this ChRM suggest that only a very small portion of an excursion or polarity episode was recorded. Rock magnetization data, TRM acquisition experiments, and the presence of components of magnetization superimposed on the ChRM suggest it to be a high–fidelity TRM. The only existing isotopic age determination on the Albuquerque Volcanoes was a whole rock K–Ar date of 190 ± 40 ka [Bachman and Mehnert, 1978]; our determinations, suggesting a younger time of eruption, are nonetheless associated with low precision and emphasize the difficulty in dating young lavas with low K_2O contents. Soil profile inspection of lava flow surfaces constrains the age of the flows as less than approximately 150 ka. We describe this geomagnetic field feature as the Albuquerque Excursion, keeping in mind the possibility that it is a portion of a polarity episode.

2. GEOLOGY OF THE ALBUQUERQUE VOLCANOES

Eruptions of alkali basaltic, olivine tholeiitic, and andesitic, lavas occurred during the Pliocene and Pleistocene within the Albuquerque–Belen Basin, Rio Grande Rift [Kelley and Kudo, 1978]. The Albuquerque Volcanoes consist of five large cinder cones, several

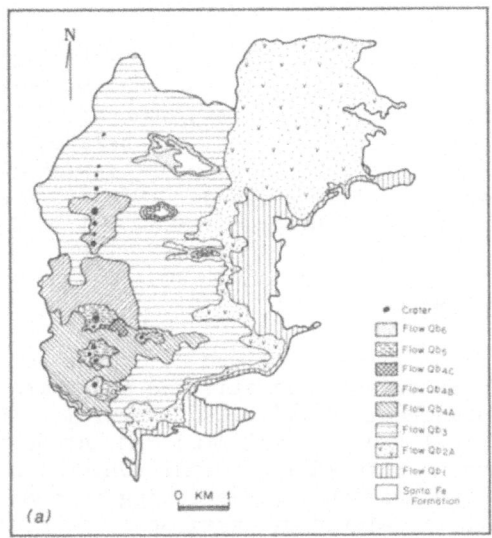

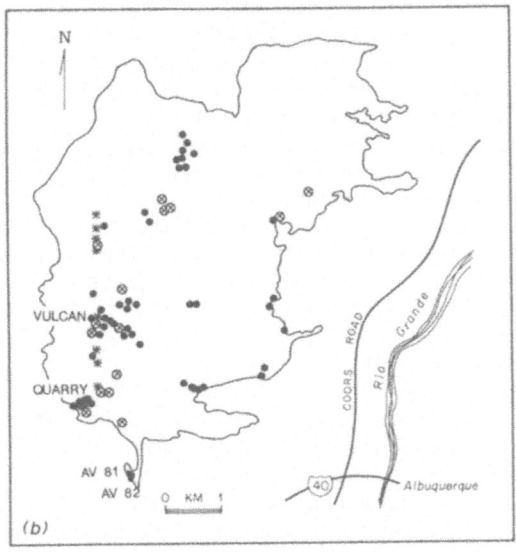

Figure 1. (a) Geologic map of the Albuquerque Volcanoes, from Kelley and Kudo [1978]. (b) Map showing locations of all palaeomagnetic sampling and soil profile investigation sites. Solid dots, UNM sites; circled crosses, UMass sites; stars, volcanic vents; open circles, soil profile investigation sites.

smaller vents, and numerous thin basalt flows (Figure 1a).
Phenocrysts consist of 3-5% euhedral, skeletal olivine and 3-10% laths
of plagioclase.

The lava field consists of at least six major flows, traceable
over several km^2 (Figure 1b). Flows Qb_1, Qb_2 and Qb_3 extend the
length of the fissure system. Younger flows erupted from one or two
localized centers. The absence of ancient soil horizons or aeolian
deposits between flows suggests a rapid rate of flow emplacement. We
have followed the mapping of Kelley and Kudo [1978] in order to assign
sites to specific flow units in Table 1.

3. FIELD AND LABORATORY METHODOLOGY

At each sampling site (Figure 1b) a minimum of 7 samples,
independently oriented using both magnetic and sun compasses,
were collected with non-magnetic drill bits. Several sites exhibited
highly variable and typically high (>10 A/m) NRM intensities
suggesting that a lightning-induced isothermal remanent magnetization
(IRM) is partially superimposed on a pre-existing remanence. Sites
AV81 and AV82 (Figure 1b) consisted of baked soil immediately
underlying flow Qb_1.

Magnetization data were obtained at both the University of New
Mexico (UNM) and the University of Massachusetts (UMASS) laboratories,
using spinner magnetometers or, in the case of the baked sediments, a
three-axis superconducting magnetometer (UNM). Alternating field (AF)
and thermal demagnetization employed commercial instrumentation. Rock
magnetic experiments were carried out to evaluate the essential
magnetic mineralogy and fidelity of the ChRM.

Interpretation of demagnetization data (Figure 2) was in most
cases straightforward. AF demagnetization was usually effective in
isolating a high coercivity component, resulting in a stable endpoint
(SEP). Vector subtraction and/or principal component analysis
techniques were used to calculate directions of magnetizations. For
sites affected by lightning, AF demagnetization data exhibited a
curvilinear decay of the NRM along a great circle segment.
Consequently, a combination of remagnetization circle and SEP [Bailey
and Halls, 1984] analysis was often used.

All measurements are expressed in SI units of induction (B,
tesla) and magnetic dipole moment per unit volume (M, ampere/metre).
An induction of 1.0 T corresponds to a cgs magnetic field H of
10^4 oersted; a magnetic dipole moment per unit volume of 1.0 A/m
corresponds to a cgs magnetization of 10^{-3} emu/cc.

Duplicate argon extractions and isotopic analyses were performed
at the Berkeley Geochronology Centre at the Institute of Human Origins
(formerly at the University of California, Berkeley). Analyses were
by standard isotope dilution methods using a 10 cm Reynolds-type
gas-source mass spectrometer according to procedures described by
Dalrymple and Lanphere [1969]. Potassium analyses were by flame
photometry using a lithium internal standard following procedures
described by Ingamells [1970]. Samples were examined for xenolith

127

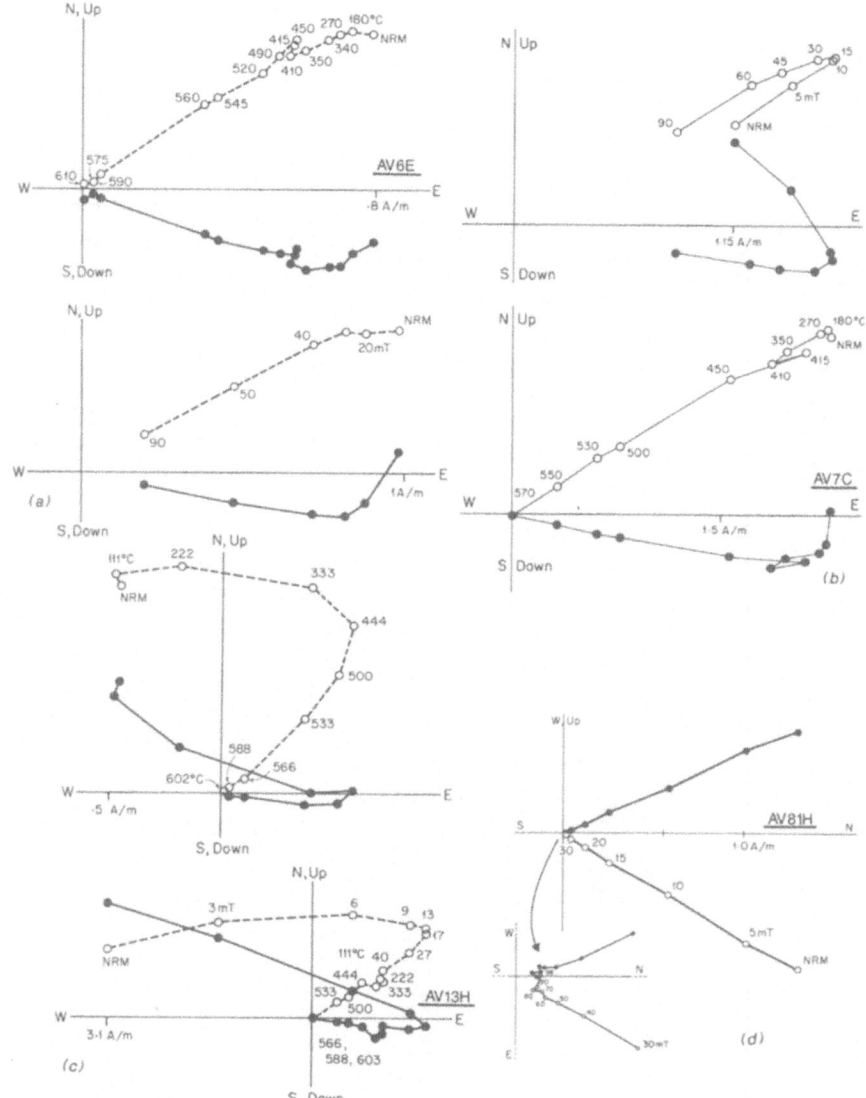

Figure 2. Representative examples of AF and thermal demagnetization
results from the Albuquerque Volcanoes. Orthogonal progressive
demagnetization diagrams showing the endpoint of the magnetization
vector plotted on to the horizontal (closed symbols) and vertical
(open symbols) planes. Peak demagnetizing fields (in mT) or
temperatures (in °C) are given along vertical projections.

TABLE 1. Palaeomagnetic data from the Albuquerque volcanoes

Flow	ChRM[1] Decl. Incl.[2]	N/No[3]	α_{95}[4]	k[6]	MAD[7]	VGP[8] Lat. Long[9]	A[10]$_{95}$	VGP[11] ASD
INDIVIDUAL FLOWS:								
Qb$_1$	99.6,−37.7	10/10	3.5 (2.8,3.3)[5] (30.7)	191.1	5.6	−19.5,355.8	3.2 (2.2,3.1) (30.7)	5.4
Qb$_{2A}$	103.4,−35.6	12/12	3.8 (2.6,4.0) (120.4)	129.8	6.8	−21.9,352.6	3.8 (1.9,4.4) (19.5)	7.0
Qb$_3$	101.0,−35.5	10/10	5.4 (3.1,5.8) (46.6)	80.7	8.5	−19.9,353.8	5.3 (2.4,6.1) (53.1)	8.9
Qb$_{4A}$	101.4,−37.2	6/6	5.6 (2.4,5.7) (161.0)	132.4	6.4	−20.8,354.6	4.6 (2.6,4.4) (158.9)	5.6
Qb$_{4B}$	101.1,−34.8	7/8	6.3 (2.0,6.8) (129.0)	92.3	7.8	−19.7,353.3	6.0 (1.5,6.7) (8.2)	8.0
Qb$_{4C}$	102.7,−36.3	3/4	2.7 (1.9,6.0) (173.9)	201.5	1.4	−21.5,353.4	5.8 (2.1,3.8) (135.5)	3.8

Qb$_5$	102.0,-35.4	7/8	5.1 (2.2,5.3) (114.9)	141.1	6.3	-20.7,353.2	5.2 (1.6,5.7) (18.5)	6.9
Qb$_6$	97.7,-36.4	4/5	5.9 (1.6,5.2) (5.2)	246.3	4.5	-17.6,356.0	4.3 (1.9,3.7) (110.0)	3.7
AV81	305.9,24.9	9/12	16.0	11.2	-	-	-	-
AV82	73.8,43.6	7/7	9.8	38.2	-	-	-	-
ALL FLOWS:	101.1,36.1	8	1.2 (0.7,1.2) (68.5)	2219.9	1.6	-20.2,354.1	1.24 (0.5,1.4) (40.5)	1.8
ALL SITES:	101.2,-36.4	63/63	2.0 (1.8,2.1) (66.9)	80.1	-	-20.5,354.0	2.07 (1.3,2.5) (40.1)	9.3

[1] Remanence characteristic of the particular site investigated. [2] In degrees east of north and positive downwards. [3] Ratio of the total number of sites accepted to the total number of sites measured, except for AV81 and AV82 (baked soils, ratio of samples). [4] Semi-angle of the cone of 95% confidence about the mean direction. [5] Bingham statistics (a^1$_{95}$, a^2$_{95}$) (oval orientation) for the population of site directions and derived VGP values. [6] Fisher's precision parameter. [7] Mean angular deviation of the mean, in degrees. [8] Virtual geomagnetic pole. [9] Degrees north and positive longitude east of 0°. [10] Semi-angle of the cone of 95% confidence about the mean VGP, using site VGP's. [11] Angular standard deviation of the VGP determination, in degrees.

contamination from the underlying Santa Fe Formation. Uncontaminated samples were crushed and sieved, then treated with dilute HCl and HF, and finally washed with distilled water in an ultrasonic bath to remove carbonate and surface alteration. Atmospheric ^{40}Ar corrections were made using the "zero–age" curve method developed at the Berkeley Geochronology Center, (Dieno and Drake, in press). The stated ± is the one sigma resolution of the particular analysis. The resolution of any one analysis, in terms of Ma, is a function of:

1) the weight percent K^+ content of the sample;
2) the mass of material used in the analyses;
3) the accuracy and precision in measuring the isotopic ratios;
4) the accuracy and precision in making the atmospheric argon correction.

Zero–age residuals for the Berkeley argon extraction systems indicate that the age resolution for 6 to 7 grams of material with a K^+ content of about 0.4 wt % is 0.051 ± 0.037 Ma. The quoted standard deviations are best estimates of the analytical precision and accuracy of each individual analysis based on a propagation of the errors listed above.

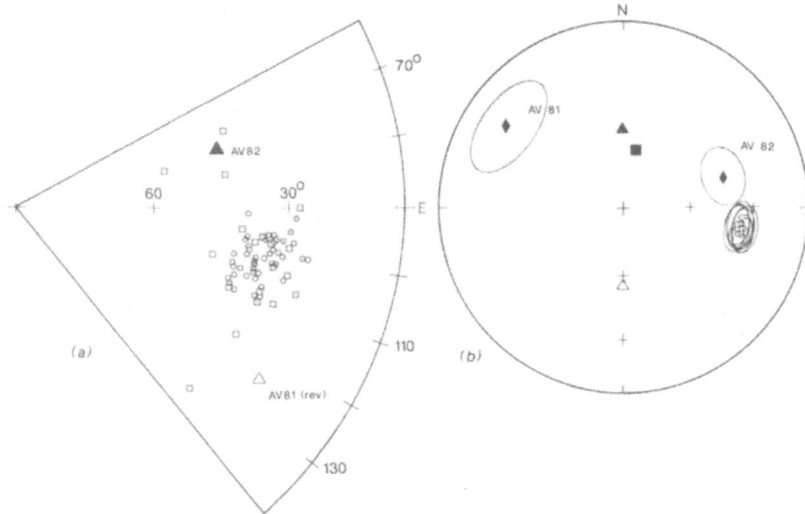

Figure 3. Equal area projections of directions in ChRM's isolated in demagnetization. (a) Individual site means (65, including two sites in baked soils, triangles). Open circles UNM sites; open squares UNM site means requiring remagnetization circle analysis; circles with dots, UMass sites; squares with dots, UMass site means requiring remagnetization circle analysis. (b) individual flow means (8) and baked soil means (diamonds, 2 sites), with projected cones of 95% confidence. Triangles are the axial geocentric dipole directions; square is the present day field.

4. PALAEOMAGNETIC DATA

Data for the eight distinct lava flows studied are provided in Table 1
and site and unit mean directions are shown in Figure 3(a,b). Data
for individual sites (65) are available from the first author upon
request. Based upon NRM intensities and directions as well as
demagnetization characteristics, the lava flows exhibit three
behaviours: (1) high (> 50 mT) median destructive field and
distributed, high laboratory unblocking temperature, single component
NRM's (12 sites) (e.g. sample AV6E, Figure 2a); (2) low intensity
secondary components superimposed on the ChRM (11 sites) (e.g. samples
AV7C and AV13H, Figure 2b,c); and (3) high NRM intensity,
lightning-induced, IRM-contaminated NRM's which required
remagnetization circle analysis of AF demagnetization data (40 sites).
With the exception of a few lightning-affected sites, all sites
yielded well-defined ChRM's. Baked soils were able to be sampled at
two sites underlying lava flow Qb_1. Samples with NRM intensities
exceeding approximately 0.1 A/m exhibit a well-defined univectorial
decay (e.g. sample AV81H, Figure 2d). At site AV81, the magnetization
isolated is of west-north-west declination and moderate positive
inclination. At site AV82, only 6 m to the south, the magnetization
is of east-north-east declination and moderate positive inclination
(Table 1).

5. ROCK MAGNETISM, PETROGRAPHIC OBSERVATIONS, AND THE ORIGIN OF CHRM

Response of the ChRM to AF and thermal demagnetization suggests that
it is carried by fine (i.e. single domain (SD) to pseudo-single-domain
(PSD)) low-Ti titanomagnetite grains. Curie temperature (J_s vs. T)
determinations, acquisition of IRM and backfield demagnetization of
saturation IRM, AF demagnetization of SIRM, and NRM, ARM acquisition
and demagnetization, and low temperature demagnetization all
corroborate demagnetization behaviour.
    Progressive TRM acquisition experiments (Figure 4) suggest the
Albuquerque Volcanoes lavas to be reliable field recorders. With
increasing temperature, the NRM of each specimen becomes aligned with
the laboratory field direction ($\pm Z$). At temperatures exceeding
approximately 550°C, a complete TRM is imparted and the RM is parallel
to Z. No systematic appreciable change in bulk susceptibility
occurred during the experiments. The experiment does not rule out the
possibility of a complex, irreversible, subsolidus reaction affecting
all flows.
    Anisotropy of magnetic susceptibility (AMS) data from three flows
(sites AV40, AV41, AV42) exposed in a quarry in the south-west corner
of the field (Figure 1b) show that all samples are characterized by
low (< 5%) anisotropy and poorly defined lineation and foliation
fabrics. The unusual ChRM in the Albuquerque lavas does not appear to
be controlled by a flow-related anisotropy.
    Response by baked soils beneath lava flow Qb_1 at sites AV81 and
Av82 is puzzling, yet not inconsistent with the ChRM in the lavas

being a reliable record of an unusual field. Perhaps the magnetic mineralogy of the baked soils, which appears to have been produced during contact thermal metamorphism with the lava flow, is often capable of partial self-reversal or some unusual coupling of an extremely local field with the geomagnetic field.

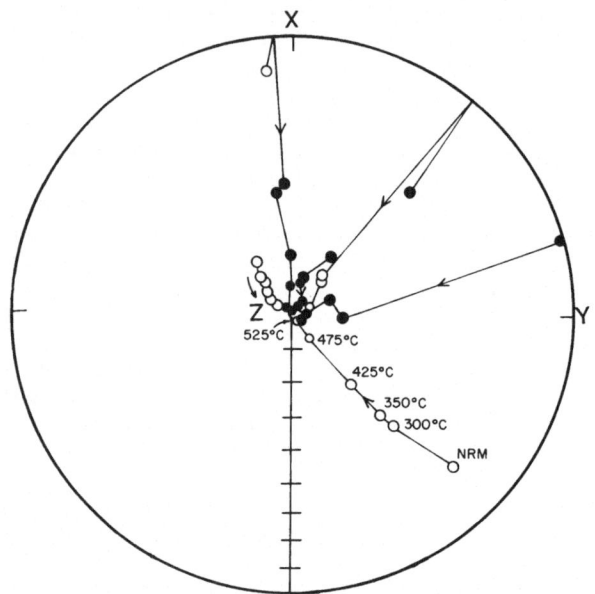

Figure 4. Equal area projection in specimen coordinates of results of incremental TRM acquisition experiments. The results show the progressive acquisition of a RM parallel to Z, as applied during the experiments, by all six specimens (three with TRM parallel to +Z; three with TRM parallel to -Z).

6. ISOTOPIC AND RELATED AGE DETERMINATIONS

6.1 K-Ar Studies

The K-Ar analytical data are presented in Table II. Sample AV-4, the oldest flow stratigraphically, yields K-Ar ages of 0.11 ± 0.38 and 0.03 ± 0.60 Ma. Sample AV-3, overlying sample AV-4, gives 0.04 ± 0.16 and 0.10 ± 0.28 Ma. The youngest flow, sample AV-1, gives -0.46 ± 0.07 and 0.17 ± 0.05 Ma. The age of -0.46 ± 0.070 Ma obtained for sample AV-1/1 is best interpreted as an age of 0.070 Ma or less.

The lack of aeolian deposition, lack of soil development between the lava flows, and the similarity in the palaeomagnetic directions of the three flows indicate that the flows were erupted over a short period of time. · Therefore the age data for all three flows are used in evaluating the age of the palaeomagnetic excursion. The arithmetic mean for all six analysis is 0.00 ± 0.23 Ma, but the age for AV-1/1 differs from the mean by two standard deviations. Using Chauvenot's criterion the value for AV-1/1 is therefore rejected.

The simple arithmetic mean of the five accepted analyses is then 0.089 ± 0.026 Ma, where the standard error is based on the scatter about the mean. Taking the weighted mean of the five accepted analyses gives 0.155 ± 0.047 Ma, where the standard error is based on the weighting standard deviations.

The weighted mean 0.16 ± 0.05 Ma is the preferred age of the lava flow because the more precise analyses are given a greater weight. However this value is determined largely by the 0.17 ± 0.07 Ma of AV1/2, which has the highest percentage radiogenic argon yield and therefore the lowest calculated standard deviation. If the higher radiogenic argon yield is due to contamination from partially degassed Santa Fe Formation xenoliths, then the age for this sample, and therefore the weighted mean, may be slightly too large. We feel, however, that this weighted mean (0.16 ± 0.05 Ma) is the most statistically defensible value.

TABLE II. K-Ar isotopic age data, Albuquerque volcanoes lava flows.

Sample Number	Sample Weight/gm	$\%K^+$	$^{40}Ar^*$ $/10^{-13}mol/gm$	$\%Ar^*$	AGE/Ma ± 1σ
AV-1/1	7.0	0.44	-3.49	-3.4	-0.46 ± 0.07
AV-1/2	5.3	0.44	1.33	1.4	0.17 ± 0.05
AV-3/1	6.9	0.41	0.28	0.2	0.04 ± 0.16
AV-3/2	5.4	0.41	0.70	0.2	0.10 ± 0.28
AV-4/1	6.3	0.44	0.84	0.1	0.11 ± 0.38
AV-4/2	6.5	0.44	0.19	<0.1	0.03 ± 0.60

Decay Constants: $\lambda_\varepsilon + \lambda_{\varepsilon'} = 0.581 \times 10^{-10}$ a^{-1}, $\lambda_\beta = 4.962 \times 10^{-10}$ a^{-1}

$^{40}K/K_{total} = 1.167 \times 10^{-4}$; see Steiger and Jager [1977].

Simple mean of all analyses, less AV-1/1:   0.089 ± 0.026 Ma

Weighted mean of all analyses, less AV-1/1:   0.155 ± 0.047 Ma

Best age for magnetic excursion: **0.155 ± 0.047 Ma.**

6.2 Soil Profile Investigations

Examination of soil profiles on several flow surfaces (Figure 1b) shows that buried soils were not encountered; aeolian material blanketed flow surfaces before soil development progressed to any significant extent. The degree of soil development in aeolian deposits on flow surfaces reflects the age of the underlying lavas [McFadden et al., 1986]. The degree of development of soils in aeolian deposits in the Albuquerque Volcanoes field can be compared with that of latest Pleistocene to early Holocene soils forming in aeolian deposits in the San Juan Basin, 150 km north-west of the study area [McFadden et al., 1983], for which absolute ages are available. The strongly developed horizons of carbonate accumulation require minimally 10 to 50 ka of accumulation, provided the soil parent materials are largely noncalcareous. A maximum age of ca. 150 ka is estimated for the flows of the Albuquerque Volcanoes on the basis of soil profile characteristics.

7. DISCUSSION

The lava flows of the Albuquerque Volcanoes recorded a non-axisymmetric geomagnetic field direction during a portion of an excursion or polarity episode in late Pleistocene time. The span of geomagnetic field time recorded by the lavas is most likely less than a century or so and severely limits the significance of correlating this Albuquerque data with, or incorporating it in, possibly contemporaneous geomagnetic features (e.g. the Blake event as recorded in terrestrial and marine sediments [Smith and Foster, 1969; Tucholka et al., 1987]). Furthermore, until the age of Albuquerque volcanism is more accurately determined, the utility of the feature for dating sections of late Pleistocene sediments is limited. Because field directions more typical of the Brunhes chron have not been recorded by any of the Albuquerque lavas, an internal comparison of magnetization intensities for relative palaeointensity evaluation is also precluded.
    The excursion record is clearly insufficiently detailed to allow comparison with morphologies of other excursions and polarity episodes [e.g. Hoffman, 1986]. Comparison of the Albuquerque result with other data, in either VGP or rotated field vector coordinates [Hoffman, 1984], by extrapolating excursion directions to Albuquerque coordinates assuming an axial geocentric dipole, is not justified. We note simply that the Albuquerque VGP is over 40° from the nearest VGP defined by TRM data (direction FZ, Lake Mungo Excursion, Barbetti and McElhinny, 1972) and approximately 30° from the same result in rotated field direction coordinates though, strictly speaking, the Albuquerque result can only be compared with features recorded in nearby units.
    The general similarity between the Albuquerque and the Skalamaelifell area results [Levi et al., 1987] is worth noting. While the directions recorded, and their associated VGP's, differ significantly for the two areas, several igneous units at each nonetheless recorded, over presumably a short period of time, constant

field directions. If a comparison with complete field transition morphologies [e.g. Prevot et al., 1985, Hoffman, 1986] can be made, perhaps these two examples are illustrating intermediate field "hang-ups", as recorded in continuous records of "multiple-style" polarity transitions.

Further identification of geomagnetic field excursions in young volcanic and sedimentary rocks will only improve basic knowledge of excursion phenomena (i.e. duration, amplitude, and field intensity). As also noted by Champion et al. [1988], the TRM-defined excursion record for the past several 100 ka shows numerous high-amplitude (deviations greater than 90°) data. Perhaps when dynamo instabilities occur, they are typically characterized by very high amplitude field variations. Accurate dating of excursion or polarity episode records, or even fragments thereof, should continue to be a critical aspect of excursion research. Several relatively recent techniques for dating young lavas (e.g. ^{40}Ar/^{39}Ar step heating; cation-ratio and accelerator radiocarbon; and ^{3}He/^{4}He, ^{36}Cl, and ^{40}Ar/^{39}Ar exposure ages might be considered to evaluate the timing of excursion phenomena. Further documentation of excursions or polarity episodes will allow us to develop a better understanding of the periodicity of short-term, yet nonetheless significant dynamo instabilities.

8. REFERENCES

Bachman, G.O., and H.H. Mehnert, 1978. 'New K-Ar dates and the late Pliocene to Holocene geomorphic history of the central Rio Grande region, New Mexico', Geol. Soc. America Bull., 89, 283-292.

Bailey, R.C. and H.C. Halls, 1984. 'Estimates of confidence in paleomagnetic direction derived from mixed remagnetization circle and direct observational data', J. Geophys., 54, 174-182.

Barbetti, M.F. and M.W. McElhinny, 1972 'Evidence of a geomagnetic excursion 30,000 yr. BP', Nature, 239, 327-330.

Champion, D.E., M. Lanphere, and M. Kuntz, 1988. 'Evidence for a new geomagnetic reversal in the Brunhes Normal Polarity Chron from lava flows at the Idaho National Engineering Laboratory: The Emperor Event retracted', J. Geophys. Res., in press.

Dalrymple, G.B. and M. Lanphere, 1969. Potassium-argon dating, 258pp, W.H. Freeman Company, San Francisco.

Denham, C.R., 1976. 'Blake polarity episode in two cores from the Greater Antilles Outer ridge', Earth Planet. Sci. Lett., 29, 422-434.

Hoffman, K.A., 1981. 'Paleomagnetic excursions, aborted reversals, and transitional fields, Nature', 294, 67-69.

Hoffman, K.A., 1984. 'A method for the display and analysis of transitional palaeomagnetic data', J. Geophys. Res., 89, 6285-6292.

Hoffman, K.A., 1986. 'Transitional field behaviour from southern
    hemisphere lavas: Evidence for two-stage reversals of the
    dynamo', Nature, 320, 6059-6061.
Ingamells, C.O., 1970. 'Lithium metaborate flux in silicate
    analysis'. Analytica Chimica Acta, 52, 323-334.
Jacobs, J.A., Reversals of the Earth's Magnetic Field, Adam Hilger,
    Bristol, 230 pp., 1983.
Kelley, V.C., and A.M. Kudo, 1978. Volcanoes and related basalts of
    Albuquerque Basin, New Mexico, Circular 156, New Mexico Bur.
    Mines and Mineral Resources, 30 pp.
Levi, S., H. Audunsson, R.A. Duncan, and L. Kristjansson, 1987. 'The
    geomagnetic excursion at Skalamaelifell, Iceland: Additional
    evidence for unstable geomagnetic behaviour circa 40 ka ago'
    [abstract], EOS, Trans. Amer. Geophys. Union, v. 68,
    p. 1249.
Liddicoat, J.C. and R.S. Coe, 1979. 'Mono Lake geomagnetic excursion',
    J. Geophys. Res., 84, 261-271.
McFadden, L.D., S.G. Wells and J.G. Schultz, 1983. 'Soil development
    on late Quaternary eolian deposits in San Juan Basin,
    northwestern New Mexico', in Wells, S.G., D.W. Love, and
    T.W. Gardner (eds.) Chaco Canyon country, A Fieldguide to the
    Geomorphology, Quaternary Geology, Palaeocology, and
    Environmental Geology of Northwestern New Mexico, Amer.
    Geomorph. Field Group 1983 Field Trip Guidebook, 167-176.
McFadden, L.D., S.G. Wells and J.C. Dohrenwend, 1986. 'Influences of
    Quaternary climatic changes on processes of soil development
    on desert loess deposits of the Cima volcanic field,
    California', Catena, 13, 361-389.
Merrill, R.T. and M.W. McElhinny, 1983. The Earth's Magnetic Field:
    Its History, Origin, and Planetary Perspective, Academic Press,
    London, 401 pp.
Prevot, M., E. Mankinen, C.S. Gromme, and R.S. Coe, 1985. 'How the
    geomagnetic field vector reverses polarity', Nature, 316,
    230-234.
Smith, J.D. and J.H. Foster, 1969. 'Geomagnetic reversal in Brunhes
    normal polarity epoch'. Science, 163, 565-567.
Steiger, R.H. and E. Jager, 1977. 'Subcommission on geochronology-
    convention on the use of decay constants in geo- and
    cosmo-chronology; Earth Planet. Sci. Lett., 36, 359-362.
Tarling, D.H., 1983. Palaeomagnetism, 379 pp., Chapman and Hall,
    London.
Tucholka, P., M. Fontugne, F. Guichard, and M. Parterne, 1987. 'The
    Blake magnetic polarity eposide in cores from the Mediterranean
    Sea', Earth Planet. Sci. Lett., v. 86, 320-326.
Verosub, K. and S. Banerjee, 1977. 'Geomagnetic excursions and their
    paleomagnetic record', Rev. Geophys. Space Phys., 15,
    145-155.
Watkins, N.D., 1976. 'Polarity group sets up guidelines', Geotimes,
    21, 18-20.

SHORT REVERSAL OF THE PALAEOMAGNETIC FIELD ABOUT 280 000 YEARS AGO AT
LONG VALLEY, CALIFORNIA

Joseph C. Liddicoat and Roy A. Bailey*

Earth Sciences Board
University of California
Santa Cruz
California 95064, U.S.A.

*U.S. Geological Survey
345 Middlefield Road
Menlo Park
California 94025, U.S.A.

ABSTRACT. A reversal of the palaeomagnetic field is recorded in
exposed lake sediments at Long Valley and Mono Basin in east-central
California. The reversal is estimated to be several thousand years
long and 280 000 years old. The chronology is based on correlation of
volcanic ash beds at Long Valley and Mono Basin with ones at Summer
Lake, Oregon, and correlation of ash beds at Summer Lake with ash beds
in a core at Tulelake, California, where age control is provided by
tephrochronology and magnetostratigraphy.

The path of the virtual geomagnetic poles forms a large clockwise trending loop in the hemisphere of Long Valley and is below 45°S
latitude for approximately 1000 years. An alternate interpretation of
the field behaviour is that it is an excursion, i.e. an aborted
reversal, as proposed by Hoffman (1981). In that possibility the
record at Long Valley is inconsistent with the examples selected in
favour of aborted reversals during times of stable normal polarity.

1. INTRODUCTION

An abundance of palaeomagnetic data from sedimentary and volcanic
rocks indicates the Earth's magnetic field has been normally directed
throughout most of the Brunhes Normal Chron (the last 730 000 years),
although brief departures from that polarity have been encountered.
The Laschamp reversed subchron (Bonhommet and Zahringer, 1969) 35 000
to 45 000 years (35–45 ka) ago (Heller and Petersen, 1982) is one that
was discussed during this conference. In this paper we describe
another departure of short duration recognized in lacustrine sequences
at two localities about 40 kilometres apart in eastern California –
one in Long Valley caldera and the other in Mono Basin (Fig. 1). We

137

F. J. Lowes et al. (eds.), Geomagnetism and Palaeomagnetism, 137–153.
© *1989 by Kluwer Academic Publishers.*

138

consider the departure a reversal, recognizing that it does not
entirely satisfy that definition (Cox et al., 1975); i.e., its
occurrence worldwide remains to be documented. Correlation of the
reversal between Long Valley and Mono Basin is possible by the
presence in both sedimentary sections of a distinctive ash bed that
formed at about 280 ka.

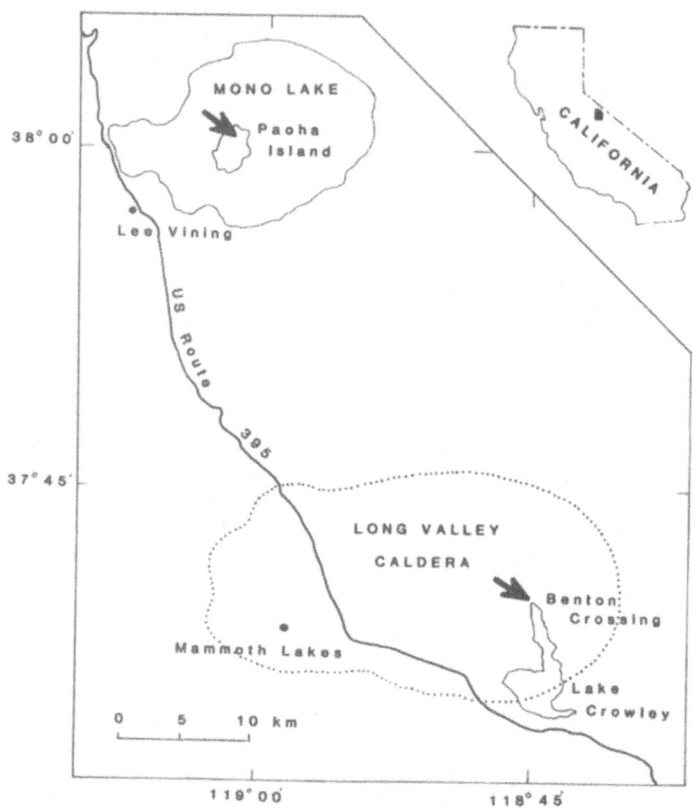

Figure 1. Location map for Long Valley caldera (dotted line) and
Paoha Island, Mono Lake, California. Arrows identify the
palaeomagnetic sample localities on the island and at Benton Crossing.

2. SAMPLE LOCATIONS

The section sampled in Long Valley caldera is in the uppermost part of
a flat-lying sequence of lacustrine silts deposited in Pleistocene
Long Valley Lake (Mayo, 1934) between 600 and 100 ka. Remnant lake
terraces and strandlines on the walls of the caldera indicate that the
lake initially formed during or shortly after the rise of the
resurgent dome (730 to 620 ka) and finally drained through a breach in
the southeastern caldera wall between 100 and 50 ka (Bailey et al.,

1976). The Long Valley sampling site is at Benton Crossing on a low gravel-capped hill, an erosional remnant formed sometime after drainage of the Pleistocene lake.

The section sampled at Mono Lake is on Paoha Island in a sequence of poorly consolidated clayey silts that have been uplifted from the lake bottom on the crest of a rhyolite cryptodome, probably about 200 years ago (Lajoie, 1968; Stine, 1984). The sequence has been gently arched, but, where collected on the crest of the island, the sediments are nearly horizontal. Their age is poorly constrained, but on the basis of radiocarbon dating of ostracods and tufa and of estimated sedimentation rates, they probably were deposited between 300 and 36 ka (Lajoie, 1968; pers. comm., 1987).

At both localities samples were collected above and below a distinctive, pale orange, ash bed. This ash bed, designated LV-III in Long Valley and PAOH-III on Paoha Island, has been identified and correlated on the basis of association with other ash beds in both sequences and on similarity in chemical and mineralogical composition (see Tephra Analysis and Age Estimate, below).

3. SAMPLING PROCEDURE

In Long Valley, oriented samples were collected at three sites, A, B, and C, spaced laterally over a distance of 15 m along the north slope of the hill at Benton Crossing. Initially, samples for alternating field (a.f.) demagnetization were taken at about 10 cm intervals at all three sites (samples 1 through 35, Table I), about 30 to 70 cm above and below ash bed LV-III. Later, samples for thermal demagnetization were collected at Site A only, at 2 cm intervals over a vertical distance of 95 cm (samples 36 through 79, Table I).

On Paoha Island, oriented samples for a.f. demagnetization were collected at two sites, A and B, 50 m apart. No samples for thermal demagnetization were taken because the clayey silt was poorly consolidated and had to be encased in plastic boxes for measurement. At Site A, samples were taken at 1 to 3 cm intervals 4 to 5 cm above and below ash bed PAOH-III; at Site B they were taken 15 cm or less intervals over an 81 cm section spanning the ash bed (Table II).

4. LABORATORY PROCEDURE

For each sample from Long Valley, a single specimen (6 cc cube cut on a band saw) was prepared for thermal demagnetization. On those specimens 400°C was used after full demagnetization of pilot specimens showed no significant change in the palaeomagnetic directions above that temperature. The pilot specimens have a maximum blocking temperature of about 590°C, which identifies magnetite as the primary source of the magnetization (Fig. 2).

For a.f. demagnetization of Long Valley samples from Site A, pilot specimens were progressively demagnetized in fields as high as 800 oersted (80 mT) (Fig. 3). The median destructive field is about

Table I.   Palaeomagnetic data following a.f. demagnetization (samples 1-35) or thermal demagnetization (samples 36-79) for Long Valley, California

SAMPLE	POSITION	N	INCL	DECL	INT	k	α_{95}	PLAT	PLONG
				SITE A (a.f.)					
19	+68.0	6	44.4	336.2	2.33	2120	1.5	66.8	128.1
18	+60.0	6	52.2	310.5	2.42	1200	1.9	49.7	159.6
17	+55.0	6	46.5	283.8	3.38	2022	1.5	26.9	166.6
16	+51.0	6	46.9	275.9	4.67	793	2.4	21.1	170.9
15	+48.0	6	47.3	273.2	4.17	1388	1.8	19.2	172.6
14	+34.0	6	9.0	217.3	4.37	1397	1.8	-35.5	193.1
13	+30.0	6	5.3	211.8	4.87	3908	1.1	-40.1	197.5
12	+20.0	6	3.1	206.9	3.72	847	2.3	-43.6	202.3
11	+13.0	6	37.1	201.6	.215	37	11.2	-28.2	218.0
10 (LV-III)	---	6	76.7	142.4	1.23	172	5.1	16.5	256.8
9	0.0	6	-13.3	165.7	1.71	801	2.4	-56.5	267.4
8	-4.0	6	16.7	154.4	.615	929	2.2	-38.0	273.9
7	-10.0	6	56.9	44.9	.118	13	12.2	54.8	317.0
6	-18.0	6	-28.8	33.2	.463	447	3.2	28.5	24.1
5	-29.0	6	45.6	20.8	.145	71	8.0	69.6	356.0
4	-36.0	6	44.0	355.1	.218	126	6.0	77.5	81.8
3	-40.0	6	54.0	5.2	.350	242	4.3	84.8	5.5
2	-49.0	6	33.6	65.0	6.97	1132	2.0	30.7	330.7
1	-59.0	6	48.3	53.9	3.28	629	2.7	44.9	325.0
				SITE A (thermal)					
79	+50.0	1	46.4	274.2	.947	---	---	19.6	171.4
78	+48.0	1	31.8	261.6	.652	---	---	4.1	169.7
77	+46.0	1	31.2	257.6	.501	---	---	0.8	171.8
76	+44.0	1	28.3	239.1	.493	---	---	-13.5	182.6
75	+42.0	1	14.5	219.6	.448	---	---	-31.8	193.0
74	+40.0	1	16.4	221.4	.495	---	---	-29.9	192.0
73	+38.0	1	13.7	223.9	.616	---	---	-29.5	188.8
72	+36.0	1	13.0	219.8	.692	---	---	-32.3	192.2
71	+34.0	1	19.1	221.0	.755	---	---	-29.0	193.4
70	+32.0	1	13.0	211.4	1.00	---	---	-36.9	200.7
69	+30.0	1	28.9	212.0	.800	---	---	-29.0	205.3
68	+28.0	1	12.1	199.4	.966	---	---	-42.6	214.3
67	+26.0	1	-0.9	197.6	1.09	---	---	-49.4	213.3
66	+24.0	1	16.0	213.1	.795	---	---	-34.7	199.9
65	+22.0	1	12.0	207.8	.568	---	---	-39.2	204.3
64	+20.0	1	7.3	205.0	.671	---	---	-42.6	206.1
63	+18.0	1	8.3	199.9	.580	---	---	-44.2	212.7
62	+15.0	1	10.7	199.2	.495	---	---	-43.3	214.2
61	+13.0	1	1.9	197.7	.458	---	---	-48.0	214.0
60	+11.0	1	5.6	199.1	.378	---	---	-45.8	213.1
59	+9.0	1	6.5	180.2	.456	---	---	-49.0	240.7
58	+7.0	1	8.9	175.6	.328	---	---	-47.6	247.5
57	+5.0	1	15.8	162.9	.314	---	---	-41.5	263.9
56	+3.0	1	-0.8	156.5	.391	---	---	-46.9	276.7
55	0.0	1	-5.2	154.0	.534	---	---	-47.6	281.5
54	-2.0	1	3.7	155.1	.511	---	---	-44.3	277.0
53	-4.0	1	13.4	149.3	.666	---	---	-37.1	280.5
52	-6.0	1	18.9	139.3	.495	---	---	-29.2	288.4
51	-8.0	1	9.2	128.5	.438	---	---	-26.2	301.4
50	-10.0	1	-2.2	113.7	.318	---	---	-19.3	316.9

Table I continued

SAMPLE	POSITION	N	INCL	DECL	INT	k	α95	PLAT	PLONG
49	-12.0	1	-11.1	86.0	.191	---	---	-0.3	337.9
48	-14.0	1	-8.6	62.0	.234	---	---	18.9	352.5
47	-16.0	1	-14.3	38.3	.245	---	---	32.6	14.1
46	-18.0	1	-5.1	37.3	.206	---	---	37.0	11.7
45	-20.0	1	-2.6	63.4	.163	---	---	19.9	349.1
44	-22.0	1	-16.5	77.6	.195	---	---	4.5	345.3
43	-29.0	1	36.8	7.4	.107	---	---	71.7	38.5
42	-31.0	1	53.7	15.1	.524	---	---	77.3	342.3
41	-33.0	1	42.3	345.7	.177	---	---	72.0	107.7
40	-35.0	1	32.4	357.2	.148	---	---	70.0	68.7
39	-37.0	1	43.2	345.5	.120	---	---	72.4	109.6
38	-39.0	1	53.0	356.0	.209	---	---	84.7	100.4
37	-41.0	1	44.6	26.4	.131	---	---	65.0	350.5
36	-45.0	1	39.6	53.5	.255	---	---	42.0	333.6

SITE B

SAMPLE	POSITION	N	INCL	DECL	INT	k	α95	PLAT	PLONG
26	+28.0	6	1.8	202.9	4.72	5798	0.9	-46.1	206.9
25	+20.0	6	-1.9	204.6	4.05	2540	1.3	-46.9	203.5
24	+12.0	6	-13.4	194.6	2.92	1084	2.0	-56.5	214.1
23	+8.0	6	-15.4	180.6	3.48	2872	1.3	-60.2	239.8
22	-6.0	6	3.8	153.2	3.57	2213	1.4	-43.4	279.3
21	-18.0	6	8.7	28.6	3.38	108	6.5	47.4	15.8
20	-29.0	6	18.3	36.5	4.02	387	3.4	46.7	2.1

SITE C

SAMPLE	POSITION	N	INCL	DECL	INT	k	α95	PLAT	PLONG
35	+27.0	6	6.5	205.8	3.20	1327	1.8	-42.7	204.8
34	+20.0	6	15.7	210.6	1.33	180	5.0	-36.2	202.4
33	+11.0	6	62.5	228.3	.120	18	16.1	2.4	208.4
32	+3.0	6	-6.9	159.4	.407	360	3.5	-51.0	274.9
31	0.0	6	-17.2	162.5	1.78	2335	1.4	-57.2	274.2
30	-6.0	6	33.3	153.4	.328	112	6.4	-28.9	270.1
29	-16.0	6	36.1	29.4	.220	512	3.0	59.0	357.3
28	-20.0	6	0.1	22.8	.260	236	4.4	47.0	26.4
27	-31.0	6	46.8	7.5	.120	27	13.1	78.6	25.5

SAMPLE:	Number of the sample
POSITION:	Centimeters above (+) or below (-) the base of ash bed LV-III
N:	Number of specimens demagnetized
INCL:	Mean inclination (degrees) after a.f. demagnetization at 300 oersteds or thermal demagnetization at 400°C
DECL:	Mean declination (degrees) after a.f. demagnetization at 300 oersteds or thermal demagnetization at 400°C
INT:	Intensity ($\times 10^{-5}$ emu/cc) of a 6-cc specimen after a.f. demagnetization at 300 oersteds or thermal demagnetization at 400°C
k:	Fisher precision parameter
α_{95}:	Alpha-95: semiangle (degrees) of cone of 95% level of confidence for paleomagnetic directions
PLAT:	North or south (-) latitude of Virtual Geomagnetic Pole (VGP)
PLONG:	East longitude of VGP

142

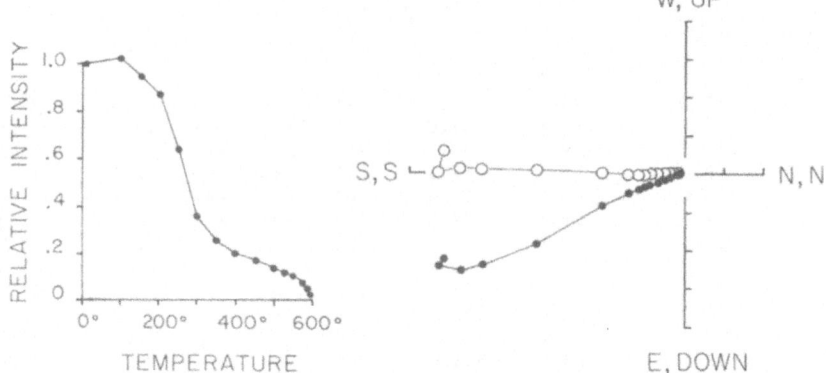

Figure 2. Vector diagram and normalized intensity for lacustrine silt immediately below ash bed LV-III at Long Valley Site A. The maximum blocking temperature for this sample is about 590°C. In the vector plot, solid circles are projections on the NS-EW plane and open circles are projections on the NS-Vertical plane; divisions on the axes are 1.00×10^{-5} emu/cc (10^{-2} A/m).

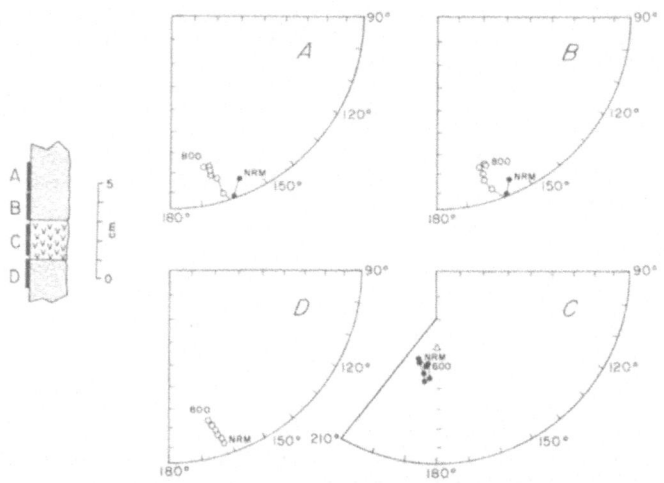

Figure 3. Equal-area plot of palaeomagnetic directions for progressive a.f. demagnetization to 600 oersted for ash bed LV-III (plot C) and to 800 oersted for lacustrine silt (plots A, B and D) for Long Valley Site A. Lettering of plots corresponds to lettering of samples on the stratigraphic column at left. In the column, the broken horizontal pattern represents silt, v's represent LV-III, and black is the interval sampled. Filled circles are plotted on the lower hemisphere, open circles on the upper hemisphere. The open triangle in plot C is the direction of a reverse axial dipole field. The silt records reverse polarity and ash bed LV-III records an anomalous direction - south declination and steep positive inclination.

300 oersted, which was used for blanket demagnetization; the within-sample scatter of palaeomagnetic directions is usually 5° or less (Table I). At all levels of demagnetization, ash bed LV-III records an anomalous palaeomagnetic direction – southerly declination and steep positive inclination (Fig. 3c). The same direction was found at Sites B and C, eliminating the possibility that samples were misoriented during collection.

For Paoha Island Site A, pilot specimens were progressively a.f. demagnetized to 600 oersted; the remaining specimens were demagnetized at 200 oersted. The scatter of palaeomagnetic directions in the clayey silts at Paoha Island is high (Table II), unlike that in the silts at Long Valley. Only ash bed PAOH-III records reversed polarity (Fig. 4). The samples from Site B produced similar results – southeasterly declination and shallow negative inclination for ash bed PAOH-III and erratic palaeomagnetic directions for the surrounding clayey silt (Fig. 5; Table II).

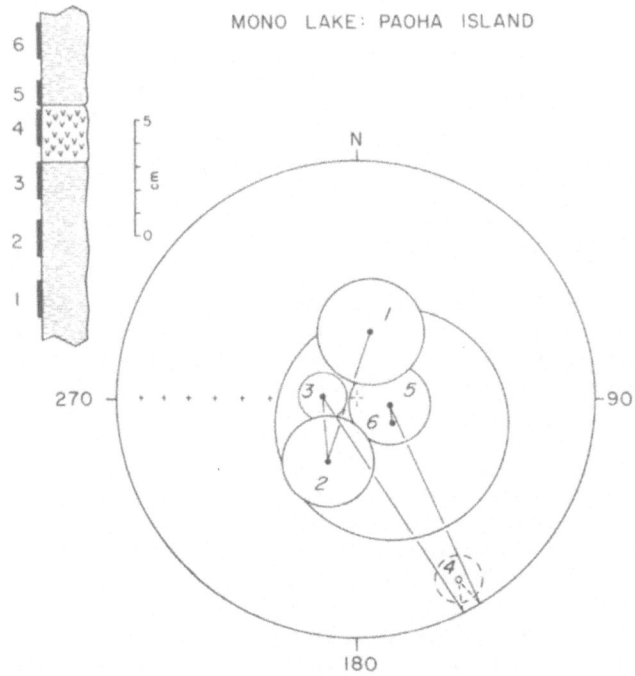

Figure 4. Equal-area plot of the mean palaeomagnetic directions and circle of confidence (alpha-95) for ash bed PAOH-III and clayey silt on Paoha Island, Mono Lake, after a.f. demagnetization at 200 oersted. The patterns in the stratigraphic column are the same as in Fig. 3. Note that only PAOH-III (sample 4) records reverse polarity and that the direction is similar to that for lacustrine silts adjacent to ash bed LV-III in Long Valley (Fig. 3). Filled circles are plotted on the lower hemisphere, open circles on the upper hemisphere.

144

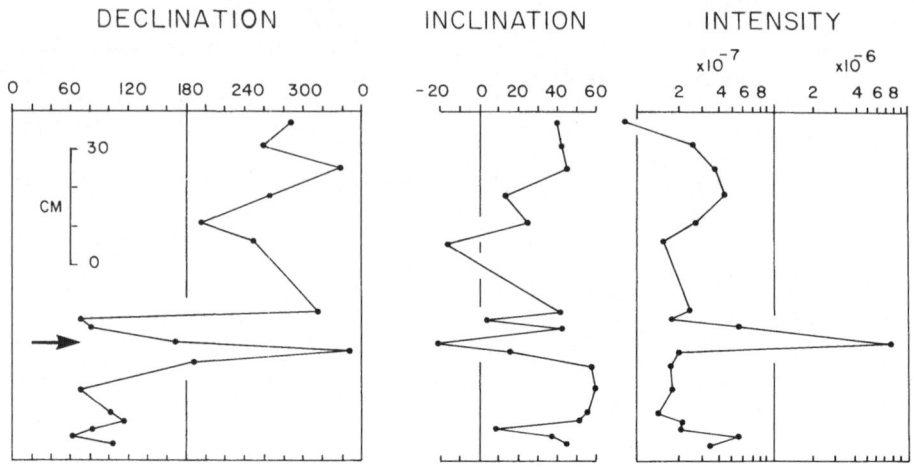

DECLINATION   INCLINATION   INTENSITY

Figure 5. Plot of mean declination, inclination, and intensity
(emu/cc) after a.f. demagnetization at 200 oersted for Paoha Island
Site B. The palaeomagnetic direction for ash bed PAOH-III is
indicated by the arrow. Note in Table II that the scatter of
directions (alpha-95) for most samples exceeds 20°.

5. TEPHRA ANALYSIS AND AGE ESTIMATE

Discovery of the polarity reversals in Long Valley and on Paoha Island
requires establishing that the intervals surrounding ash bed LV-III
and PAOH-III are indeed the same. Correlation of the ash beds would
prove that the record arose from the palaeomagnetic field and is not
an artifact of soft-sediment deformation in the lacustrine silts
(Verosub, 1975).
     Volcanic glass separates from the ash beds were analysed by
electron-microprobe (EM), energy-dispersive X-ray fluorescence (XRF),
and instrumental neutron-activation (INA) for 31 major, minor, and
trace elements. The analytical and data-evaluation methods used have
been described by Sarna-Wojcicki et al. (1979, 1984 and 1985), and our
data are available upon request). Correlative ash beds at several
other localities in California and Oregon, previously analyzed by EM
(Davis, 1985; Rieck et al., in press; Sarna-Wojcicki et al., in
press), provide age estimates for the Long Valley and Paoha Island
sections, as described below.
     Results of the chemical analyses indicate that ash beds LV-III
and PAOH-III are chemically identical within the limits of accuracy of
the analytical methods and with precisions equivalent to those
obtained upon replicate analyses of glass from the same ash bed. The
minerals present in the ash bed from both localities and the shapes of
glass shards are also the same (Table III). Another ash bed that
stratigraphically underlies the reversal at both localities also is

Table II. Palaeomagnetic data following a.f. demagnetization for
Paoha Island, Mono Lake, California.

SAMPLE	POSITION	N	INCL	DECL	INT	k	α_{95}	PLAT	PLONG
				SITE A					
6	+3.5	6	75.1	123.7	.48	4	40.8	19.8	265.5
5	+1.0	5	78.7	100.0	.20	35	13.1	31.4	266.4
4 (PAOH-III)	---	6	-13.7	151.5	1.95	44	10.2	-49.6	288.0
3	0.0	5	78.2	273.6	.083	84	8.4	36.0	212.6
2	-2.5	5	66.1	206.2	.37	21	17.2	-0.5	224.0
1	-5.0	4	67.0	11.1	.18	21	20.3	75.9	271.7
				SITE B					
24	+57.0	6	39.6	288.4	.090	6	31.4	27.7	158.9
23	+51.0	6	42.5	259.1	.24	6	31.2	7.0	176.9
22	+45.0	6	45.2	337.8	.37	16	17.3	68.2	126.4
21	+37.0	6	13.4	265.7	.41	2	71.2	0.8	159.0
20	+31.0	6	24.4	197.0	.27	2	84.8	-36.8	220.2
19	+26.0	6	-17.3	248.2	.17	2	66.9	-22.6	157.5
18	+8.0	6	40.9	313.0	.23	2	63.5	47.6	144.9
17	+6.0	6	4.7	70.9	.19	1	>90.0	16.4	341.1
16	+4.0	6	42.7	81.5	.55	2	69.0	21.3	315.6
15 (PAOH-III)	---	6	-21.2	168.8	7.87	14	18.5	-61.2	264.3
14	0.0	4	15.1	346.3	.21	3	69.3	57.3	86.7
13	-3.0	6	57.5	187.9	.18	2	72.3	-13.5	234.6
12	-10.0	6	59.8	69.8	.19	2	63.7	37.4	304.7
11	-16.0	6	54.8	101.0	.14	8	24.7	13.5	296.5
10	-18.0	6	50.9	115.0	.23	8	25.8	2.2	291.6
9	-20.0	6	8.2	81.5	.22	4	40.8	9.2	333.0
8	-22.0	6	37.4	62.1	.54	5	34.1	34.4	330.2
7	-24.0	6	44.3	103.1	.36	4	40.7	6.3	302.7

SAMPLE: Number of the sample
POSITION: Centimeters above (+) or below (-) the base of ash bed PAOH-III
N: Number of specimens demagnetized
INCL: Mean inclination (degrees) after a.f. demagnetization
DECL: Mean declination (degrees) after a.f. demagnetization
INT: Intensity (x10^{-6} emu/cc) of a 6-cc specimen after a.f. demagnetization
k: Fisher precision parameter
α_{95}: Alpha-95: semiangle (degrees) of cone of 95% level of confidence for
              paleomagnetic directions
PLAT: North or south (-) latitude of Virtual Geomagnetic Pole (VGP)
PLONG: East longitude of VGP

correlative – ash bed LV-I at Long Valley and PAOH-I at Mono Lake
(Fig. 6). That ash bed and another, PAOH-II, which lies between
PAOH-I and PAOH-III, are significantly different in chemical
composition from each other and from the ash bed within the reversal
(LV-III and PAOH-III).

An estimate of the age of the reversal is based on correlation of
ash bed PAOH-II with Ash Bed JJ (Davis, 1985) in a stratigraphic
section containing multiple ash beds at Summer Lake, Oregon, and
correlation of two other ash beds in the latter section (Ash Bed V and

Table III.   Heavy mafic mineral frequencies of tephra samples from
Long Valley and Paoha Island, Mono Lake. Mineral separates are from
the magnetite +0.6 amp. fraction, density greater than 2.85 g/cm^3.
Frequencies were determined by the line counts method. Analyst: P.C.
Russell, U.S. Geological Survey, Menlo Park, California.

Mineral	1 LV-III (100-200) mesh	2 LV-III (200-350) mesh	3 PAOH-III (200-350) mesh	1 2 3 Percentages		
Green to brown hornblende	220	121	219	66	36	80
Hypersthene	89	133	45	27	39	16
Clinopyroxene	1	2	1	1	1	1
Oxyhornblende	1	p[1]	-	-	-	-
Biotite	p[1]	-	-	-	-	-
Opaque minerals	23	82	10	7	24	4
Totals	334	338	275	100	100	100
Unidentified transparent minerals	6	5	5			

[1] Mineral present but not counted.

Ash Bed KK; Davis, 1985) with two ash beds (T1193 and T2023) in a core
of lacustrine sediments that contains multiple ash beds at Tulelake,
California (Rieck et al., in press). Ages of the ash beds in the
Tulelake core have been estimated from a sedimentation-rate curve
based on tephrochonology and magnetostratigraphy (Rieck et al., in
press). The core interval containing ash beds T1193 and T2023 is
underlain by the Rockland ash bed, about 400 ka old (Sarna-Wojcicki et
al., 1985), and overlain by late Pleistocene ash beds and a
palaeomagnetic reversal believed to be the Blake reversed subchron
(Rieck et al., in press). Ash beds T1193 and T2023 at Tulelake are
estimated to be about 250 ka and 370 ka old, respectively. Using
250 ka and 370 ka as estimated ages for correlative Ash Beds V and KK,
respectively, at Summer Lake, Ash Bed JJ, by extrapolation, and
PAOH-II, by correlation, are then about 300 ka old. The age of ash
bed PAOH-III is then estimated to be about 280 ka by extrapolation
between ash beds PAOH-I and PAOH-II (Fig. 6).

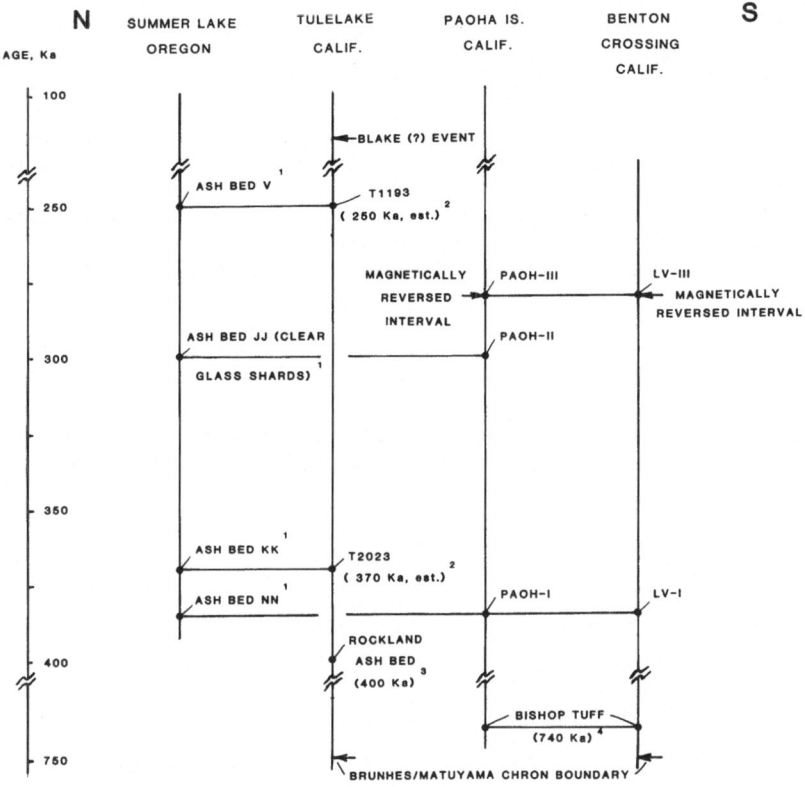

Figure 6. Correlation chart of ash beds from the locations in the western United States used in this investigation. Note that ash beds LV-III and PAOH-III are correlative at the magnetically reversed interval and are assigned the age 280 ka. (Written communication, A.M. Sarna-Wojcicki, 1988). Superscripts on some ash beds apply to references [1]Davis, 1985; [2]Reick et al., in press; [3]Sarna-Wojcicki et al., 1985; and [4]Bailey et al., 1976 and Izett et al., in press.

Thus the age of the Long Valley reversal appears to be about 280 ka. However, this estimate may be in error by as much as 50,000 years (about 20%), as a constant sedimentation rate was assumed between age control points at the three localities (Paoha Island, Summer Lake, and Tulelake), and such an assumption is probably not justifiable for most continental depositional basins. In addition, an aggregate error may have been accrued from the two successive interpolations and extrapolations. Nevertheless, an error larger that plus or minus about 20% is unlikely because unrealistically rapid sedimentation rates would have had to exist for certain intervals within these sections. There is no sedimentological or stratigraphic evidence at the three localities for such an unusually rapid sedimentation rate. Indeed, the lacustrine sediments at all four localities shown in Fig. 6 are fine grained and well bedded. Nor did we detect evidence of disconformities or hiatuses within those parts of the stratigraphic sections under consideration. A firm maximum age for the reversal is about 400 ka, based on the age of the Rockland ash bed (Sarna-Wojcicki et al., 1985; Fig. 6). That ash bed, although not found at Summer Lake or Tulelake, is present a few tens of kilometres north of Mono Basin near Bridgeport, California, and it very likely fell in the Mono Lake area. Although not exposed on Paoha Island, the Rockland ash bed probably is buried deeper in the section.

6. DISCUSSION

In this discussion of the palaeomagnetic field at Long Valley, only the data for specimens subjected to a.f. demagnetization are considered. However, the thermal demagnetization data from Site A are in general agreement with the a.f. demagnetization data (Fig. 7).

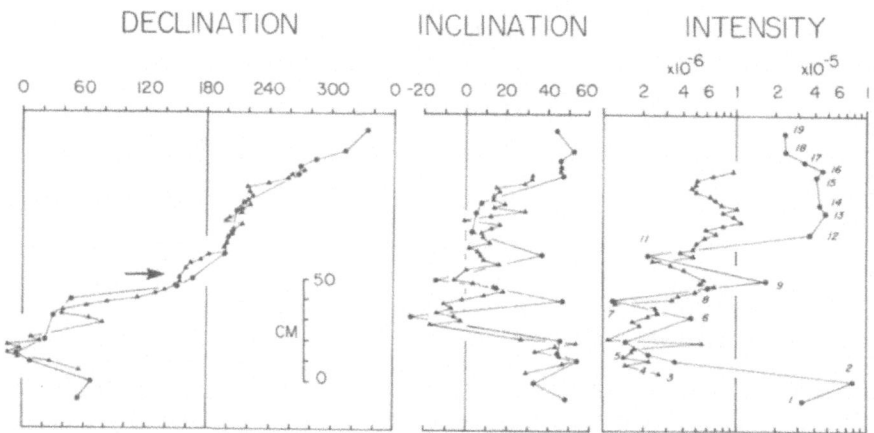

Figure 7. Plot of mean declination, inclination, and intensity (emu/cc) for Long Valley Site A after a.f. demagnetization at 300 oersted (circles, Table I, samples 1-19) and thermal demagnetization at 400°C (triangles, Table I, samples 36-79). The anomalous palaeomagnetic direction (D=142.4°, I=76.7°, Table I, sample 10) for ash bed LV-III is not plotted, but the position of the ash bed is indicated by the arrow.

The single sample that records reversed polarity (virtual geomagnetic pole [VGP] descends to 56.5°S latitude; Table I) is immediately below ash bed LV-III. The mean palaeomagnetic directions for the underlying sample and four successively overlying ones have VGPs that also plot in the southern hemisphere, and one sample (number 12) reaches 43.6°S (Fig. 8a; Table I). Those and subsequent VGPs

form a path that is about 90° west of the locality and trends from
southeast to northwest in the midlatitudes when traced from old to
young (Fig. 9). Following a pause, the traverse of poles from the
southern to northern hemisphere is rapid and favours a "near" field
geometry as proposed by Hoffman and Fuller (1978) and Fuller et al.
(1979). Similar behaviour is observed for Sites B and C (Figs. 8b and
8c). Ash bed PAOH-III in Mono Basin also has a VGP of reversed
polarity (samples 4 and 15, Table II) that is close to the VGP for the
sample adjacent to ash bed LV-III in Long Valley (Fig. 9).

Note that for Long Valley the declination changes systematically
from east to west when followed from old to young and that the pattern
is repeated at each site (Table I). Inclination, however, is quite
variable, and for two of the sites (A and C), has fluctuations of up
to 80°. Accompanying the directional changes is a relative decrease
in field intensity (unnormalized) at Site A; that behaviour is matched
somewhat in the thermal curve (Fig. 7) and at Site C, but not at Site
B (Table I). The variability of the records for inclination and
relative intensity between the three Long Valley sites demonstrates
the need for multiple records because different conclusions might be
drawn depending on which of the curves is examined. Additionally, our
ability to locate the reversal in Long Valley, rather than only on
Paoha Island, means we can discount the reversal as having been

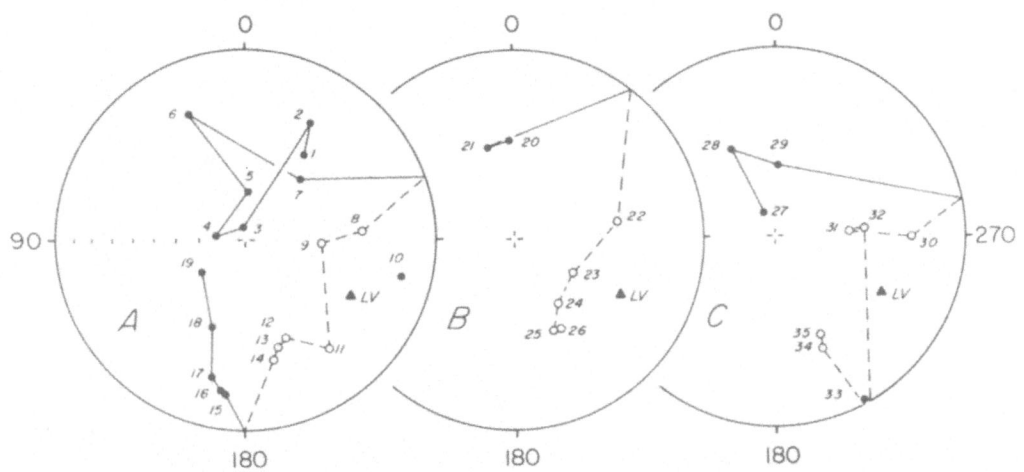

Figure 8. Equal-area plot of VGPs for Long Valley Sites, A, B and C.
All but one pole at Site A for the younger segment of the reversal
(samples 11-19) plot in the "near" hemisphere, as they do at Sites B
(samples 23-26) and C (samples 32-35). The filled triangle is the
locality position (37.6°N, 241.3°E), and the numbering of data is the
same as in Table I. Note that samples 12, 25 and 34 from Sites A, B
and C, respectively, have almost the same pole; they are all from
20 cm above ash bed LV-III and attest to the excellent reproducibility
of the record in silt at Benton Crossing.

150

extremely brief; that is, it lasted considerably longer than the
volcanic eruption that produced ash bed LV-III and PAOH-III.

Is the Long Valley reversal one that has been previously reported
in the literature, or is it a new one? The reversal cannot be the
Lake Mungo (Barbetti and McElhinny, 1972) or Laschamp reversed
subchrons, which are established as no older than 45,000 years,
because of age constraints provided by the tephra correlation shown in
Fig. 6. In addition, the ash beds on Paoha Island are
stratigraphically well below the base of deposits that correlate with
sediments dated about 12 to 36 ka on the northwest shore of Mono Lake
(Lajoie et al., 1980) that contain the Mono Lake Excursion (Denham and
Cox, 1971; Denham, 1974; Liddicoat and Coe, 1979).

If the age of the Emperor reversed subchron is about 450 ka
(Champion et al., 1981) then the 400 ka maximum age for the Long
Valley reversal imposed by the Rockland ash bed excludes the Emperor
from consideration. The tephra correlations shown in Fig. 6 indicate
that the Long Valley reversal also cannot be the Blake reversed
subchron, which is interpreted to be present high in the Tulelake core
(Rieck et al., in press). A more likely correlation would be the
short reversal Biwa 2, recorded in Lake Biwa, Japan, and estimated to
be about 290 ka old (Yaskawa et al., 1973). However, until the age of
280 ka for the Long Valley reversal can be confirmed by direct dating,
the correlation with Biwa 2 must remain conjectural.

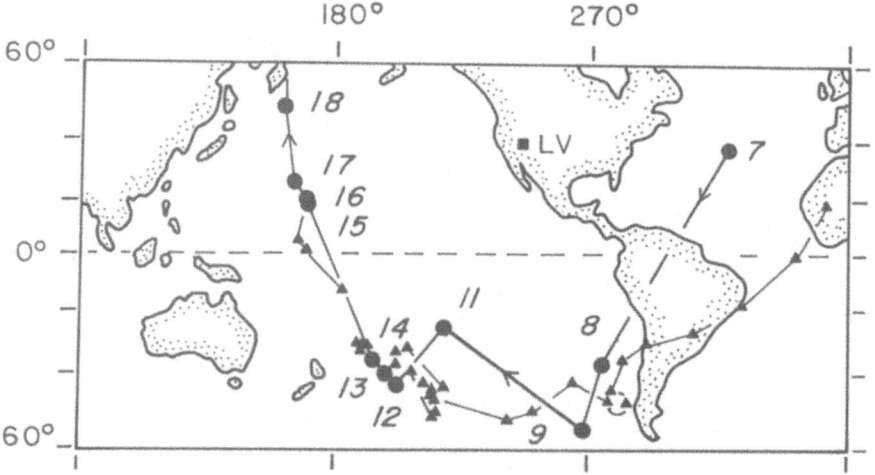

Figure 9. Plot of the VGPs for samples 7-18 for Long Valley Site A
(filled circles). Sample 9 records a reversal as defined by Cox et
al. (1975); samples 23 and 24 for Site B and 31 for Site C, not
plotted but shown in Fig. 8, also satisfy that definition, as do the
poles for samples 55, 56, 58-61, and 67 (Table I) that were thermally
demagnetized (filled triangles). The open circle between poles 8 and
9 that contains three triangles near the tip of South America is the
VGP for ash bed PAOH-III (sample 4, Table II) on Paoha Island.
Numbering of the poles is the same as in Table I.

Finally, if the Long Valley palaeomagnetic data are interpreted as a geomagnetic excursion (Cox et al., 1975), or an aborted reversal in a period of stable normal polarity, they counter the observation by Hoffman (1981) that the VGPs will follow a narrow longitudinal band in the hemisphere away ("far side") from the locality, suggestive of a quadrupolar intermediate field geometry. Most of the VGPs in Table I are confined to the hemisphere of Long Valley (i.e., are "near-sided"), and there is a broad zone in which they fall. As well, a large clockwise-trending loop is formed when the VGPs are followed from old to young (Fig. 8). Separating the field behaviour at Long Valley Site A into normal-to-reverse (N–R; older) and reverse-to-normal (R–N; younger) segments further shows that some N–R poles lie in the "far" hemisphere, and all but the youngest pole (number 19, Table I) during the R–N transition is in the "near" hemisphere. That pattern is consistent with the suggestion by Fuller et al. (1979) that there is a bias toward the "near" hemisphere during R–N transitions, such as the 730 ka old Matuyama/Brunhes. Regardless of how the succession of VGPs is interpreted, the Long Valley (and Mono Basin) palaeomagnetic data provide evidence of a large and rapid departure from the usual field geometry that should serve as a useful stratigraphic marker when dated more precisely.

7. CONCLUSIONS

1. Anomalous behaviour of the palaeomagnetic field resulting in a VGP of reversed polarity is recorded in dry lake sediments at two localities 40 km apart in the western Great Basin of the United States. At Long Valley, the palaeomagnetic directions are in silt above and below a distinctive ash bed (LV-III). In Mono Basin, that ash bed (PAOH-III) also records a VGP that descends below 45°S latitude. The clayey silts on Paoha Island do not record the reversal, but the mean direction for samples above and below PAOH-III is substantially different from that of an axial dipole field. At Long Valley, there is evidence of a relative decrease of the intensity of the palaeomagnetic field during the reversal.
2. The palaeomagnetic record in Long Valley and Mono Basin occurs within the Brunhes normal chron. Because a definitive age cannot be estimated for the ash bed within the reversal, the field behaviour cannot be positively correlated with other recorded subchrons or excursions of Brunhes age. However, the reversal is older than 36 ka and younger than 400 ka; on the basis of long-distance tephra correlations, it is estimated to be about 280 ka old. The reversal was of short duration, perhaps a few thousand years, assuming that the sedimentation rate in Long Valley was similar to that of other pluvial lakes in the Great Basin (Smith et al., 1983).

152

8. ACKNOWLEDGEMENTS

We thank K.R. Lajoie and Andrei M. Sarno-Wojcicki of the U.S.
Geological Survey for their assistance in many aspects of this
investigation. W.D. Bridge and J.M. Glen were also helpful in the
field and made some of the palaeomagnetic measurements.

9. REFERENCES

Bailey, R.A., Dalrymple, G.B. and Lanphere, M.A., 1976. 'Volcanism,
    structure, and geochronology of Long Valley caldera, Mono County,
    California'. Journal of Geophysical Research, 81, 725-744.
Barbetti, M.F. and McElhinny, M.W., 1972. 'Evidence of a geomagnetic
    excursion 30,000 yr B.P.'. Nature, 239, 327-330.
Bonhommet, N. and Zahringer, J., 1969. 'Paleomagnetism and potassium
    argon age determinations of the Laschamp geomagnetic polarity
    event'. Earth and Planetary Science Letters, 6, 43-46.
Champion, D.E. and Dalrymple, G.B., 1981. 'Radiometric and
    paleomagnetic evidence for the Emperor reversed polarity event
    at 0.46 ± 0.05 m.y. in basalt lava flows from the eastern Snake
    River Plain, Idaho'. Geophysical Research Letters, 8,
    1055-1058.
Cox, A., Hillhouse, J. and Fuller, M., 1975. 'Paleomagnetic records
    of polarity transitions, excursions, and secular variation'.
    Reviews of Geophysics and Space Physics, 13, 185-189.
Davis, J.O., 1985. 'Correlation of Late Quaternary tephra layers in
    a long pluvial sequence near Summer Lake, Oregon'. Quaternary
    Research, 23, 38-53.
Denham, C.R., 1974. 'Counter-clockwise motion of paleomagnetic
    directions 24,000 years ago at Mono Lake, California'. Journal
    of Geomagnetism and Geoelectricity, 26, 487-498.
Denham, C.R. and Cox, A., 1971. 'Evidence that the Laschamp polarity
    event did not occur 13,300-30,400 years ago'. Earth and
    Planetary Science Letters, 13, 181-190.
Fuller, M., Williams, I. and Hoffman, K.A., 1979. 'Paleomagnetic
    records of geomagnetic field reversals and the morphology of the
    transitional fields'. Reviews of Geophysics and Space Physics,
    17, 179-203.
Heller, F. and Petersen, N., 1982. 'The Laschamp excursion'.
    Philosophical Transactions of the Royal Society of London. A306,
    169-177.
Hoffman, K.A., 1981. 'Paleomagnetic excursions, aborted reversals
    and transitional fields'. Nature, 294, 67-68.
Hoffman, K.A. and Fuller, M., 1978. 'Transitional field
    configurations and geomagnetic reversal'. Nature, 273, 715-718.
Izett, G.A., Obradovich, J.D. and Mehnert, H.H., in press. 'The
    Bishop ash bed (middle Pleistocene) and some older (Pliocene and
    Pleistocene) chemically and mineralogically similar ash beds in
    California, Nevada, and Utah'. U.S. Geological Survey Bulletin
    1675.

Lajoie, K.R., 1968. 'Quaternary stratigraphy and geologic history of Mono Basin, eastern California'. Ph.D. Dissertation, University of California (Berkeley), 271 p.

Lajoie, K.R., Liddicoat, J.C. and Robinson, S.W., 1980. 'Dating and refinement of the Mono Lake paleomagnetic record'. Transactions of the American Geophysical Union (EOS), 61, 215.

Liddicoat, J.C. and Coe, R.S., 1979. 'Mono Lake geomagnetic excursion'. Journal of Geophysical Research, 81, 261-271.

Mayo, E.B., 1934. 'The Pleistocene Long Valley Lake in eastern California'. Science, 80, 95-96.

Rieck, H.J., Sarna-Wojcicki, A.M. and Adam, D.P. in press. 'Magnetostratigraphic and tephrochronology of upper Pliocene and Quaternary lake sediments, Tulelake, northern California', U.S. Geological Survey Open-File Report.

Sarna-Wojcicki, A.M., Bowman, H.R. and Russell, P.C., 1979. 'Chemical correlation of some late Cenozoic tuffs of northern and central California by neutron activation analysis of glass and comparison with X-ray fluorescence analysis'. U.S. Geological Survey Professional Paper 1147, 15p.

Sarna-Wojcicki, A.M., Bowman, H.R., Mayer, C.E., Russell, P.C., Woodward, M.J., McCoy, G., Rowe, J.J., Baedecker, P.A., Asaro, F. and Michael, H., 1984. 'Chemical analyses, correlations, and ages of upper Pliocene and Pleistocene ash layers of east-central and southern California'. U.S. Geological Survey Professional Paper, 1293, 40p.

Sarna-Wojcicki, A.M., Meyer, C.E., Bowman, H.R., Hall, N.T., Russell, P.C., Woodward, M.J. and Slate, J.L., 1985. 'Correlation of the Rockland ash bed, a 400,000-year-old stratigraphic marker in northern California and western Nevada, and implications for middle Pleistocene paleogeography of central California'. Quaternary Research, 23, 236-257.

Sarna-Wojcicki, A.M., Lajoie, K.R., Meyer, C.E., Adam, D.P. and Rieck, H.J. in press. 'Tephrochronologic correlation of upper Neogene sediments along the Pacific margin, conterminous United States'. in R.B. Morrison, ED., The Quaternary of the Unglaciated United States, Geological Society of America, Decade of North America Geology Volume K2.

Smith, G.I., Barczka, V.J., Moulton, G.F. and Liddicoat, J.C., 1983. 'Core KM-3, a surface-to-bedrock record of late Cenozoic sedimentation in Searles Valley, California'. U.S. Geological Survey Professional Paper 1256, 24p.

Stine, S., 1984. 'Late Holocene lake level fluctuations and island volcanism at Mono Lake, California', in Holocene paleoclimatology and tephrochronology east and west of the central Sierran crest, Field Trip Guidebook for the Friends of the Pleistocene, Pacific Cell, 21-49.

Verosub, K.L., 1975. 'Paleomagnetic excursions as magnetostratigraphic horizons: a cautionary note'. Science, 190, 48-50.

Yaskawa, K., Nakajima, T., Kawai, N., Torii, M., Natsuhara, N. and Horie, S., 1973. 'Paleomagnetism of a core from Lake Biwa (1)'. Journal of Geomagnetism and Geoelectricity, 25, 447-474.

MAGNETIC POLARITY TIME SCALES AND REVERSAL FREQUENCY

W. Lowrie
Institut fur Geophysik
ETH-Hönggerberg
8092 Zürich
Switzerland

ABSTRACT. The interpretation of oceanic magnetic anomaly sequences
provides the longest continuous record of geomagnetic polarity
available for analysis of reversal frequency. The polarity sequence
since the middle Jurassic has been largely confirmed and dated as a
result of coordinated magnetostratigraphic and palaeontological
investigations. Different polarity time scales for the same interval
reflect successive improvements in the definition of the polarity
sequence, the number of correlated tie-levels used, the optimum ages
accepted for the tie-levels, and different critical evaluations of the
radiometric age data base. The polarity sequence since the Late
Cretaceous is dominated by a steady increase in reversal rate, on
which is superposed a cyclicity of 30 Myr and its more distinctive 15
Myr harmonic. The distribution of polarity chron lengths is modelled
as a Poissonian distribution but because of the lack of short polarity
chrons in actual time scales a gamma renewal process has been
proposed. However, despite differences in analytical methods, the
available data do not justify rejecting the hypothesis of a Poissonian
model. The existence of incompletely resolved, very short polarity
chrons in the reversal record merits further investigation. Even if
all suggested short chrons are real, an apparent asymmetry in
stability of the normal and reversed polarity states is probably an
artefact of the method of describing the stability, and there is again
no statistical justification for supposing that one polarity state is
more stable than the other.

1. MAGNETIC POLARITY REVERSAL SEQUENCES

Numerous time scales based upon the history of geomagnetic field
polarity have been published in the past 25 years. These resulted
from successive refinements of a data base derived from three sources:
radiometrically dated lavas, interpretation of marine magnetic
anomalies and magnetostratigraphic studies in sedimentary rocks. The
progressive improvements introduced in major papers on the history of
geomagnetic field polarity written in the 1960's and 1970's were

F. J. Lowes et al. (eds.), Geomagnetism and Palaeomagnetism, 155–183.
© 1989 by Kluwer Academic Publishers.

reviewed critically by Ness et al. (1980) and only the main steps will
be summarised briefly here.

1.1 Radiometrically dated time scales

The most modern radiometrically dated time scale was developed by
Mankinen and Dalrymple (1979), who used revised decay and abundance
constants to recalculate the K-Ar ages of 354 data in the range 0-5
Myr. The optimum ages of polarity chron boundaries were determined
with the aid of chronograms, a statistical technique to find the
estimated boundary age which minimizes apparent dating inconsistencies
(Cox and Dalrymple, 1967). The tabulated results confirm that
radiometrically dated time scales are only practicable for about the
last 5 Myr of geomagnetic polarity history, beyond which the
experimental uncertainty of K-Ar dating (about 2%) becomes comparable
to the durations of many known short polarity chrons. The
radiometrically dated polarity time scale has been extended to over
6.5 Myr in stratigraphically related samples from Icelandic lava
sequences (McDougall et al., 1976, 1977).

1.2 The marine magnetic polarity record

The interpretation of marine magnetic anomalies gives a continuous
geomagnetic reversal record covering about the last 150 Myr. This
data base consists of two reversal sequences which span respectively
(a) the Cenozoic and Late Cretaceous and (b) the Early Cretaceous and
Late Jurassic. These polarity sequences are separated by the
Cretaceous Superchron of normal polarity. The Cenozoic and Late
Cretaceous sequence is the more intensively studied, and several
magnetic polarity time scales have been proposed for this interval.
      From comparative analyses of distances from ridge crests to
individual magnetic anomalies in the North and South Pacific, Indian
and South Atlantic oceans, Heirtzler et al. (1968) selected the South
Atlantic Vema-20 anomaly sequence as most likely to have been formed
during conditions of constant sea-floor spreading. The most
recognisable anomalies were numbered in order of increasing age from 1
to 32. Cande and Kristoffersen (1977) selected North Pacific marine
magnetic profiles as the global standard for the Late Cretaceous and
extended the sequence of lineated anomalies to anomaly 34. Large
areas of the Atlantic and Pacific oceans older than this polarity
sequence are free of correlatable lineated magnetic anomalies. These
"magnetic quiet zones" were interpreted to have formed during an
extended period of constant normal geomagnetic polarity, the
Cretaceous Normal Polarity Superchron (Cox, 1982).
      The oldest parts of the Atlantic and Pacific ocean basins were
also found to be characterised by lineated magnetic anomalies, the
M-sequence. Larson and Pitman (1972) identified the M-sequence
anomalies analogously to the younger sequence, numbering them in order
of increasing age M1 to M22. A younger anomaly, M0, was identified by
Larson and Hilde (1975) who also extended the sequence to M25. The
transition to an older, Jurassic quiet zone is indistinct. Cande et

al. (1978) described low amplitude, lineated anomalies M25–M29 in the Pacific ocean, which were attributed to short polarity reversals which occurred during a period of low field intensity. These anomalies appear as tiny perturbations on the marine magnetic record and their interpretation as polarity chrons is not unique; small scale marine magnetic anomalies could be due to geomagnetic field intensity fluctuations (Cande and LaBrecque, 1974). However, the Pacific set of polarity chrons for M25–M29 has found its way into more recent magnetic polarity time scales.

The resolution of the marine magnetic record is not limited by being a fixed proportion of the absolute age. The resolvable anomaly amplitude at the ocean surface depends on the water depth, crustal magnetization, thickness and width of the crustal block. Typically a block about 1 km wide can be resolved, which, depending on the spreading rate, is equivalent to a resolution of about 0.02–0.10 Myr. The relative ages of chrons in marine-based magnetic time scales are determined by the anomaly spacings, and are usually given to 0.01 Myr. The absolute ages of these chrons are not nearly so well known. The absolute accuracy of Late Cretaceous time scales is about 1 Myr, and age estimates for the M-sequence polarity chrons differ by as much as 10 Myr.

1.3 Magnetostratigraphic correlation and dating of polarity chrons

Because of the global nature of geomagnetic field reversals, magnetic polarity stratigraphy is a powerful method for correlating one stratigraphic section with another. Coordinated magnetostratigraphic and biostratigraphic studies provide a more reliable and finer correlation of stratigraphic sections with each other than is possible with one data set alone. They also tie major biostratigraphic events such as major stage boundaries to the geomagnetic polarity sequence and form the basis for calibration of magnetic polarity time scales.

The most common method of identifying a sequence of magnetozones in a stratigraphic section is to find the equivalent pattern in the oceanic record. Consequently, the temporal resolutions of magnetostratigraphic studies and the marine polarity record are comparable, about 0.02–0.10 Myr. If a distinctive polarity "fingerprint" is not present, an unambiguous correlation or identification may not be possible. The pattern recognition is usually done visually, but an alternative quantitative method is by the use of correlograms (Langereis et al., 1984). The cross-correlation function of polarity interval lengths is computed at successive matching positions as one polarity sequence is moved progressively past the other, and the maximum value corresponds to the optimum matching position.

1.3.1 The Palaeogene and Cretaceous.
The magnetic polarity chrons of the Palaeogene and Late Cretaceous have been confirmed independently by magnetostratigraphic results in pelagic marine carbonate sections from the Tethyan realm (Fig. 1) now exposed on land (Alvarez et al., 1977; Lowrie et al., 1982) or sampled in Deep Sea Drilling Project

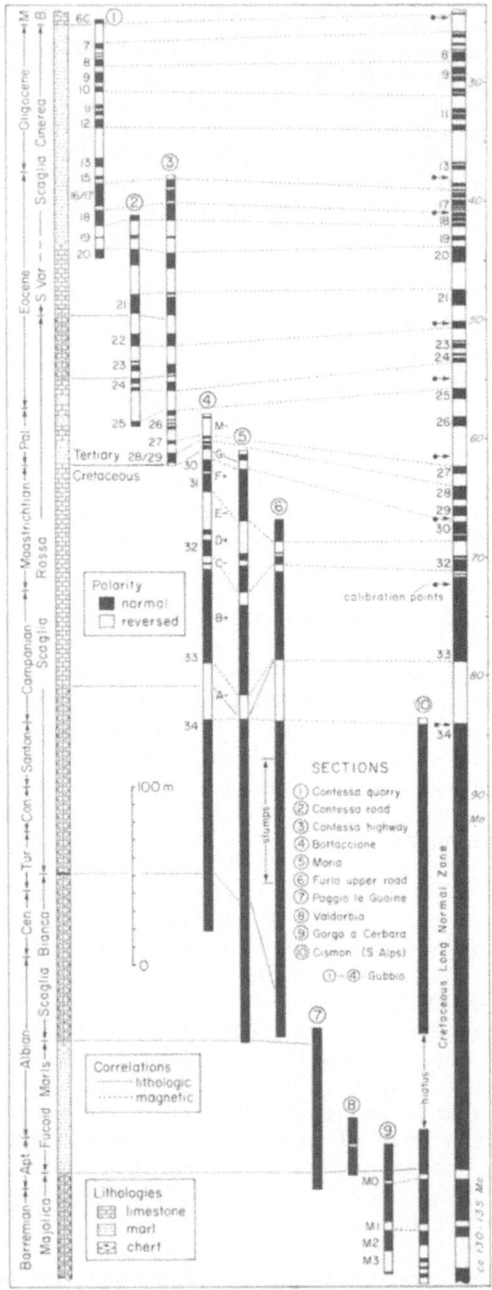

Figure 1.   Correlations of Cenozoic, Late and Middle Cretaceous magnetostratigraphic sections in Umbria and the Southern Alps with the oceanic magnetic polarity record (Lowrie and Alvarez, 1981).

cores (LaBrecque et al., 1983). The positions of major palaeontological stage boundaries in these sections have been tied securely to the geomagnetic polarity sequence. For example, the Oligocene/Miocene boundary falls near the young edge of chron 6Cr and the Eocene/Oligocene boundary lies within chron 13r (Lowrie et al., 1982). The position of the Cretaceous/Tertiary boundary is found within chron 29r close to its younger margin, and the end of the Cretaceous normal superchron (CNS) is just older than chron 33r (Alvarez et al., 1977; Lowrie and Alvarez, 1977).

The Cretaceous normal superchron starts after the youngest polarity chron of the M-sequence (MO), which has been dated as very early Aptian, just younger than the Aptian-Barremian boundary, in pelagic carbonate sections in the Southern Alps (Channell et al., 1979) and in Umbria (Lowrie et al., 1980). Occasional reports of negative polarity zones within the CNS have not been verified by repetition in other magnetostratigraphic sections.

The M-sequence oceanic magnetic anomalies span the Lower Cretaceous and Upper Jurassic. Most of the Lower Cretaceous polarity chrons were confirmed in magnetic stratigraphy investigations in Umbria (Lowrie and Alvarez, 1984; Cirilli et al., 1984), with the exception of chrons M11r to M13n (Fig. 2). The limestone magnetization in this stratigraphic interval was too weak to be measureable. However, recent studies in the Italian Southern Alps have independently confirmed and dated the polarity chrons from M8 to M23, including the interval M11r-M13n (Channell et al., 1987).

1.3.2 <u>The Jurassic/Cretaceous Boundary</u>. The location and correlation of the Jurassic/Cretaceous boundary to the magnetic polarity time scale illustrates how magnetostratigraphy can help with the synchronisation of different palaeontological dating schemes. Palaeontologists do not agree on the definition of this boundary, due in part to the difficulty of correlating one palaeontological dating scheme with another and to the standard ammonite zonation. Results from Umbria, the Southern Alps and DSDP sites (Fig. 3) give different relative positions of the boundary, depending on which palaeontological scheme is used (Ogg, 1984). It has been correlated as young as the old edge of chron M17r, on the basis of calpionellids in the Umbrian Maiolica limestone section (Lowrie and Channell, 1984a), and as old as the middle of the chron M19n, based on nannofossils, dinoflagellates and calpionellids in DSDP site 534 (Roth, 1983). Channell and Grandesso (1987) reinterpreted the Umbrian calpionellid zonation and suggested a correlation within M18n. Observing that most correlations place the Jurassic/Cretaceous boundary near reverse polarity chron M18r, Ogg and Lowrie (1986) have suggested resolving the uncertainty of palaeontological definitions of the boundary by defining it at the base of chron M18r, which is a readily identifiable event. Correlations consistent with this definition have been found in magnetostratigraphic sections in the Southern Alps (Channell et al., 1987).

160

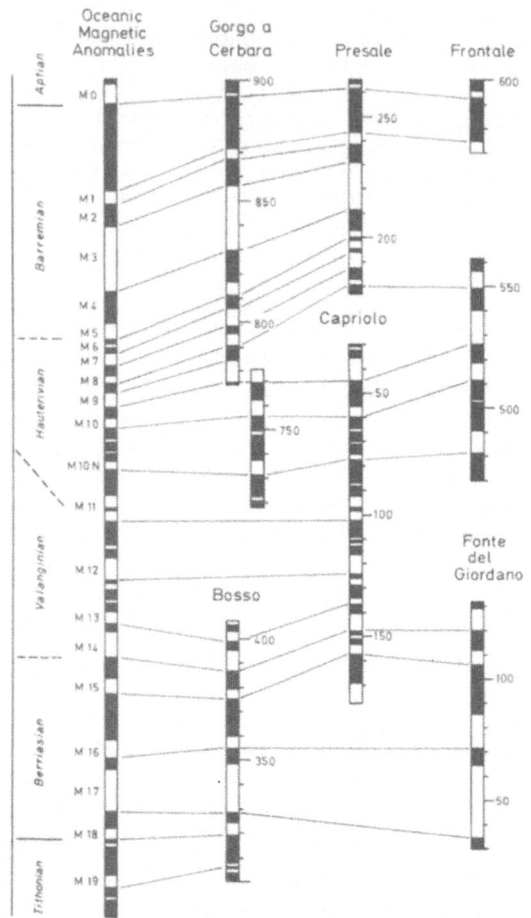

Figure 2. Correlations of Lower Cretaceous M-sequence magnetostratigraphic sections in Umbria and the Southern Alps, Italy. Section locations: <u>Umbria</u>: Bosso (Lowrie and Channell, 1984a), Gorgo a Cerbara, Presale, Frontale (Lowrie and Alvarez, 1984), Fonte del Giordano (Cirilli et al., 1984); <u>Southern Alps</u>: Capriolo (Channell et al., 1987).

1.3.3 <u>The Jurassic</u>. The M-sequence magnetic polarity for the Late Jurassic (Fig. 4) has been confirmed in the Foza and San Giorgio magnetostratigraphic sections in northern Italy (Ogg, 1981) and in the Carcabuey and Sierra Gorda sections in southern Spain (Ogg et al., 1984). An important result from these studies is the correlation of the Oxfordian/Kimmeridgian boundary with polarity chron M25n.

Magnetic stratigraphy research in older Jurassic sections has been less successful in confirming the marine record. Magnetozones with alternating polarity were found in four overlapping sections of

Middle—Late Oxfordian pelagic limestone near Aguilon in northern Spain
(Steiner et al., 1985), but the sequence did not correlate well with
the Pacific oceanic pattern for M25-M29.

For earlier Jurassic polarity history there is no oceanic record
as reference. Magnetic polarity sequences, dated with ammonites in
some cases, have been reported for early Jurassic sections in Italy,
Hungary and Spain but coeval polarity records do not agree well with
each other (Channell et al., 1982). Contemporaneous sections from
different sedimentary environments with different histories might not
be expected to show the same polarity sequence. However, overlapping
sections from nearby localities within Umbria (Channell et al., 1982)
and in northern Spain (Steiner et al., 1987) show poor consistency.
The breakdown of consensus between early Jurassic magnetostratigraphic
sections may be a result of variable sedimentation rate within and
between the sections. The lack of mutual confirmation, in the absence
of a marine polarity sequence, presents a serious obstacle to
establishing a standard polarity time scale for the middle and Early
Jurassic.

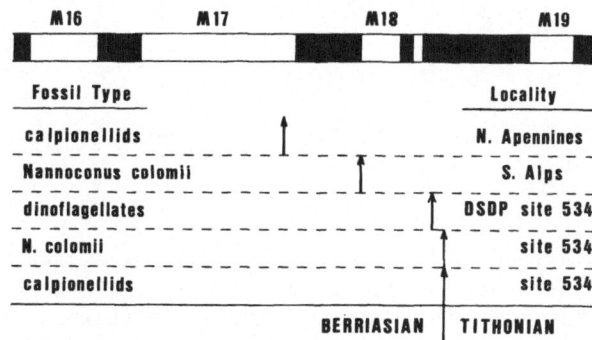

Figure 3. Comparison of correlations of the Jurassic/Cretaceous
boundary based on different fossil dating schemes (Lowrie and
Channell, 1984b).

1.4 Biostratigraphy and the magnetic polarity record

The most significant results which link the oceanic magnetic record to
a framework of biostratigraphic ages are (a) the Palaeogene stage
boundary correlations, especially the Cretaceous/Tertiary boundary
correlation with the upper part of chron 29r, (b) the end of the
Cretaceous "quiet interval" close to the Santonian Campanian boundary,
(c) the dating of chron MO as earliest Aptian, very close to the
boundary with the Barremian, (d) the association of the
Jurassic/Cretaceous boundary with chron M18r, and (e) the correlation
of the Oxfordian/Kimmeridgian boundary with polarity chron M25n. By
providing calibration points for the oceanic magnetic reversal
sequences these magnetostratigraphic correlations have most strongly
influenced the development of successive magnetic polarity time scales.

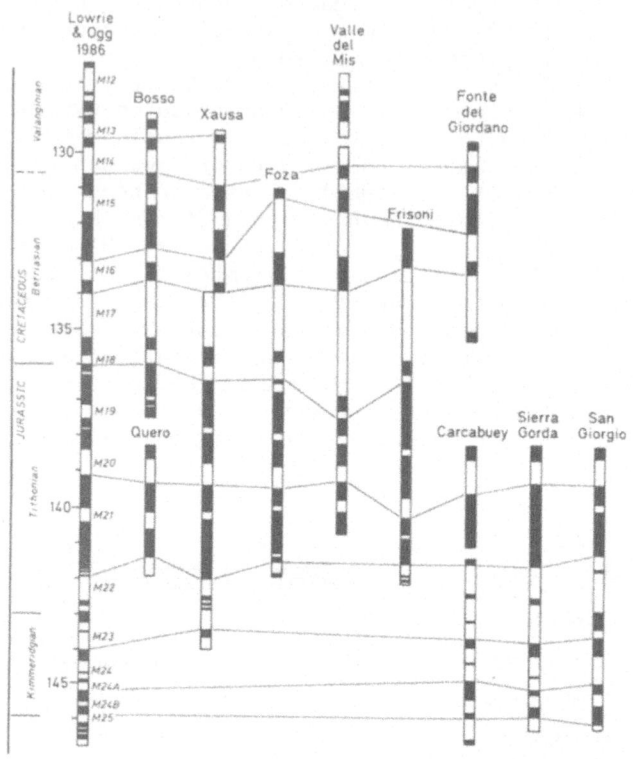

Figure 4. Correlations of Early Cretaceous and Late Jurassic
magnetostratigraphic sections in Umbria, the Southern Alps and Spain
with the M-sequence magnetic polarity time scale (Lowrie and Ogg,
1986). Locations and sections: Umbria: Bosso (Lowrie and Channell,
1984a), Southern Alps: Xausa, Valle del Mis, Quero, Frisoni (Channell
and Grandesso, 1987; Channell et al., 1987), Foza and San Giorgio (Ogg
et al., 1984); Spain: Carcabuey and Sierra Gorda (Ogg et al., 1984).

2. MAGNETIC POLARITY TIME SCALES

Of the many magnetic polarity time scales that have been constructed
for the time interval since the middle Jurassic, only those which have
made dramatic departures from their predecessors are summarized here.
These can be divided into time scales for the Cenozoic and Late
Cretaceous sequence, and the M-sequence, respectively.

2.1 Cenozoic and Late Cretaceous time scales

2.1.1 Heirzler et al. (1968). The first magnetic polarity time scale
for the entire Cenozoic and Late Cretaceous was constructed by
Heirtzler et al. (1968), on the basis of the long Vema-20 magnetic

anomaly profile in the South Atlantic. The old edge of the Gauss normal polarity chron was identified by comparison with the known radiometric time scale. Knowing its distance from the ridge and using an age of 3.35 Myr an average ridge-crest spreading rate was calculated. The ages of all other polarity chron boundaries were calculated by dividing their distances from the ridge by the ridge-crest spreading rate. Subsequent studies have confirmed that the assumption of constant sea-floor spreading in the South Atlantic was well founded. The polarity sequence dated in this way was found to extend back into the Late Cretaceous; the Cretaceous/Tertiary boundary was estimated to be just older than anomaly 26. The magnetostratigraphic results from the Gubbio Bottaccione section established directly that the correct correlation is with the top of chron 29r (Alvarez et al., 1977; Lowrie and Alvarez, 1977). It was later discovered that the feature identified as anomaly 14 in the polarity sequence did not occur on other marine profiles.

2.1.2 LaBrecque et al. (1977). The magnetostratigraphic correlation of the Cretaceous/Tertiary boundary was adopted as a tie-point with an age of 65 Myr (Van Hinte, 1976a) in a revised polarity time scale (LaBrecque et al., 1977), which also incorporated several other improvements. The anomaly 14 reversals were omitted from the polarity sequence, and the relative lengths of the Late Cretaceous chrons 29 to 34 were revised, in accordance with the results of Cande and Kristoffersen (1977). The ages of polarity chron boundaries between anomalies 2A and 29 were calculated by interpolation and between 29 and 34 by extrapolation. The sequence of polarity in the LaBrecque et al. (1977) time scale has been used with only minor modifications in all subsequent time scales, which have stretched or contracted the sequence in linear segments between selected tie-points.

2.1.3 Ness et al. (1980). Using the revised decay and abundance constants for K/Ar dating, Ness et al. (1980) recalculated the ages of palaeontological stage boundaries in the Cenozoic and Late Cretaceous. They adopted the improvements to the polarity sequence made by LaBrecque et al. (1977) and further adjusted the age of anomaly 24 to be close to 55 Myr as a compromise agreement with other time scales. They then recomputed the ages of polarity chrons in the modified Heirtzler et al. (1968) time scale. This time scale did not include any substantial improvements in correlation between the polarity sequence and biostratigraphic history.

2.1.4 Lowrie and Alvarez (1981). The magnetostratigraphic correlations of the major stage and sub-stage boundaries of the Late Cretaceous and Palaeogene, gave 11 tie-points which were used by Lowrie and Alvarez (1981) as a grid for a further revision of the magnetic polarity time scale. The tie-points were allocated the revised absolute ages calculated by Ness et al. (1980) and chron boundary ages were determined by interpolation. A comparison of palaeontological ages of bottom sediments in DSDP holes with ages predicted from this time scale shows that it is better tied to the biostratigraphy than the earlier time scale of LaBrecque et al. (1977), which had large discrepancies in the Late Palaeocene and early Eocene. However, the use of estimated ages for all correlated stage and sub-stage boundaries was not sufficiently discriminating, and resulted in apparent sudden changes in sea-floor spreading rate.

2.1.5 Cox (1982). To minimize sea-floor spreading accelerations, while utilizing as many as possible of the magnetostratigraphic correlations, Cox (1982) dropped the sub-stage boundaries as tie-points. He used the recomputed boundary ages of Ness et al. (1980), but adjusted the absolute ages of the tie-points within their error limits so as to obtain a time-scale which was more consistent with a model of nearly constant sea-floor spreading.

2.1.6 Berggren et al. (1985). A more radical approach to revising the polarity time scale for the Cenozoic was adopted by Berggren et al. (1985) who considered the absolute ages associated with magnetostratigraphically correlated stage boundaries to be unreliable. They divided the Cenozoic and Late Cretaceous polarity record into three segments, with intersection points at the old edges of anomalies 5 and 24, where the marine polarity record indicates major adjustments in spreading rate may have taken place. Three selected radiometric age dates were used to calibrate the crucial middle segment. Two of these age dates were for volcanic ash beds in an Oligocene magnetostratigraphy for continental sediments (Prothero et al., 1982). The quality of the palaeomagnetic data in this magnetostratigraphic section is poor, and the interpreted polarity sequence is not unique. The magnetostratigraphic correlations of the dated ash beds are inadequate, and moreover, the errors associated with the age dates are unknown. Because of these inadequacies the Berggren et al. (1985) magnetic polarity time scale is no more securely dated than its predecessors.

2.2 M-sequence time scales

The polarity sequence used in all M-sequence time scales is that obtained from the interpretation of magnetic anomalies in the North Pacific and Atlantic oceans (Larson and Pitman, 1972; Larson and Hilde, 1975). The several magnetic polarity time scales that have been proposed fall in two categories: earlier time scales which were dated solely using DSDP data, and later time scales which incorporate magnetostratigraphic correlations of dated stage boundaries.

2.2.1 Time scales calibrated by DSDP data. The original calibration of the M-sequence anomalies (Larson and Pitman, 1972; Larson and Hilde, 1975) was based upon the palaeontological ages of the sediment in contact with igneous basement in DSDP holes drilled near the young and old ends of the M-sequence. Assuming absolute ages for these tie points, the intervening polarity chrons were dated by linear interpolation.

The same method was used by Vogt and Einwich (1979), who added an
additional early Valanginian calibration point for DSDP site 387
within the M-sequence. This calibration technique was the only one
available until recently, but it has serious drawbacks. It is often
uncertain whether a DSDP hole has indeed reached igneous basement, or
whether it has merely encountered an intrusion. Also, the length of
time between formation of the lava and deposition of the first datable
sediments is not known. Finally, the palaeontological dates of the
bottom sediments often have large errors, amounting to an entire stage
or more, and ages determined with different fossil types can differ
appreciably.

2.2.2 <u>Magnetostratigraphically calibrated time scales</u>. The
correlation of chron MO with the very earliest Aptian, close to the
Barremian boundary in pelagic limestone magnetostratigraphic sections
in Italy (Channell et al., 1979; Lowrie et al., 1980) was incorporated
as the younger of two tie-points in a magnetic polarity time-scale for
the M-sequence by Cox (1982; in Harland et al., 1982). However, the
older tie-point was a palaeontological age from DSDP site 105.

The available radiometric ages are inadequate to date the Early
Cretaceous and Jurassic stage boundaries adequately. Harland et al.
(1982) decided that the chronogram determinations of optimum ages
could not be used from the Triassic Ladinian/Asinian boundary to the
mid-Cretaceous Aptian/Albian boundary. They interpolated the ages of
all intervening stage boundaries on the hypothesis that each stage had
an average equal duration. However, closer inspection shows that this
interval is better represented by at least two linear segments,
intersecting at the Jurassic/Cretaceous boundary.

Kent and Gradstein (1985) used the magnetostratigraphic tie for
MO and the correlation of the Oxfordian/Kimmeridgian boundary with
chron M25n (Ogg et al., 1984). However, their time scale was based
upon age estimates for the Barremian/Aptian and Oxfordian/Kimmeridgian
stage boundaries taken from the Harland et al. (1982) time scale.

Lowrie and Ogg (1986) used the correlation of the
Jurassic/Cretaceous boundary with the base of polarity chron M18r as
an additional tie-point. They re-evaluated the chronogram evidence
and found an acceptable age of 136 Myr for this tie-point. They
adopted the age estimates given by Hallam et al. (1985) for the MO and
M25n correlations in preference to those of Harland et al. (1982). A
magnetic polarity time scale for the M-sequence polarity chrons was
then constructed by linear interpolation on the two segments MO-M18
and M19-M25; ages for M25-M29 were extrapolated from the M19-M25
segment. As a result of the additional Jurassic/Cretaceous tie point
and the different ages assumed for the three calibration points, the
ages of all polarity chrons in the time scale of Lowrie and Ogg (1986)
are 5-10 Myr younger than in the time scale of Kent and Gradstein
(1985), although they agree quite closely with earlier estimates of
these stage boundaries by Van Hinte (1976a, 1976b).

2.3 Optimum magnetic polarity time scales

The differences between current versions of the Late Cretaceous and
Cenozoic magnetic polarity time scale are small but are especially
noticeable in the Eocene and Palaeocene. For example, different time
scales have given a broad range of estimates for the age of the old
boundary of chron 24 (Table I). On balance, the most appropriate
magnetic polarity time scale for the Late Cretaceous and Cenozoic is
currently that of Cox (1982). It makes optimum use of firmly
established magnetostratigraphic correlations and age dates, while
minimizing drastic changes in sea-floor spreading rate.

TABLE I

Authors of Time Scale	Chron 24 Age (Myr)
Heirtzler et al., 1968	60.5
LaBrecque et al., 1977	56.6
Ness et al., 1980	54.9
Lowrie and Alvarez, 1981	53.6
Cox, 1982	53.1
Berggren et al., 1985	56.1

The construction of a reliable magnetic polarity time scale for
the M-sequence polarity chrons is handicapped by the paucity of well
dated tie-levels. Although several Lower Cretaceous and Upper
Jurassic stage boundaries have been correlated satisfactorily to the
M-sequence, the absolute ages for these boundaries are not well known.
The M25-M29 sequence of low amplitude anomalies has not been confirmed
as short polarity intervals, yet have been included (perhaps
prematurely) in recent M-sequence time scales. The different values
accepted for the absolute ages of calibration points result in large
differences between M-sequence time scales (Fig. 5). The estimated
durations of the Cretaceous normal superchron (CNS) and the M-sequence
of alternating polarity are quite varied, lasting 27-35 Myr and 36-47
Myr, respectively, depending on the time scales used (Table II).
Although coordinated magnetostratigraphic and palaeontological
investigations have confirmed and dated both polarity sequences fairly
well, there is no definitive version of either time scale. This will
remain the case until there are significant improvements in the number
and quality of absolute age estimates for the stage boundaries used as
magnetostratigraphic tie-levels.

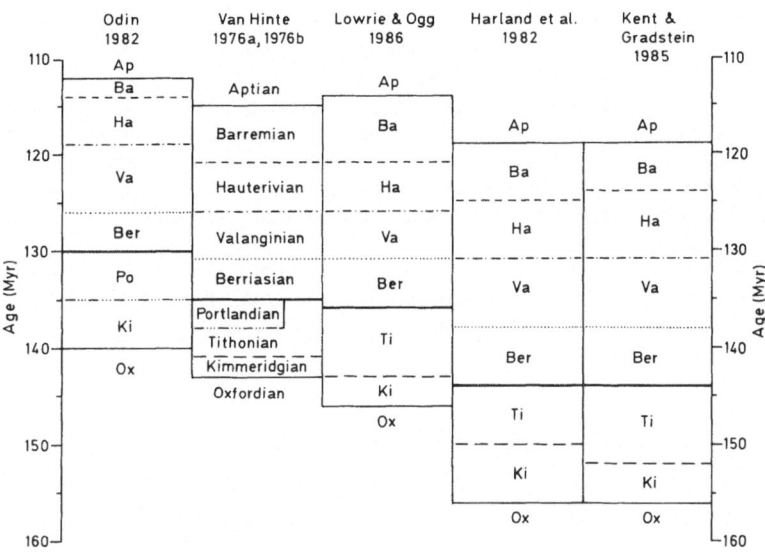

Figure 5. Absolute ages accorded to Lower Cretaceous and Upper Jurassic palaeontological stage boundaries in some recent time scales.

TABLE II

Time Scales	Length of CNS (Myr)	Length of M-sequence (Myr)
LaBrecque et al., 1977; Larson and Hilde, 1975	27	45
Harland et al., 1982	35	47
Kent and Gradstein, 1985	34	42
Lowrie and Alvarez, 1981; Lowrie and Ogg, 1986	29	36

3. ANALYSIS OF GEOMAGNETIC REVERSAL FREQUENCY

When the number of reversals per million years is calculated for successive narrow intervals a few million years in width, every modern magnetic polarity time scale for the Cenozoic-Late Cretaceous sequence has similar characteristics: the reversal frequency since the end of the CNS is dominated by a monotonic increase on which a slowly fluctuating component is superposed (Fig. 6). A histogram window width of about 8-10 Myr is enough to smooth out the fluctuating component, leaving a linear trend (Lowrie, 1982). The M-sequence shows a decreasing trend in reversal frequency towards the CNS, but it is less clear that the deviations from the trend are cyclical. Some authors have interpreted these two trends together with the CNS as a very long period fundamental harmonic frequency. Mazaud et al. (1983) suggested that the Cenozoic-Late Cretaceous long term trend may represent a core process which controls the overall frequency of reversals. The monotonic trends have no satisfactory theoretical explanation.

Studies of reversal frequency have focussed on two main themes: (1) the periodicity of the long term fluctuation, and (2) the statistical distribution of polarity chron lengths.

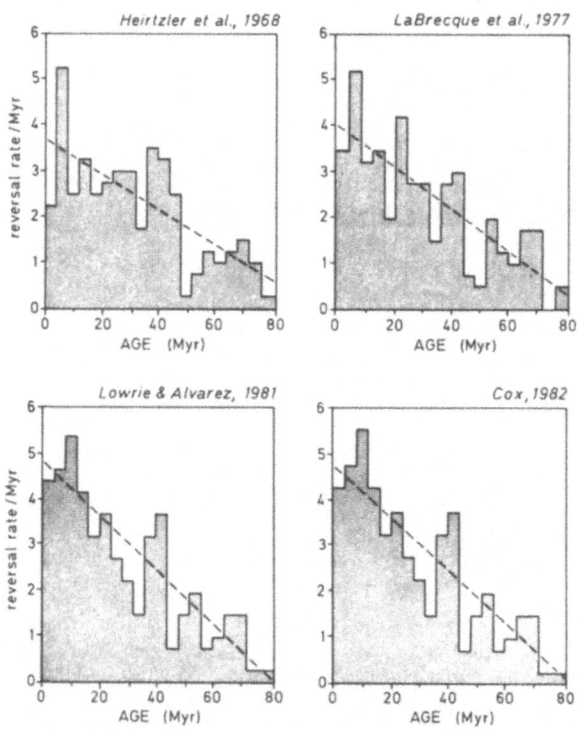

Figure 6. Variation with age of 4 Myr averages of reversal frequency in some magnetic polarity time scales. A best-fit linear trend is shown for each time scale.

3.1 The periodic component

In Fig. 7a the average reversal frequency has been computed for fixed windows with a width of 4 Myr for the Cenozoic-Late Cretaceous time

scale of Cox (1982). When the trend, represented by a third order
polynomial, is subtracted, the cyclical nature of the residual
component is clearly evident (Fig. 7b). By averaging the times
between in-phase zero-crossings a period of 14.5 Myr is obtained.

Mazaud et al. (1983) examined the reversal data with a 4 Myr
width moving window and modelled the long-term trend with a Cauchy
function; autocorrelation of the residual cyclical signal gave a
period of approximately 15 Myr. McFadden (1984a) disputed this result
and proposed that the periodicity was an artefact of the fixed-length
sliding window method of analysis. However, Mazaud et al. (1984)
found that for all window widths from 1 to 10 Myr the periodicity was
confined to the narrow range of 13-17 Myr. They applied their method
to synthetic computer-generated time series and found that, although
some produced a periodic behaviour, this was not the general case.
They did not exclude the possibility that the observed periodicity
could result from the probabilistic process controlling polarity
interval lengths.

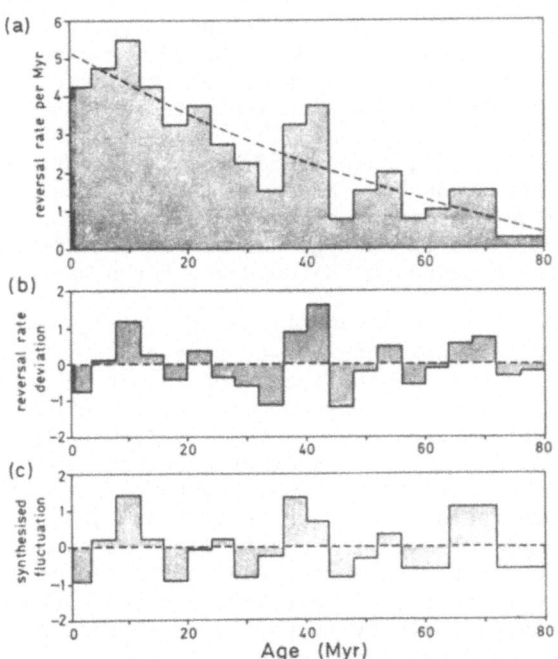

Figure 7. (a) The reversal rate averaged for successive 4 Myr
intervals in the Cenozoic-Late Cretaceous magnetic polarity time scale
of Cox (1982). (b) Reversal rate deviations from the third order
polynomial trend. (c) The sum of in-phase fluctuations with periods
29 Myr and 14.5 Myr.

When the Cenozoic-Late Cretaceous reversal sequence is filtered through a 15 Myr sliding window to remove the cyclicity described by Mazaud et al. (1983), Fourier and maximum entropy spectral analysis of the residual signal give dominant peaks near to 30 Myr (Pal and Creer, 1986). This period could not be securely established for the M-sequence, which was too short for this type of analysis.

In most studies of the fluctuation in reversal frequency the entire Harland et al. (1982) time scale has been analysed as a single continuous time series. Raup (1985) analysed the reversal record of the last 165 Myr in the Harland et al. (1982) polarity time scale. He used a nonparametric test method developed by Stothers (1979) which does not assume that the time series contains a sinusoidal component. However, spectral peaks at several periods were found, the strongest corresponding to a periodicity of 30 Myr. A significance test of this spectral peak was devised using a bootstrap method. A synthetic time scale was simulated by rearranging the sequence of chrons in the Harland et al. (1982) time scale with a Monte Carlo technique, so that the simulation contained the same chrons (including the CNS), but in a different order. The time series analysis was repeated for 200 simulations to determine the spectral intensity below which 95% of the simulated spectra were contained. Only the 30 Myr spectral peak in the origin analysis was higher than this 95% confidence limit. Raup (1985) concluded that the peak was unlikely to be an accident or an artefact of the method of analysis. In a separate test of the Cenozoic-Late Cretaceous sequence (0-83 Myr) the 30 Myr period was again found to be the strongest, although in this case only 83% of simulations gave a poorer goodness of fit test. All periods in the range 22-23 Myr tested in the M-sequence showed excellent fit, possibly because the M-sequence is too short for a search for long periods.

Instead of the Stothers (1979) linear prediction model Lutz (1985) used a circular model with a "wrapping effect" correction, which compensated for the length of the record not being an exact multiple of the trial period. When applied to the Harland et al. (1982) time scale this method again gave a 30 Myr periodicity. However, Lutz (1985) found that the wrapping effect correction changed the significance of all spectral peaks. Systematic truncation of the record from the recent end caused the periods and phases of all peaks to change in a way consistent with their origin as harmonics of the long-term variation between the high reversal frequency at present and at the old end of the record. Lutz (1985) concluded that there is no indication of a true periodicity in the reversal record.

Stothers (1986) did not agree with this conclusion. By progressively truncating the Harland et al. (1982) record from the young end he investigated the robustness of the spectral peaks. The dominant peak was always near to 30 Myr, independently of the record length. Separate spectral analyses of the M-sequence and Cenozoic-Late Cretaceous sequence also gave peaks near to 30 Myr, which showed that the length of the CNS was not responsible for this periodicity. Starting with very short chrons lasting only 0.04 Myr he systematically pruned ever longer chrons from the record and found

that this simulation of the effect of possible undetected short
reversals caused no major change in the 30 Myr period. On the basis
of this statistical modelling Stothers concluded that the robust 30
Myr period was a real feature of the magnetic reversal record, and
that the 15 Myr period (Mazaud et al., 1983) was probably a harmonic
of the 30 Myr period. The sum of a 14.5 Myr periodicity and an
in-phase 29 Myr periodicity with 60% amplitude is shown for comparison
with the residual fluctuation in Fig. 7c.

The use of the entire Harland et al. (1985) time scale for
analysis of reversal frequency is inappropriate, because part of the
database is too insecure for this purpose. The M-sequence reversals
are inadequately calibrated and dated, thus the durations of the very
long normal polarity interval (CNS) and the M-sequence are uncertain
(Table II). The unconfirmed M25-M29 chrons introduce 20 reversals in
the oldest 5 Myr of the time scale, producing a strong, possibly
spurious peak in reversal frequency which dominates the old end of the
Harland et al. (1985) 0-165 Myr polarity record. The CNS and older
part of the polarity record should not be incorporated in analyses of
geomagnetic reversal frequency. The 30 Myr periodicity deduced from
analyses of the entire Harland et al. (1985) polarity record (Raup,
1985; Lutz, 1985; Stothers, 1986) are suspect, as they are derived
from an insecure database. However, the 30 Myr period has been found
by different techniques in separate analyses of the more securely
established Cenozoic-Late Cretaceous reversal sequence (Raup, 1985;
Pal and Creer, 1986; Stothers, 1986), and the 15 Myr cyclicity (Mazaud
et al., 1983) may be a harmonic of the 30 Myr period.

3.2 The statistical distribution of polarity chron lengths

A stochastic model proposed by Cox (1968) is widely accepted as a
likely explanation of the process governing the occurrence of
geomagnetic reversals. The model assumes that a reversal is triggered
when strong non-dipole field behaviour coincides with weak dipole
intensity. Assuming about 0.01 Myr as the time constant for dipole
processes and about 0.20 Myr as the average length of an interval of
constant polarity in the last few million years, this suitable
configuration has happened only once in 20 cycles of dipole intensity.
A stochastic model is attractive, because it can explain a wide range
of polarity interval lengths without requiring long time constants for
core fluid motions. The probability of a new reversal at any time is
independent of events that preceded the previous reversal; the core
has no memory of past fluid motions beyond the decay time of
magnetohydrodynamic processes. The model predicts a Poissonian
distribution for the lengths of polarity intervals. The validity of
this statistical model has been tested in several studies of different
time scales.

Cox (1969) tested his model with the known radiometric time scale
for the last 4 Myr and with the Heirtzler et al. (1968) time scale.
For reversal history younger than 10 Myr, the distribution of polarity
chrons agreed well with a Poissonian distribution, but the agreement
was much poorer for older reversals because of a lack of short

polarity chrons. As discussed above, the marine polarity record
measured at the ocean surface has a resolution limited to about 0.02
Myr under optimum conditions. It may represent a filtered version of
true polarity history, in which many short polarity events which have
occurred have simply not been detected. The underlying process
governing reversals may in this case be Poissonian, and the
probability of a new reversal at any time after the previous one is
constant. On the other hand, if the short events which apparently are
missing from the record never occurred, the core renewal process may
be described better by a gamma distribution. The probability of a new
reversal is now not constant; from a low value immediately after a
reversal the probability recovers with time to somewhat above the
Poissonian value.

For a gamma distribution, the probability P(T) dT of observing a
polarity chron with length between T and T dT is given by

$$P(T) \; dT = \frac{(k/\mu)^k}{\Gamma(k)} \; T^{k-1} \; \exp\left(- \frac{k}{\mu}T\right) dT$$

where μ is the average chron length, and k is the gamma index, which
are independent parameters of the distribution. The Poisson
distribution is a special case of the gamma distribution with index
k = 1.

The optimum data base for analysis of the distribution of
polarity chron lengths in the Cenozoic and Late Cretaceous is the Cox
(1982) time scale, which is very similar to the Lowrie and Alvarez
(1981) time scale. Both are correlated well with the biostratigraphic
ages in DSDP holes. The LaBrecque et al. (1977) and Ness et al.
(1980) time scales are poorly tied to the biostratigraphy in the
Eocene and Palaeocene, the only time interval where they differ
appreciably from the later time scales. Most recent analyses of the
statistical distribution of polarity chron lengths have used the
LaBrecque et al. (1977) or Ness et al. (1980) time scales. The
characteristics of all recent Cenozoic-Late Cretaceous time scales are
so similar that the choice of time scale as data basis is not
critical. However, there are major differences with analyses based on
the earlier Heirtzler et al. (1968) time scale, which has been
superseded.

3.2.1 Analyses of the Heirtzler et al. (1968) time scale. A
discontinuity in reversal rate marked by an abrupt threefold increase
at about 45-50 Myr was observed by Heirtzler et al. (1968). This time
scale is appreciably different from its successors because of the
effects of the spurious anomaly 14, which consisted of four short
normal and four short reverse chrons. The eight additional reversal
boundaries artificially increased the average reversal frequency in
the 35-45 Myr age range, and contributed to the 45 Myr discontinuity
(Fig. 6a).

Naidu (1970) divided the Heirtzler et al. (1968) time scale into
nine adjacent windows, each 8 Myr wide, and showed that the parts

younger and older than the discontinuity were stationary time series, with different average polarity interval lengths of 0.33 and 0.60 Myr, respectively. He calculated the gamma index (k) with a maximum likelihood method, and found that the best values were k = 2 for chrons younger than 48 Myr, and k = 4 for chrons older than 56 Myr. He also applied a Kolmogorov-Smirnov statistical test to the younger stationary series to establish that the observed cumulative chron lengths fitted a gamma distribution with μ = 0.33 Myr and k = 2 within the 95% confidence limits.

A different statistical approach was used by Phillips (1977), who analysed the Heirtzler et al. (1968) time scale through a moving window extending over a fixed number of polarity chrons (25). The average chron length was computed as a Gaussian weighted mean for each window, and the optimum gamma index was calculated with the maximum likelihood method. The discontinuity near 45-50 Myr showed distinctly, and divided the time scale into two periods of apparently stationary behaviour, in which the mean chron lengths were 0.33 Myr during the last 45 Myr, and 0.90 Myr in the early Cenozoic. In contrast to Naidu (1970), no discontinuity was found in k, which had an average value of 1.72 for the entire time scale. The distributions of normal and reverse polarity chrons were also analysed separately. The moving window means behaved similarly, and no significant difference between the mean lengths of normal and reverse polarity chrons were found in the Cenozoic. However, the values of k were significantly different, with values estimated from normal chrons (around 3) consistently larger than estimates from reverse chrons (slightly larger than 1). Phillips (1977) attributed the asymmetry in k for normal and reverse chrons to processes in the core, which made the geodynamo more stable after a transition to normal polarity than it is after a transition to reverse polarity.

3.2.2 Analyses of magnetostratigraphically calibrated time scales.
The reversal rate in later polarity time scales for the Cenozoic and Late Cretaceous is not discontinuous but is characterised by a strong, nearly linear trend (Fig. 6, Fig. 7) which makes the time series non-stationary. The estimation of mean, variance, probability distribution function, etc. by using the entire sequence is meaningless unless the sequence is stationary. Naidu (1970) was careful to establish stationarity before making a statistical analysis of the 0-48 and 56-72 Myr parts of the Heirtzler et al. (1968) time scale. Lowrie and Kent (1983) analysed the LaBrecque et al. (1977) and Lowrie and Alvarez (1981) time scales using a fixed width, 8 Myr wide averaging window. This was moved 1 Myr at a time, so that every eighth point was independent. Linear regression of the independent points showed that the time scales in their entirety, and also the 0-40 Myr segments separately, were not stationary. The distribution of polarity chron lengths resembled a gamma distribution with index k = 1.5 (Fig. 8a), but the non-stationarity denies this observation much meaning.

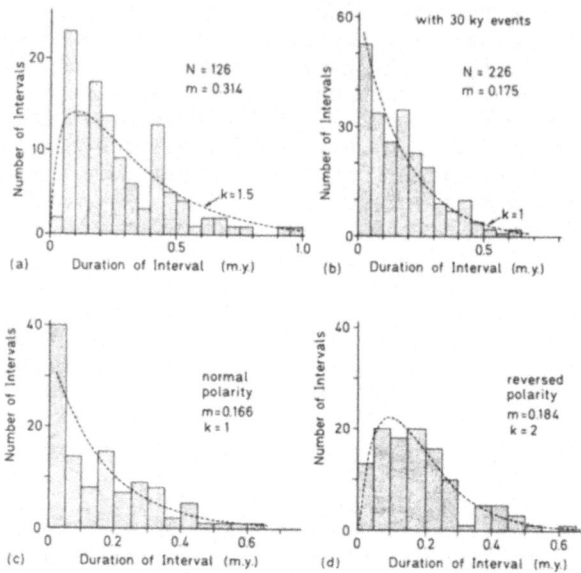

Figure 8. Histograms of polarity chron lengths in the 0–40 Myr part of the LaBrecque et al. (1977) time scale. (a) The polarity chron lengths are best fitted by a gamma distribution (k = 1.5). (b) After addition of 50 short polarity events each lasting 30 kyr the distribution is Poissonian (k = 1). The short events induce an asymmetry in which (c) the normal chrons have a Poissonian distribution and (d) the negative chrons have a gamma distribution.

McFadden (1984b) discussed the possible incompleteness of the magnetic polarity record due to the filtering out of short chrons and their incorporation into adjacent chrons. He showed that the concatenation of equal numbers of successive chrons had the effect of creating a gamma distribution from the original Poissonian distribution. Although in the observed polarity sequence many chrons are the original ones, some will be the concatenation of two or more of the original chrons, and the resultant distribution approximates a gamma distribution. Thus McFadden interpreted k as the effective ratio of the true number of chrons in the original sequence to the number in the filtered sequence. This contrasted with Phillips (1977) who interpreted k in terms of the recovery time of the dynamo process after a reversal.

The problem of non-stationarity of the observed polarity sequences has been handled by analysing the polarity chron statistics for a short sliding window 25 polarity chrons wide (Phillips, 1977; McFadden, 1984b). The k index was obtained by a maximum likelihood method, akin to that used by Naidu (1970). McFadden (1984b) demonstrated how the sliding window representation could give the visual impression of apparent differences between values of k for sequences of normal and reverse polarity chrons which had in fact been

drawn from the same population. He concluded that the relative
stabilities of the normal and reverse polarity states were better
judged by the values of the mean lengths of the normal and reverse
polarity chrons. McFadden and Merrill (1984) demonstrated the extreme
sensitivity of the maximum likelihood estimation procedure for k, and
concluded that the apparent differences in k for the normal and
reverse states in the time scale of Ness et al. (1980) were not
significant. They investigated the nature of the non-stationarity in
reversal rate and found that the normal and reverse polarity chrons
had the same statistical properties and could be assumed to be drawn
from the same populations. This implied that there was no discernible
difference in the relative stabilities of the two polarity states.

3.2.3 Effect of undetected short polarity chrons. The geomagnetic
field spends about 5 kyr in a transitional configuration during a
polarity reversal. The duration of a polarity zone that can be
readily identified in marine magnetic anomaly or magnetostratigraphic
records is about 50 kyr. Within the range 5-50 kyr the detailed
behaviour of the Earth's geomagnetic field is not well known. Low
amplitude, lineated marine magnetic anomalies have ambiguous
interpretations; they may be due to intensity fluctuations or short
polarity chrons lasting about 30 kyr (Cande and LaBrecque, 1974).
Using a stacking procedure for closely spaced, parallel marine
profiles, small coherent anomalies were interpreted as short polarity
chrons (Blakely and Cox, 1972; Blakely, 1974). Because of their
ambiguous nature, very short wavelength magnetic anomalies were not
included as polarity chrons in the LaBrecque et al. (1977) time scale,
although the times of occurrence of 57 were indicated.
    It is perhaps unlikely that all 57 of the reported short events
represent true polarity chrons. Nevertheless, Lowrie and Kent (1983)
assumed a uniform duration of 30 kyr for these chrons and added them
to the LaBrecque et al. (1977) time scale at the indicated positions.
Most of the short chrons have normal polarity and 50 of them are
younger than 40 Myr. The time series in the 0-40 Myr interval was now
stationary, and the distribution of polarity chron lengths in this
time interval was Poissonian (Fig. 8a, 8b). However, a strong
asymmetry developed between the distributions of normal and reverse
polarities (Fig. 8c, 8d). The normal distribution was Poissonian (k =
1) but the reverse distribution was gamma distributed with k = 2.
This asymmetry is opposite to that observed by Phillips (1977) in the
Heirtzler et al. (1968) time scale without very short chrons.
    The interpretation of the ambiguous features as short polarity
chrons accepts the hypothesis that they are indeed the result of field
reversals which were missing from the original record because of
detection problems. It is not surprising that the addition of so many
short chrons converts a gamma distribution into a Poissonian
distribution; because they are dominantly normal chrons, the normal
distribution also becomes Poissonian. Few short negative chrons are
added, so the distribution of negative polarity chrons remains lacking
in short chrons, i.e. a gamma distribution. The statistical
properties of the new reverse polarity distribution depend strongly on

drawn from the same population. He concluded that the relative
stabilities of the normal and reverse polarity states were better
judged by the values of the mean lengths of the normal and reverse
polarity chrons. McFadden and Merrill (1984) demonstrated the extreme
sensitivity of the maximum likelihood estimation procedure for k, and
concluded that the apparent differences in k for the normal and
reverse states in the time scale of Ness et al. (1980) were not
significant. They investigated the nature of the non-stationarity in
reversal rate and found that the normal and reverse polarity chrons
had the same statistical properties and could be assumed to be drawn
from the same populations. This implied that there was no discernible
difference in the relative stabilities of the two polarity states.

3.2.3 <u>Effect of undetected short polarity chrons</u>. The geomagnetic
field spends about 5 kyr in a transitional configuration during a
polarity reversal. The duration of a polarity zone that can be
readily identified in marine magnetic anomaly or magnetostratigraphic
records is about 50 kyr. Within the range 5–50 kyr the detailed
behaviour of the Earth's geomagnetic field is not well known. Low
amplitude, lineated marine magnetic anomalies have ambiguous
interpretations; they may be due to intensity fluctuations or short
polarity chrons lasting about 30 kyr (Cande and LaBrecque, 1974).
Using a stacking procedure for closely spaced, parallel marine
profiles, small coherent anomalies were interpreted as short polarity
chrons (Blakely and Cox, 1972; Blakely, 1974). Because of their
ambiguous nature, very short wavelength magnetic anomalies were not
included as polarity chrons in the LaBrecque et al. (1977) time scale,
although the times of occurrence of 57 were indicated.

It is perhaps unlikely that all 57 of the reported short events
represent true polarity chrons. Nevertheless, Lowrie and Kent (1983)
assumed a uniform duration of 30 kyr for these chrons and added them
to the LaBrecque et al. (1977) time scale at the indicated positions.
Most of the short chrons have normal polarity and 50 of them are
younger than 40 Myr. The time series in the 0–40 Myr interval was now
stationary, and the distribution of polarity chron lengths in this
time interval was Poissonian (Fig. 8a, 8b). However, a strong
asymmetry developed between the distributions of normal and reverse
polarities (Fig. 8c, 8d). The normal distribution was Poissonian (k =
1) but the reverse distribution was gamma distributed with k = 2.
This asymmetry is opposite to that observed by Phillips (1977) in the
Heirtzler et al. (1968) time scale without very short chrons.

The interpretation of the ambiguous features as short polarity
chrons accepts the hypothesis that they are indeed the result of field
reversals which were missing from the original record because of
detection problems. It is not surprising that the addition of so many
short chrons converts a gamma distribution into a Poissonian
distribution; because they are dominantly normal chrons, the normal
distribution also becomes Poissonian. Few short negative chrons are
added, so the distribution of negative polarity chrons remains lacking
in short chrons, i.e. a gamma distribution. The statistical
properties of the new reverse polarity distribution depend strongly on

the positions of the added short chrons. The addition of a short positive chron divides a longer negative chron into two unequal shorter chrons. For the reverse polarity chron distribution, an exact knowledge of the durations of the short chrons is less critical than the accuracy of their locations (McFadden et al., 1987).

The question of the relative stabilities of the normal and reverse polarity states, with and without the additional short polarity chrons, was further investigated by McFadden et al. (1987). The statistical methods of McFadden (1984b) were employed for a sliding window of width 25 polarity chrons. Addition of the short intervals to the LaBrecque et al. (1977) time scale produced the asymmetry observed by Lowrie and Kent (1983); the normal polarity chrons were satisfied by a value of k close to 1, while generally larger values of k around 2 were observed for the reverse polarity chrons. However, taking into account the sensitivity of the method of calculation of k, McFadden et al. (1987) concluded that this difference is probably not real; the data give no reason for rejecting the hypothesis of a common k for the normal and reverse sequences. They showed also that the data, with or without the 57 short chrons, are compatible with the normal and reverse sequences having a common reversal rate. The addition of the 57 short chrons does, however, alter the structure of the non-stationarity in the reversal sequence, because only 7 of them are older than 40 Myr. The 0-40 Myr interval becomes stationary after addition of the short chrons (Lowrie and Kent, 1983), while the older part of the time scale remains non-stationary.

4. CONCLUSIONS

The ages of the M-sequence polarity chrons in the Harland et al. (1982) time scale are uncertain. The interpretation of a 30 Myr periodicity in reversal frequency from time series analysis of the 0-165 Myr polarity sequence in its entirety is therefore unwarranted. However, this period shows up in separate analyses of the better dated Cenozoic-Late Cretaceous polarity sequence, and is probably real. A better defined cyclicity with period close to 15 Myr is possibly a harmonic of the 30 Myr period.

The use of a moving window of fixed time length to analyse the reversal frequency has been criticised (McFadden, 1984a) because a rectangular box filter can introduce spurious frequencies. However, the 15 Myr cycle is apparent even when the reversal data are averaged for stationary windows only 4 Myr wide (Fig. 7a) and is unlikely to be an artefact of the method of analysis.

The most exhaustive statistical analyses of reversal frequency in the Cenozoic-Late Cretaceous sequence and in the M-sequence have been carried out by using analytical methods applied within a moving window defined by a fixed number of polarity chrons (Phillips, 1977; McFadden, 1984b). This technique may circumvent the problem that neither reversal sequence is a stationary time series, but it introduces other undesirable features; it gives a false picture of the

amount of independent information in the polarity sequence and of the
temporal variation of the statistical properties.

As noted by McFadden (1984b), the sliding window method gives a
strong visual impression that there is more information than really
exists, and that differences between parameters for the two polarity
states are persistent. In fact when a 25 point sliding window is
used, only every 25th analysis is independent. Only four independent
estimates of the statistical parameters of each polarity are obtained
for the Cenozoic-Late Cretaceous polarity sequence and exact
statistical tests of differences between the two states are difficult
to perform.

A further problem with analysing a constant number of chrons is
that the time represented by a fixed number of intervals is quite
variable along the polarity sequence. Thus, the Pliocene polarity
data are averaged with a filter only about 5 Myr wide; beyond the
Miocene (24 Myr) the filter is more than 10 Myr wide, and beyond the
Lower Eocene it is more than 20 Myr wide. The smoothed data are
plotted against the mid-point of a window of variable width, which
together with the variable smoothing gives a false visual impression
of the temporal variations of the statistical parameters. For
example, the reversal rate determined by a sliding window 25 chrons
wide for the Cenozoic-Late Cretaceous polarity sequence (McFadden and
Merrill, 1984; McFadden et al., 1987) shows almost no evidence of the
cyclical fluctuations shown in Figs. 6 snd 7, suggesting that the
cyclicity has been suppressed by the heavy smoothing.

The generally accepted stochastic model of reversals proposed by
Cox (1968) has been tested in several studies. The model predicts a
Poissonian distribution for the lengths of polarity chrons, but some
studies have indicated that a gamma distribution may be more apt.
Analyses of the LaBrecque et al. (1977) and later time scales differ
on the meaning that should be accorded to the gamma index k. If the
incidence of reversals is governed by a gamma renewal process, the
probability per unit time of a new reversal is described by the hazard
function (Phillips, 1977) which is constant for a Poissonian process
(k = 1) but which for any other gamma process increases with time from
a low value immediately after a reversal. Phillips (1977) interpreted
the meaning of the gamma k index in terms of this postreversal
stability, which if the suspected short polarity chrons do exist, is
asymmetric with respect to polarity (Lowrie and Kent, 1983).

McFadden (1984b) proposed a different interpretation of the
meaning of k which assumes that a Poissonian model governs reversals
and that some very short chrons have not been resolved because they
have concatenated with their neighbours. The gamma index k then
represents the fraction of very short polarity chrons which have been
lost from the Poissonian model by concatenation. He showed that the
method of estimation of k is ultrasensitive to the data set. Thus,
functions based on the value of k may not offer the most suitable
criterion for judging the stability of a polarity state. McFadden
(1984b) proposed instead that the best criterion of stability was the
mean length of a polarity state. With or without the addition of
suspected short polarity chrons, the mean lengths of normal and

reversed polarity chrons are not significantly different, and there is therefore at present no reason to reject Cox's hypothesis that geomagnetic reversals are governed by a Poissonian process (McFadden and Merrill, 1984; McFadden et al., 1987).

5. REFERENCES

Alvarez, W., M.A. Arthur, A.G. Fischer, W. Lowrie, G. Napoleone, I. Premoli Silva, and W.G. Roggenthen, 1977. 'Upper Cretaceous-Paleocene magnetic stratigraphy at Gubbio, Italy. V: Type section for the Late Cretaceous-Paleocene geomagnetic reversal time scale'. Geol. Soc. Amer. Bull., 88, 383-389.

Berggren, W.A., D.V. Kent, J.J. Flynn, and J.A. Van Couvering, 1985. 'Cenozoic geochronology'. Geol. Soc. Amer. Bull., 96, 1407-1418.

Blakely, R.J., 1974. 'Geomagnetic reversals and crustal spreading rates during the Miocene'. J. Geophys. Res., 79, 2979-2985.

Blakely, R.J., and A. Cox, 1972. 'Evidence for short geomagnetic polarity intervals in the Early Cenozoic'. J. Geophys. Res., 77, 7065-7072.

Cande, S.C., and Y. Kristoffersen, 1977. 'Late Cretaceous magnetic anomalies in the North Atlantic'. Earth Planet. Sci. Lett., 35, 215-224.

Cande, S.C., and J.L. LaBrecque, 1974. 'Behaviour of the earth's palaeomagnetic field from small scale marine magnetic anomalies. Nature, 247, 26-28.

Cande, S.C., R.L. Larsen, and J.L. LaBrecque, 1978. 'Magnetic lineations in the Pacific Jurassic quiet zone'. Earth Planet. Sci. Lett., 41, 434-440.

Cirilli, S., P. Marton, and L. Vigliotti, 1984. 'Implications of a combined biostratigraphic and palaeomagnetic study of the Umbrian Maiolica Formation'. Earth Planet. Sci. Lett., 69, 203-214.

Channell, J.E.T., and P. Grandesso, 1987. 'A revised correlation of Mesozoic polarity chrons and calpionellid zones'. Earth Planet. Sci. Lett., 85, 222-240.

Channell, J.E.T., W. Lowrie and F. Medizza, 1979. 'Middle and Early Cretaceous magnetic stratigraphy from the Cismon section, northern Italy'. Earth Planet. Sci. Lett., 42, 153-166.

Channell, J.E.T., J.G. Ogg, and W. Lowrie, 1982. 'Geomagnetic polarity in the Early Cretaceous and Jurassic'. Phil. Trans. R. Soc. London, Ser. A. 306, 137-146.

Channell, J.E.T., T.J. Bralower, and P. Grandesso, 1987. 'Biostratigraphic correlation of Mesozoic polarity chrons CM1 to CM23 at Capriolo and Xausa (Southern Alps, Italy)'. Earth Planet. Sci. Lett., 85, 203-221.

Cox, A., 1968. 'Lengths of geomagnetic polarity intervals'. J. Geophys. Res., 73, 3247-3260.

Cox, A., 1969. 'Geomagnetic reversals'. Science, 163, 237-245.

Cox, A., 1982. 'Magnetostratigraphic time scale'. In: W.B. Harland et al. (eds.), A Geologic Time Scale, Cambridge Univ. Press, Cambridge, 63–84.

Cox, A., and G.B. Dalrymple, 1967. 'Statistical analysis of geomagnetic reversal data and the precision of potassium–argon dating'. J. Geophys. Res., 72, 2603–2614.

Hallam, A., J.M. Hancock, J.L. LaBrecque, W. Lowrie, and J.E.T. Channell, 1985. 'Jurassic and Cretaceous geochronology and Jurassic to Paleogene magnetostratigraphy'. In: N.J. Snelling (ed.), Geochronology and the Geological Record, Geol. Soc. London Memoir 10, 118–140.

Harland, W.B., A.V. Cox, P.G. Llewellyn, C.A.G. Pickton, A.G. Smith, and R. Walters, 1982. A Geologic Time Scale, Cambridge Univ. Press, Cambridge, 131 p.

Heirtzler, J.R., G.O. Dickson, E.M. Herron, W.C. Pitman III, and X. Le Pichon, 1968. 'Marine magnetic anomalies, geomagnetic field reversals and motions of the ocean floor and continents'. J. Geophys. Res., 73, 2119–2136.

Kent, D.V., and F.M. Gradstein, 1985. 'A Cretaceous and Jurassic geochronology'. Geol. Soc. Am. Bull., 96, 1419–1427.

LaBrecque, J.L., D.V. Kent, and S.C. Cande, 1977. 'Revised magnetic polarity time scale for Late Cretaceous and Cenozoic time'. Geology, 5, 330–335.

LaBrecque, J.L., K.J. Hsu, M.F. Carman, A.-M. Karpoff, J.A. McKenzie, S.P. Percival, N.P. Petersen, K.A. Pisciotto, E. Schreiber, L. Tauxe, P. Tucker, H.J. Weissert, and R. Wright, 1983. 'DSDP Leg 73: Contributions to Paleogene stratigraphy in nomenclature, chronology and sedimentation rates'. Palaeogeogr., Palaeoclimatol., Palaeoecol., 42, 91–125.

Langereis, C.G., W.J. Zachariasse and J.D.A. Zijderveld, 1984. 'Late Miocene magnetobiostratigraphy of Crete'. Marine Micropaleontol., 8, 261–281.

Larson, R.L., and T.W.C. Hilde, 1975. 'A revised time scale of magnetic reversals for the Early Cretaceous and Late Jurassic'. J. Geophys. Res., 80, 2586–2594.

Larson, R.L., and W.C. Pitman III, 1972. 'World-wide correlation of Mesozoic magnetic anomalies, and its implications'. Geol. Soc. Amer. Bull., 83, 3645–3662.

Lowrie, W., 1982. 'A revised magnetic polarity timescale for the Cretaceous and Cainozoic'. Phil. Trans. R. Soc. Lond. A, 306, 129–136.

Lowrie, W., and W. Alvarez, 1977. 'Upper Cretaceous–Paleocene magnetic stratigraphy at Gubbio, Italy. III. Upper Cretaceous magnetic stratigraphy'. Geol. Soc. Amer. Bull., 88, 374–377.

Lowrie, W., and W. Alvarez, 1981. 'One hundred million years of geomagnetic polarity history'. Geology, 9, 392–397.

Lowrie, W., and W. Alvarez, 1984. 'Lower Cretaceous magnetic stratigraphy in Umbrian pelagic limestone sections'. Earth Planet. Sci. Lett., 71, 315–328.

Lowrie, W., and J.E.T. Channell, 1984a. 'Magnetostratigraphy of the Jurassic-Cretaceous boundary in the Maiolica limestone (Umbria, Italy)'. Geology, 12, 44-47.

Lowrie, W., and J.E.T. Channell, 1984b. 'Reply to Comments on: Magnetostratigraphy of the Jurassic-Cretaceous boundary in the Maiolica limestone (Umbria, Italy)'. Geology, 12, 702.

Lowrie, W., and D.V. Kent, 1983. 'Geomagnetic reversal frequency since the Late Cretaceous'. Earth Planet. Sci. Lett., 62, 305-313.

Lowrie, W., and J.G. Ogg, 1986. 'A magnetic polarity time scale for the Early Cretaceous and Late Jurassic'. Earth Planet. Sci. Lett., 76, 341-349.

Lowrie, W., W. Alvarez, I. Premoli Silva, and S. Monechi, 1980. 'Lower Cretaceous magnetic stratigraphy in Umbrian pelagic carbonate rocks'. Geophys. J.R. astr. Soc., 60, 263-281.

Lowrie, W., W. Alvarez, G. Napoleone, K. Perch-Nielsen, I. Premoli Silva, and M. Toumarkine. Paleogene magnetic stratigraphy in Umbrian pelagic carbonate rocks: The Contessa sections, Gubbio'. Geol. Soc. Amer. Bull., 93, 414-432.

Lutz, T.M., 1985. 'The magnetic reversal record is not periodic'. Nature, 317, 404-407.

Mankinen, E.A., and G.B. Dalrymple, 1979. 'Revised geomagnetic polarity time scale for the interval 0-5 m.y.B.P.'. J. Geophys. Res., 84, 615-626.

Mazaud, A., C. Laj, L. de Sèze, and K.L. Verosub, 1983. '15-Myr periodicity in the frequency of geomagnetic reversals since 100 Myr'. Nature, 304, 328-330.

Mazaud, A., C. Laj, L. de Sèze, and K.L. Verosub, 1984. '15-Myr periodicity in the frequency of geomagnetic reversals since 100 Myr: Reply'. Nature, 311, 396.

McDougall, I., N.D. Watkins, G.P.L. Walter, and L. Kristjansson, 1976. 'Potassium-argon and paleomagnetic analysis of Icelandic lava flows: limits on the age of anomaly 5'. J. Geophys. Res., 81, 1505-1512.

McDougall, I., K. Saemundsson, H. Johannesson, N.D. Watkins, and L. Kristjansson, 1977. 'Extension of the geomagnetic polarity time scale to 6.5 M.y.: K-Ar dating, geological, and paleomagnetic study of a 3500 m lava section in western Iceland'. Geol. Soc. Amer. Bull., 88, 1-15.

McFadden, P.L., 1984a. '15-Myr periodicity in the frequency of geomagnetic reversals since 100 Myr'. Nature, 311, 396.

McFadden, P.L., 1984b. 'Statistical tools for the analysis of geomagnetic reversal sequences'. J. Geophys. Res., 89, 3363-3372.

McFadden, P.L., and R.T. Merrill, 1984. 'Lower mantle convection and geomagnetism'. J. Geophys. Res., 89, 3354-3362.

McFadden, P.L., R.T. Merrill, W. Lowrie, and D.V. Kent, 1987. 'The relative stabilities of the reverse and normal polarity states of the earth's magnetic field'. Earth Planet. Sci. Lett., 82, 373-383.

Naidu, P.S., 1970. 'Statistical structure of geomagnetic field reversals'. J. Geophys. Res., 76, 2649–2662.

Ness, G., S. Levi, and R. Couch, 1980. 'Marine magnetic anomaly timescales for the Cenozoic and Late Cretaceous: a précis, critique and synthesis'. Rev. Geophys. Space Phys., 18, 753–770.

Ogg, J.G., 1981. 'Sedimentology and paleomagnetism of Jurassic pelagic limestones: "Ammonitico Rosso" facies'. Ph.D. thesis, University of California at San Diego, 212 pp.

Ogg, J.G., 1984. 'Comment on: Magnetostratigraphy of the Jurassic-Cretaceous boundary in the Maiolica limestone (Umbria, Italy)'. Geology, 12, 701.

Ogg, J.G., and W. Lowrie, 1986. 'Magnetostratigraphy of the Jurassic/Cretaceous boundary'. Geology, 14, 547–550.

Ogg, J.G., M.B. Steiner, F. Oloriz, and J.M. Tavera, 1984. 'Jurassic magnetostratigraphy, 1. Kimmeridgian–Tithonian of Sierra Gorda and Carcabuey, southern Spain'. Earth Planet. Sci. Lett., 71, 147–162.

Pal, P.C., and K.M. Creer, 1986. 'Geomagnetic reversal spurts and episodes of extraterrestrial catastrophism'. Nature, 320, 148–150.

Phillips, J.D., 1977. 'Time variation and asymmetry in the statistics of geomagnetic reversal sequences'. J. Geophys. Res., 82, 835–843.

Prothero, D.R., C.R. Denham, and H.G. Farmer, 1982. 'Oligocene calibration of the magnetic polarity time scale'. Geology, 10, 650–653.

Raup, D.M., 1985. 'Magnetic reversals and mass extinctions'. Nature, 314, 341–343.

Roth, P.R., 1983. 'Jurassic and Lower Cretaceous calcareous nannofossils in the western North Atlantic (Site 534): Biostratigraphy, preservation, and some observations on biogeography and paleoceanography'. In: R.E. Sheridan, E.M. Gradstein, et al., Initial Reports of the Deep Sea Drilling Project, Vol. 76, 578–621, U.S. Govt. Printing Office, Washington, D.C.

Steiner, M.B., J.G. Ogg, G. Melendez, and L. Sequeiros, 1985. 'Jurassic magnetostratigraphy, 2. Middle–Late Oxfordian of Aguilon, Iberian Cordillera, northern Spain'. Earth Planet. Sci. Lett., 76, 151–166.

Steiner, M., J. Ogg, and J. Sandoval, 1987. 'Jurassic magnetostratigraphy, 3. Bathonian–Bajocian of Carcabuey, Sierra Harana and Campillo de Arenas (Subbetic Cordillera, southern Spain)'. Earth Planet. Sci. Lett., 82, 357–372.

Stothers, R.B., 1979. 'Solar activity cycle during classical antiquity'. Astronomy and Astrophysics, 77, 121–127.

Stothers, R.B., 1986. 'Periodicity of the Earth's magnetic reversals'. Nature, 322, 444–446.

Van Hinte, J.E., 1976a. 'A Cretaceous time scale'. Amer. Assoc. Petrol. Geologists Bull., 60, 498–516.

Van Hinte, J.E., 1976b. 'A Jurassic time scale'. <u>Amer. Assoc.</u>
    <u>Petrol. Geologists Bull.</u>, **60**, 489-497.
Vogt, P.R., and A.M. Einwich, 1979. 'Magnetic anomalies and sea-
    floor spreading in the western North Atlantic, and a revised
    calibration of the Keathley (M) geomagnetic reversal
    chronology'. In: B.E. Tucholke, P.R. Vogt et al., <u>Initial</u>
    <u>Reports of the Deep Sea Drilling Project</u>, **Vol. 43**,
    857-876, U.S. Govt. Printing Office, Washington, D.C.

OBSERVATIONS AND MODELS OF REVERSAL TRANSITION FIELDS

C. Laj
Centre des Faibles Radioactivités
Laboratoire Mixte CNRS-CEA
Avenue de la Terrasse
Gif-sur-Yvette
France

R. Weeks and M. Fuller
Department of Geological Sciences
University of California
Santa Barbara
CA. 93108
USA

ABSTRACT.   Palaeomagnetic records of reversals are now available from ocean sediments, sedimentary rocks, lavas and intrusions. From these records, it is evident that
    (1) the reversal in direction of the field takes place over a few thousand years.
    (2) the reversal in direction is accompanied by a decrease in intensity of approximately an order of magnitude.
    (3) the reversal defines a period of rapid fluctuation of intensity and direction of the field.
    (4) zonal harmonics were a dominant aspect of the last reversal. Other aspects of the reversal transition fields are suggested by the data, but are not yet unambiguously established.
    The field was not dipolar during the last reversal, but was dominated by zonal harmonics. Models of the transition field geometry which include zonal harmonics and a drifting non-dipole field are able to simulate some aspects of reversals, but we still do not have a satisfying understanding of transition field geometries.

1.   INTRODUCTION

There are now numerous palaeomagnetic records of reversals, so that some impression of the transition fields can be gained. The records have been obtained from ocean sediments, sedimentary rocks, lavas and intrusions. However, because the nature of a reversal record is dependent upon the type of material in which it is recorded, the various records give different impressions of the transition fields.

185

F. J. Lowes et al. (eds.), Geomagnetism and Palaeomagnetism, 185–203.
© 1989 by Kluwer Academic Publishers.

In the first part of this paper, we review the various records paying particular attention to the effect of the recording process upon the fidelity of the record. This then leads to a review of our knowledge of the transition fields. In the second part of the paper, we describe the various models of transition fields which have been proposed.

2. REVERSAL RECORDS

2.1 Nature of Recording Process

The magnetization process by which a reversal is recorded has an important effect upon the nature of the record. We therefore now consider the recording processes involved in the various types of records beginning with lavas and proceeding to sediments and intrusions.

Records from lavas have the great advantage that they are carried by thermo-remanent magnetization (TRM) acquired by the lava as it initially cooled on the earth's surface. TRM has been studied extensively, and there is good evidence that in isotropic material it is acquired parallel to the ambient field and that its magnitude is linearly related to the field. Each lava thus gives a vector record of the field at the time it cooled through the blocking temperature of the magnetic phases carrying its Natural Remanent Magnetization (NRM). Because the cooling time of a lava was thought to be short compared with significant changes in the geomagnetic field, it was assumed that each individual flow gave a spot reading of the field and that no averaging of field changes was involved. In general, this still appears to be a good assumption, but the possibility of very rapid changes during a reversal introduces an element of doubt (Coe and Prevot, 1988).

The chief disadvantage of lava records arises from our inability to establish the chronology of the flows within the reversal: it is impossible to determine the absolute ages with sufficient resolution to establish the time frame for the spot readings provided by the flows. One is at the mercy of the vagaries of the extrusion process and does not know whether a sequence of lavas was erupted over a period of years, or thousands of years.

Kono (1972) used historical records of eruption of various volcanoes to show that the eruption process could be well modelled by a Poisson process. An attempt (Weeks, 1988) has been made to simulate sampling of transition fields by lava sequences, using such a Poisson process in order to gain some understanding of the possible effects on the reversal records. To illustrate the effect, a VGP path generated with a model of a reversal (Figure 1a) is sampled according to a Poisson scheme. The various records obtained for different sampling parameters are illustrated in Figure 1b-c. The result is to degrade the record and to give it an appearance of rapid fluctuations which were not in the model. Hence when such fluctuations are seen in lava records they should be interpreted with some caution. In some cases, the nature of the path is completely altered and the apparent mean

longitudinal location of the path is moved by 90° or more. Random
noise is not likely to be a severe problem in lavas; measuring about 6
samples from each flow should ensure an α_{95} of a few degrees. Whether
there are effects which may give systematic errors is not known.

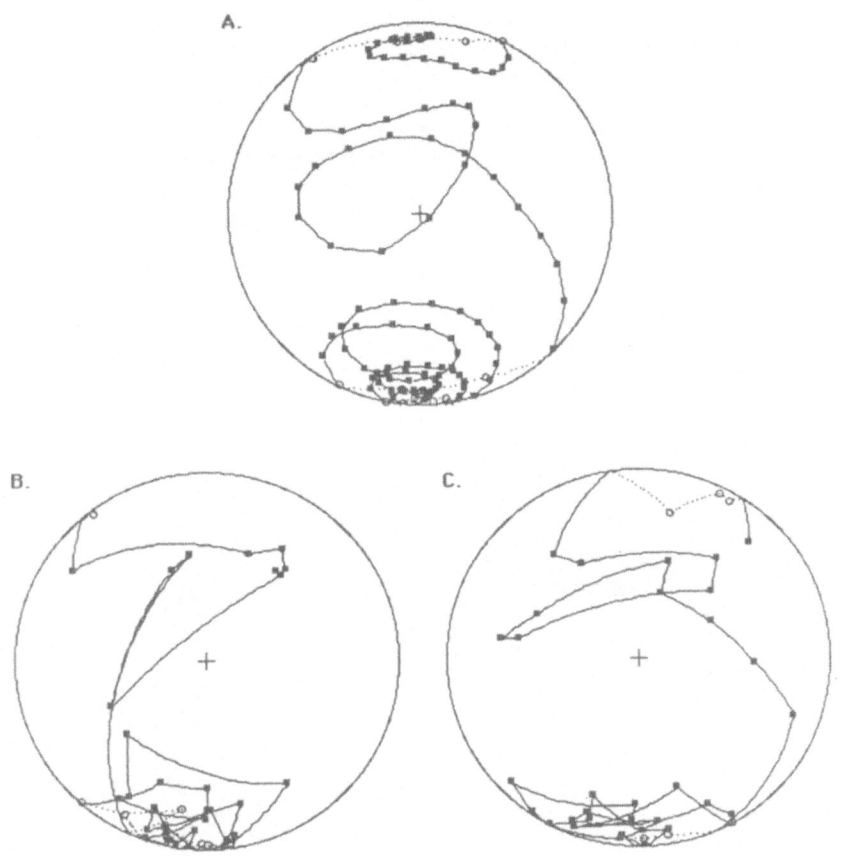

Figure 1. Effect of sampling a reversal by lava sequences; (A) VGP
path for reversal generated from g_2^0, g_3^0, g_4^0 and westward drift
(Williams et al., 1988), (B) and (C) records generated by Poisson .
sampling with 40 samples.

In contrast to records from lavas, those from both sedimentary
and intrusive rocks suffer from the problem of time averaging of the
field. The acquisition of magnetization takes place over a period of
time during which significant changes in the field occur. Thus even in
the most favourable circumstances of a record obtained from the margin
of a rapidly cooled intrusion with a rock which has a very narrow
blocking temperature range, field variations over times of the order
of hundreds of years are likely to be averaged (e.g. Dodson et al.,
1978). The effect of this time averaging can be simulated by sampling

a reversal model according to the criteria appropriate for the process
of magnetization of intrusive rocks. The results for such a sampling
of the same reversal model (Figure 2a) analysed above are shown in
Figure 2b-c. The results from intrusives are not however like this.
Figure 3 illustrates a record from the Tatoosh intrusion (Dodson et
al., 1978). Part of the problem may arise from high temperature
viscous magnetization (VRM), but the principal difference probably
arises from noise contributions. Indeed, subsequent work using thermal
demagnetization with measurements at high temperature (Dunn and
Fuller, 1980) has provided records more like the synthetic. The
effect of adding random noise to the smoothed model records is

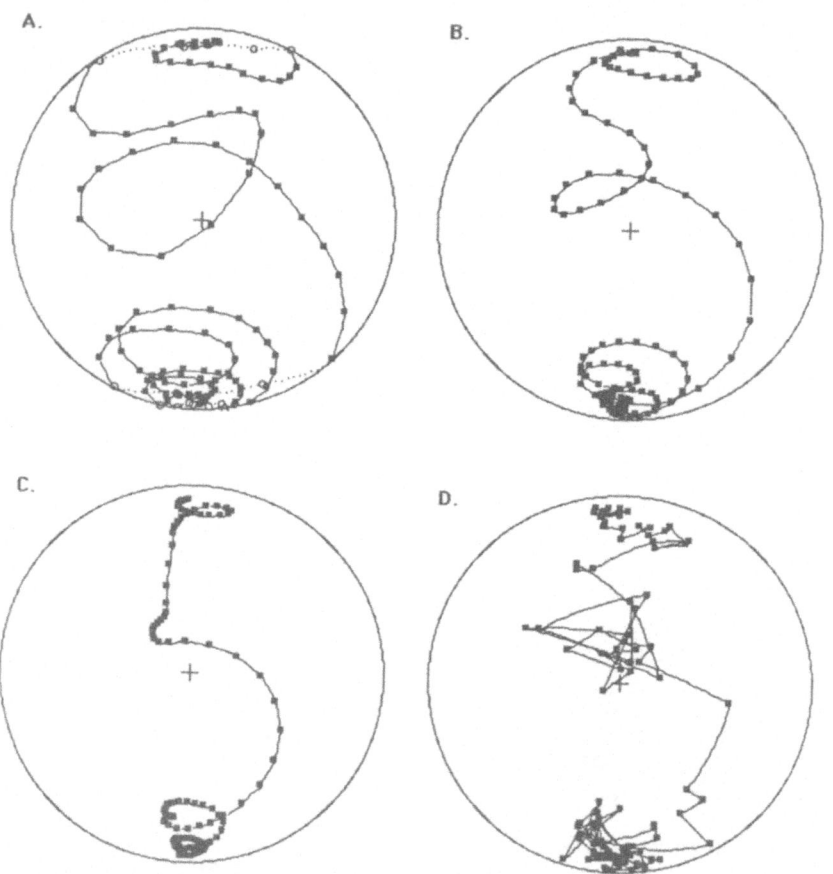

Figure 2. Effect of sampling transition fields by intrusions; (A) VGP
path for reversal generated from g_2^0, g_3^0, g_4^0 and westward drift
(Williams et al., 1988), (B) record generated by intrusion simulated
by a 7 point running average, (C) record generated by an 11 point
running average, (D) record generated by an 11 point running average
plus random noise.

illustrated in Figure 2d. However, the noise present in the Tatoosh record seems more systematic; north-south oscillations in the VGP path probably result from varying degrees of contamination of the normal polarity component. These results suggest that even in the most favourable circumstances intrusion records should be used only to gain an impression of the longer term features of the reversal record.

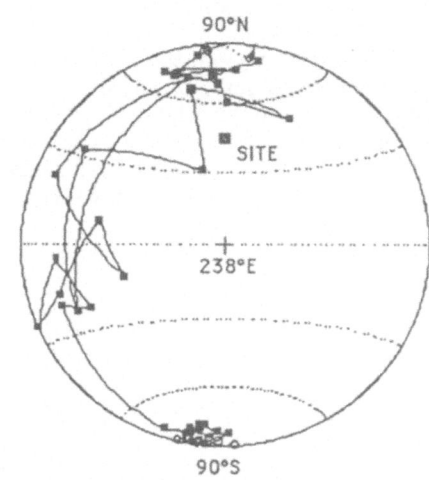

Figure 3. VGP path from the Tatoosh intrusion (Dodson et al., 1978).

Sedimentary reversal records, whether they be from ocean sediments, or from sedimentary rocks on land, may be able to provide very detailed records of field changes provided that the magnetization process has a high fidelity and the sedimentation rate is high and constant. Moreover, the possibility of obtaining multiple records of the last few reversals from piston cores makes the potential importance of ocean core records very great. The resolution of a sedimentary record is limited by a combination of the ability to measure very thin slices of sediment reliably and the thickness of sediment, which acts as a homogeneous time record of the field. The fidelity depends upon the constancy of the sedimentation rate and the absence of sedimentary hiatuses. The model can once again be analysed, this time with the smoothing representative of the sedimentation processes. Again, the synthetic record differs from the sedimentary records, so that we are left with the puzzling high frequency features. These may be a true record of the field, in which case it must exhibit very rapid fluctuations in direction and intensity, or they may be noise. They are not simply noise due to poor cleaning, because difference vectors between successive vectors appear to be random rather than systematically related to the present field direction (Valet et al., 1988).

2.2 Principal Features of Transition Fields

The reversal in direction has been shown to take place over a period
of a few thousand years from ocean sediment records (Harrison and
Somayajulu, 1966). Clement and Kent (1984) analysed how the duration
of the Brunhes/Matuyama reversal varied with the observation site
latitude using deep ocean sediment cores taken mostly from the Pacific
(Figure 4). Using the magnetostratigraphy of the cores to estimate
sedimentation rates they deduced that the reversal duration varied
from about 10 kyr to 4 or 5 kyr between 45.3°N and 33.4°S. The
assumption that deep-sea sediments have a slow and constant
sedimentation rate is probably good on these timescales and though the
records do not give very good directional resolution the ages
interpolated from magnetostratigraphy are thus likely to be reliable.
Recently much shorter directional reversals have been reported from
sequences of more rapidly deposited marine clays (Linssen, 1988; Laj
et al., 1987, 1988). Durations of reversal between 0.5 and 3.0 kyr are
reported in these records.

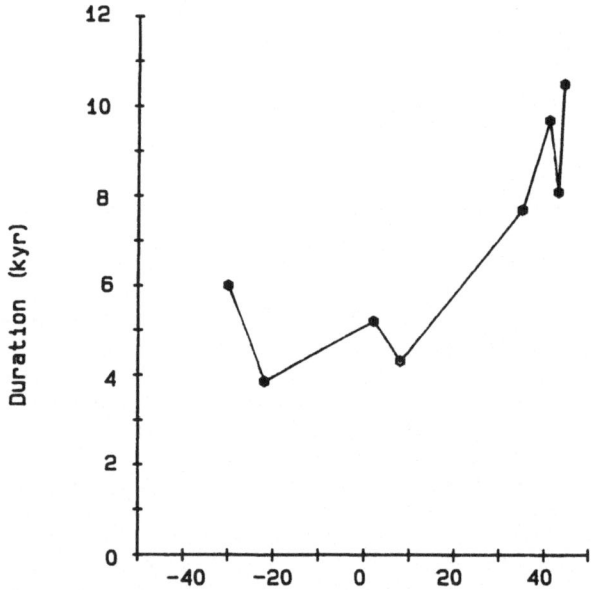

Figure 4.   Estimates of the time taken for a reversal from ocean cores
at various latitudes (Clement and Kent, 1984).

    That the reversal in direction is accompanied by a decrease in
intensity is seen qualitatively in all records, and the application of
the Koenigsberger-Thellier-Thellier intensity method to the lava
sequences has demonstrated the magnitude of the decrease. For example,
in the Steen's Mountain record, a decrease from around 31.5 to 10.9 µT,
with a recovery to 48 µT, was described by Prevot et al. (1985). Bogue

and Coe (1984) demonstrated decreases from about 40 to 10 T. These are somewhat smaller than the indications of about an order of magnitude decrease inferred from intrusive records (e.g. Dodson et al., 1978) and some sedimentary records (Linssen, 1988; Clement et al., 1987; Laj et al., 1987, 1988; Valet, 1988).

All records of reversals reveal greater fluctuations in direction and intensity than during times when the field is not reversing. These fluctuations are not only greater but more rapid. For example, Valet et al. (1988) showed, in sedimentary records from Crete, a substantial increase (3-10 times) in the angular changes between two successive vectors upon entering the reversal. Whether to attribute these fluctuations to field variation or to a worsening of the signal to noise ratio during recording of a weakened geomagnetic field is not certain. However, even in the lava records, a substantial increase in the vector differences is seen during the directional transition (e.g. Mankinen et al., 1985).

Multiple records of the Brunhes-Matuyama reversal indicate that zonal (axisymmetric) harmonics dominate the harmonic contents of transitional fields. Hoffman and Fuller (1978) pointed out that the VGP paths of R-N reversals from mid-latitudes in the northern hemisphere tend to be predominantly in the hemisphere centred on the site longitude while N-R paths show a tendency to lie in the hemisphere opposite to the site. This tendency is still particularly marked when the eight presently available paths for the Brunhes-Matuyama reversal are considered. If plotted in geographic coordinates no obvious correlations are observed among them. If, however, each path is plotted with respect to the site longitude, the paths all group within the hemisphere centred on the site meridian. Since this result seemed contentious at the meeting, a summary of all available data for the Brunhes-Matuyama reversal is presented in Figure 5. Virtual geomagnetic poles (VGP's) were calculated from each observation of declination and inclination within the records. The VGP distributions are then compared in geographic coordinates, and in coordinates with respect to site longitude. It is quite evident that the VGP's plot dominantly in the site hemisphere, requiring a dominance of zonal harmonics.

2.3 Additional Features of Transitional Fields

In addition to the features of transition fields discussed above, upon which there is general agreement, there are a number of other features, less well established, but still worthy of discussion. Several of these are related to the features discussed above.

The intensity decrease may take longer than the directional change by a factor of about three. This is inferred principally from intrusion and sedimentary records (Dodson et al., 1978; Fuller et al., 1979; Valet et al., 1988; Theyer, 1985). A small number of sedimentary records show intensity and direction changes that are simultaneous. For example, in an ocean core from the Indian Ocean, Opdyke et al. (1973) showed that the intensity change was coincident with the onset of the directional change in three successive reversals while the

intensity recovery took longer. However, recently, more detailed work
on the same core (Clement, 1984) seems to contradict the earlier
report, and shows a long intensity decrease relative to the
directional change. Coincidence of directional change and main
intensity decrease also appears to be the view of Prévot et al.
(1985). However, the Steen's Mountain record exhibits a lower
intensity at the beginning of the studied record than after the
reversal. This might be interpreted as an indication that the low
intensity associated with the reversal had already begun by the time
of the extrusion of the first lava in the main section.

There have been a number of suggestions that immediately prior
to, and after the reversal, there are periods of anomalously high
field intensity (Van Zijl et al., 1962; Hoffman and Morse, 1987). In
sedimentary and intrusion records such a feature would be hard to
distinguish from the usual fluctuation in field intensity prior to and
following reversals. However, at least in the case reported by Hoffman
and Moore (1987), there does appear to be an anomalously high absolute
intensity.

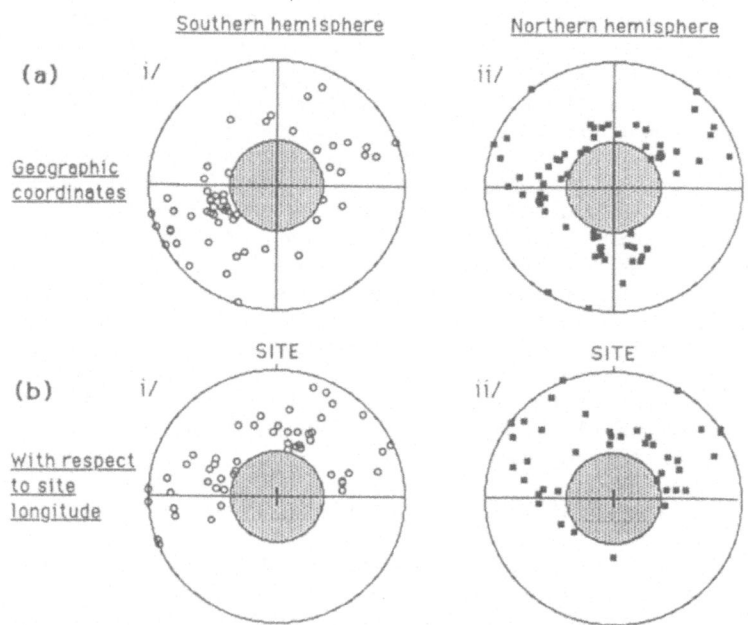

Figure 5. VGP's computed from values of declination and inclination
of eight records of the last reversal viewed from above the geographic
North Pole;
      (a) in geographic coordinates, (i) southern hemisphere VGP's,
(ii) northern hemisphere VGP's,
      (b) plotted with respect to the site longitude, (i) southern
hemisphere VGP's, (ii) northern hemisphere VGP's.

Valet et al. (1986) have suggested that secular variation, similar to that seen when the field is not reversing, may persist through the reversal. Slight smoothing of a reversal record from Crete revealed periodic oscillations in the angular displacement from the pre-reversal dipole field direction (Figure 6). In the figure we see the continuation throughout the reversal of field fluctuations which have time constants of between 1000 and 2000 years, similar to secular variation. The magnitude of fluctuation increases during the directional transition.

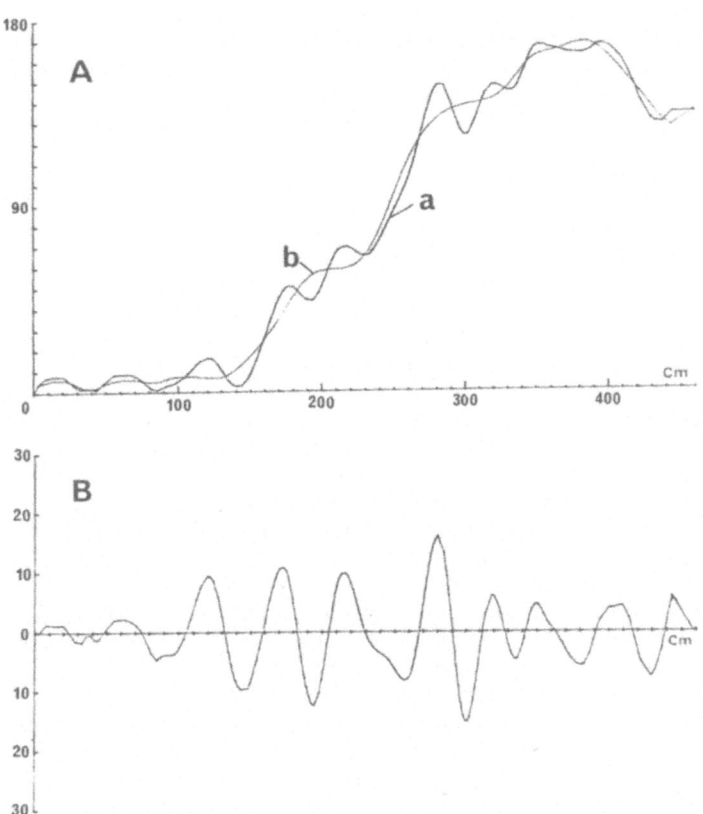

Figure 6. The presence of secular variation-like behaviour in a reversal record from Crete (Valet et al., 1988), (A) Angular deviations from the pre-reversal directions plotted against stratigraphic height (a) after smoothing with cubic splines (27 knots), (b) same with 15 knots. (B) Periodic oscillations obtained by subtracting curve (b) from curve (a) in (A).

3. MODELS

3.1 Standing and Flooding Field Mechanisms

The recognition of the non-dipolar nature of the transitional field
has been a major step in the study of geomagnetic reversals. It led
naturally to investigation of whether the transitional field resembles
the usual non-dipole field, after collapse of the main dipole field,
or whether it arises from time dependent harmonics related to the
reversal.
        Hillhouse and Cox (1976) assumed that the non-dipole field can be
described as the sum of a westward drifting and a standing component,
and accepted the estimate of 4600 years for the duration of a
reversal, obtained from records in deep-sea cores. They calculated
that, if the usual non-dipole field were basically unchanged during a
polarity reversal, its westward drifting component should cause large
longitudinal swings of the VGP which would be clearly visible in the
records. The absence of such characteristics in the Lake Tecopa record
where transitional directions are, on the contrary, rather confined
along a great circle, led Hillhouse and Cox to suggest that a standing
component of the field is present during the main part of the
transition, while the dipole field decreases to zero and regenerates
with opposite polarity. This has been called the 'standing field'
model for reversal.
        A completely different conclusion was reached by Hoffman (1977),
who suggested that a reversal originates in a localised region of the
core and subsequently 'floods' into other regions. The transitional
field originates from time dependent components associated with this
process in the core. An important aspect of this approach was the
recognition that this process would lead to zonal transition fields.
These zonal harmonics could account for multiple records of
transitional fields which involved simultaneous vertical inclinations
in successive bands of latitude. In terms of the VGP paths, the path
passes near to the site, or near to the antipode of the site ('near'
or 'far' sided path). This is illustrated in Figure 7, where a
quadrupolar transitional field is produced during a reversed to normal
transition by flooding of normal flux out from the initiation region
at the South Pole. A band of latitude at which the inclination is +90°
travels north from 90°S to 90°N. Later, in 1981, Hoffman pointed out
that the standing field and flooding field models predict different
field behaviour associated with sequential field reversals at the same
site that is theoretically distinguishable. Specifically, if a
standing field predominates during the transition and remains
unchanged for times longer than the interval separating two successive
reversals, the intermediate field geometries should be identical
(independent of the sense of the transition). Widely different
intermediate directions are, on the contrary, expected if a flooding
field controls the reversal process.
        In this respect, one may, for instance, examine the VGP paths
associated with four different reversals, three of which are
successive, recorded in an upper Miocene marine section in western

Crete, which have been reported by Valet and Laj (1981, 1983, 1984).
The main result, shown in Figure 8, is that the north-VGP paths
associated with the two R-N transitions and the south-VGP path for the
middle N-R transition are identical within the noise limits. Thus the
directional field variation at the site for this N-R transition is
identical to that associated with the R-N's, except for a change in
sign. The south-VGP path for the upper N-R reversal is different,
about 60° of longitude away from the other ones. This situation is
exactly opposite to the standing field situation, which requires
identical transition fields including the sign. It is consistent with
the flooding model even though the paths are not confined to the site
meridian or its antipode. An investigation of two sequential R-N and
N-R reversals of early Pliocene age from lava sequences in Kauai,
reported by Bogue and Coe (1982), was initially interpreted in terms
of a standing field. Indeed the two paths, although ill-defined
because of the very limited number of transitional directions, seem
identical and close to the site. However, the palaeointensity record
for the R-N reversal obtained more recently is quite asymmetric, with
an abrupt decay and smooth recovery of the field intensity. A plot of
field intensity versus angular deviation for any particular standing
field model should be symmetrical. Bogue and Coe (1984) thus

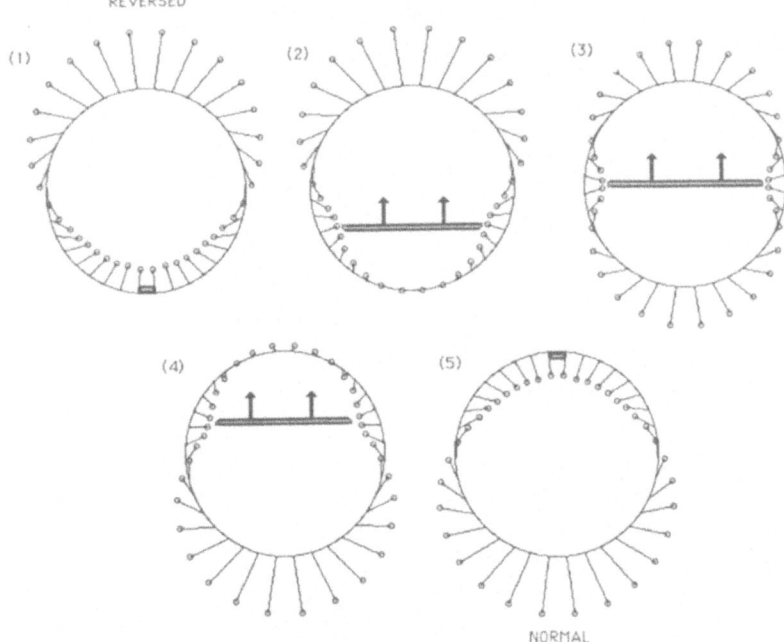

Figure 7. Progress of a latitudinal band of 90° inclination northward
across lines of latitude during a R-N reversal in which the
transitional field is quadrupolar and reversal initiates at the South
Pole.

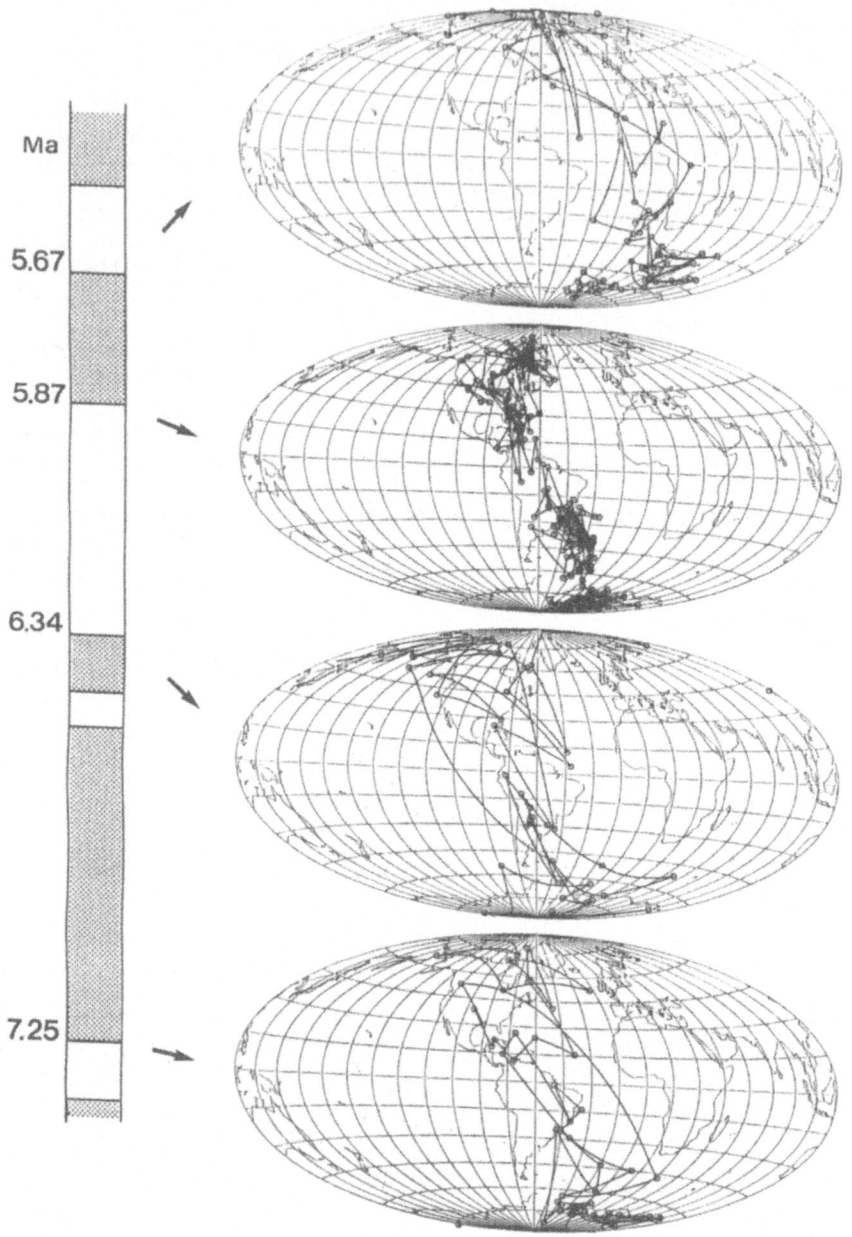

Figure 8. VGP paths of four polarity transitions in Crete (north VGP's for the R-N and south VGP's for the N-R transitions) (Valet and Laj, 1984).

reinterpreted their data in terms of a particular zonal flooding model.

If these two studies clearly favour the flooding model, on the other hand, conflicting evidence has been obtained from two independent studies of the Olduvai and Jaramillo transitions. Clement and Kent (1984) studied the upper Olduvai and lower Jaramillo transitions in the southern Indian Ocean. The two paths were slightly far sided, and consistent with a standing field model. Herrero-Bervera and Theyer (1986) studied both the lower and upper reversals of the Olduvai and Jaramillo chrons in deep-sea sediments of the north-central Pacific. Although the records are characterised by a large scatter, the back-to-back reversals have well separated paths, a situation which cannot be reconciled with a standing field model. The paths are not exactly antipodal, but can readily be described with a flooding model. These two studies cannot, therefore, be reconciled, illustrating one of the difficulties in these studies.

Summarising, although the application of Hoffman's test to sequential reversals leads in certain cases to a somewhat confusing situation, the clearest results favour the flooding model. The additional observation that in many volcanic and sedimentary records the VGP paths undergo large directional deviations might require that there be drifting components or superimposed standing and drifting components of the field during reversals.

3.2 Transitional Field Geometry

Along with studies of field reversal mechanisms, the symmetry of the transitional field, which is basic to any interpretation of the reversal records, has been the object of many investigations: the observation that zonal harmonics seemed to dominate transitional fields (Fuller et al., 1979) led Williams and Fuller (1981) to develop a mathematical model in which the dipole energy is partitioned among the low order zonal terms. In this zonal harmonic model geomagnetic reversals were simulated by allowing the dipole to decay to zero and grow in the opposite sense, with the variation in dipole field energy controlled by a tangent function. The energy lost from the dipole field was fed into the axial quadrupole, octupole, and hexadecapole in the ratio of 20%, 30%, and 50% with the energy in the field outside the core being conserved.

The model demonstrated that individual records of a reversal would vary considerably, depending upon the latitude of the observation site. However, the axisymmetry of the model predicted that records would be invariant with longitude. The zonal model was partially successful in simulating reversal records, but it was too simple to explain the variety of phenomena presently seen during reversals. The principal failing was that it could not account for records which did not include inclinations of + or -90°. In presentations of records in terms of virtual geomagnetic pole (VGP) paths, the zonal model predicts that the pole will always pass immediately under the site, or under its antipode (Figure 9A). However, records were observed without the necessary high inclination,

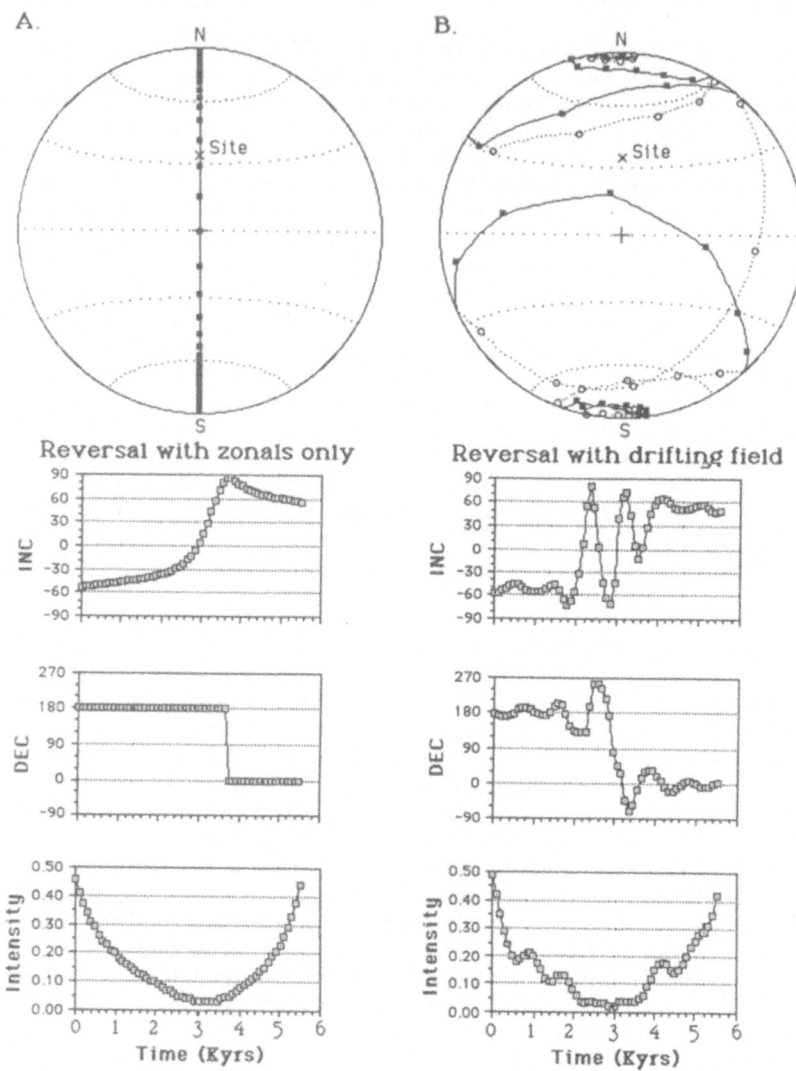

Figure 9. Synthetic palaeomagnetic records of a reverse to normal transition of the geomagnetic field observed at 35°N. The uppermost row of plots are virtual geomagnetic pole (VGP) paths viewed from above the equator where it intersects the site meridian. The lower profiles give the succession of inclination, declination and field intensity values as a function of time during the reversal.
(A) Zonal model.
(B) Drifting non-dipole field model.

so that the VGP path was neither close to the site, nor to its antipode. An additional complication was that such paths were frequently confined to a small band of longitude but displaced from the site, or the antipodal meridian. This also indicated that a drifting non-dipole field could not alone be responsible for the observed paths; such a field geometry defines a path which sweeps around the surface of the Earth (Figure 9B).

There was a paradox, the importance of axisymmetric terms was demonstrated by multiple reversal records of the last reversal (Figure 5), while, in detail, many individual paths showed departures from

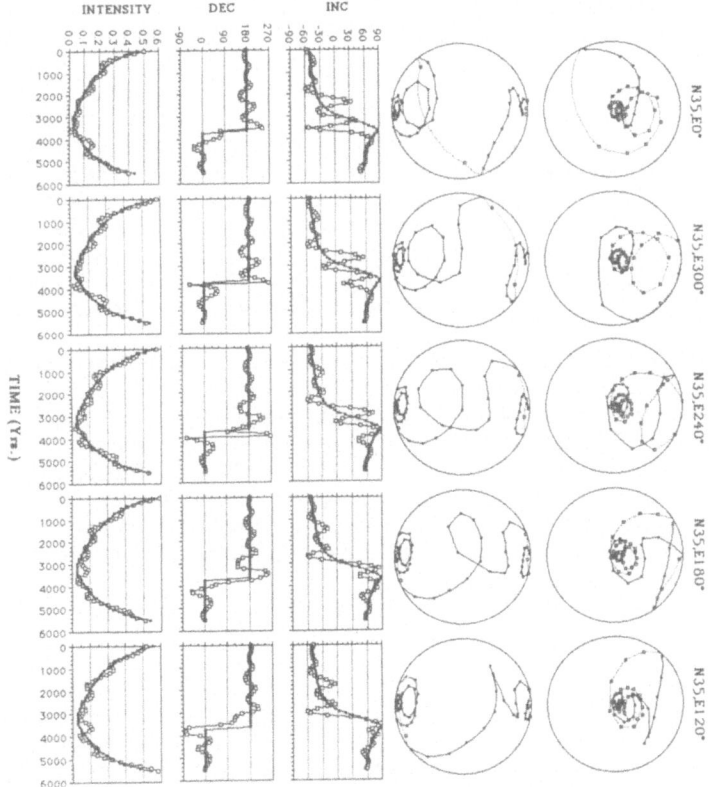

Figure 10. Synthetic palaeomagnetic records of a reverse to normal transition of the geomagnetic field according to the zonal + drifting non-dipole field model. Observation sites are at 35°N at longitudes 0°, 300°, 240°, 180°, and 120°.

The uppermost row of plots are polar plots of VGP paths. The second row gives the same VGP paths viewed from above the intersection of the equator and the site meridian.

The lower profiles give inclination, declination and intensity as a function of time throughout the reversal.

axisymmetry. An extension of the zonal model was, therefore, developed which included a westward drifting non-dipole field (Williams, Weeks and Fuller, 1988). Figure 10 illustrates the results for observation sites on a single line of latitude in the northern hemisphere, in the form of VGP paths and declination, inclination, and intensity plots. Note that, as in the case of the last reversal, there are simple VGP paths confined to the site meridian, which would be interpreted as evidence of the dominance of zonals, but there are also others revealing the importance of non-zonals. The paths are controlled by the phase relationship of the variation of strength of the zonals and the drifting non-dipole field. Hence we see a wide variety in the paths along a line of latitude as a function of longitude.

It is clear that the synthetic reversal paths include a number of features seen in the observed records. The preference for near sidedness is due to the control of the path by the zonals even in the absence of a dipole. The drifting non-dipole field can displace the VGP path in longitude away from the site meridian and, providing that the main directional change takes place when the non-dipole field is not changing rapidly, the path will still be confined in longitude. One sees the evolution of loops and hang-ups in the path as the longitude of the observation site varies. These are consequences of the phase relationships at the various sites between the growth and decay of the zonals and the passages of features of the drifting non-dipole field.

The model demonstrates that drifting fields superimposed upon a zonal field can account for many of the features of both individual palaeomagnetic reversal records and multiple records of the same reversal. This model has additional appeal since it makes use of features observed in the present geomagnetic drift: The decay of the dipole while the next higher order terms are growing, and the westward drift of the non-dipole field. However, the problem is non-unique, and undoubtedly these results could be obtained using other geometries such as varying standing fields.

4. CONCLUSION

The ultimate resolution of the problem of the geometry of transition fields during reversals requires sufficient records of a single reversal to permit spherical harmonic inversion of the data. In particular we still need good southern hemisphere records of the last reversal. Meanwhile, our knowledge of transitional fields will grow with the accumulation of records, but it is important to recognise the limitations of the type of data we study.

5. REFERENCES

Bogue, S.W. and R.S. Coe, 1982. 'Successive palaeomagnetic reversal records from Kauai', Nature, **295**, 399–401.

Bogue, S.W. and R.S. Coe, 1984. 'Transitional paleointensities from Kauai, Hawaii, and geomagnetic reversal models', J. Geophys. Res., **89**, B2, 1049–1055.

Clement, B.M., D.V. Kent, and N.D. Opdyke, 1982. 'Brunhes–Matuyama polarity transition in three deep-sea cores', Phil. Trans. R. Soc. London, **A306**, 113–119.

Clement, B.M. and D.V. Kent, 1984. 'A detailed record of the lower Jaramillo polarity transition from a Southern hemisphere, deep-sea sediment core'. J. Geophys. Res., **89**, 1049–1058.

Clement, B.M. and D.V. Kent, 1984. 'Latitude dependency of geomagnetic polarity transition durations', Nature, **310**, 488–491.

Clement, B.M. et al., 1987. 'Geomagnetic polarity transition records from five hydraulic piston core sites in the North Atlantic. In Ruddiman, W.F., Kidd, R., Thomas, E. et al., Init. Repts, DSDP, 94: Washington (U.S. Printing Office).

Coe, R.S. and M. Prevot, 1988. 'Evidence of very rapid field variation during reversals', NATO Advanced Study Institute on Geomagnetism and Palaeomagnetism – Book of Abstracts.

Dodson, R., J.R. Dunn, W. Fuller, I. Williams, H. Ito, V.A. Schmidt, and M. Wu Yu, 1978. 'Paleomagnetic record of a late Tertiary field reversal', Geophys. J., **53**, 373.

Dunn, J.R. and M. Fuller, 1984. 'Thermal demagnetization with measurements at high temperature using SQUID magnetometer', EOS Trans. Am. Geophys. Union, **65**, 863.

Fuller, M., I. William, K.A. Hoffman, 1979. 'Paleomagnetic records of geomagnetic field reversals and the morphology of the transitional fields'. Rev. Geophys. Space Phys., **17**, 179–203.

Harrison, C.G.A. and B.L.K. Somayajulu, 1966. 'Behaviour of the Earth's magnetic field during a reversal', Nature, **212**, 1193–1195.

Herrero-Bervera, E. and F. Theyer, 1986. 'Non-axisymmetric behaviour of Olduvai and Jaramillo polarity transitions recorded in north-central Pacific deep-sea sediments', Nature, **322**, 159–162.

Hillhouse, J. and A. Cox, 1976. 'Brunhes–Matuyama polarity transition', Earth Planet. Sci. Lett., **29**, 51–64.

Hoffman, K.A., 1977. 'Polarity transition records and the geomagnetic dynamo'. Science **196**, 1329–1332.

Hoffman, K.A. and M. Fuller, 1978. 'Transitional field configurations and geomagnetic reversal', Nature, **273**, 715–718.

Hoffman, K.A., 1981. 'Quantitative description of the geomagnetic field during the Matuyama–Bhrunes polarity transition'. Phys. Earth Planet. Int., **24**, 229–235.

Hoffman, K.A. and D.L. Morse, 1987. 'Strong dipole paleointensities following the Olduvai-Matuyama polarity transition', EOS, 68, Vol. 44, 1249.

Kono, M., 1973. 'Geomagnetic polarity changes and the duration of volcanism in successive flows', J. Geophys. Res., 78, 5972-5982.

Laj, C., S. Guitton and C. Kissel, 1987. 'Rapid changes and near stationarity of the geomagnetic field during a polarity reversal', Nature, 330, 145-148.

Laj, C., S. Guitton, C. Kissel and A. Mazaud, 1988. 'Complex behaviour of the geomagnetic field during three successive polarity reversals, 11-12 M.Y.B.P.', J. Geophys. Res., in press.

Linssen, J.H., 1988. 'Preliminary results of four successive sedimentary geomagnetic reversal records from the Mediterranean (Upper Thvera, Lower- and Upper-Sidufjall, Lower-Nunivak)', Phys. Earth and Planet. Int., in press.

Mankinen, E.A., M. Prevot, C.S. Gromme and R.S. Coe, 1985. 'The Steen's Mountain geomagnetic polarity transition, 1, Directional history, duration of episodes and rock magnetism', J. Geophys. Res., 90, 10,379-10,417.

Opdyke, N.D., D.V. Kent, W. Lowrie, 1973. 'Details of magnetic polarity transitions recorded in a high deposition rate deep-sea core'. Earth Planet. Sci. Lett., 20, 215-324.

Prevot, M., E.A. Mankinen, R.S. Coe, and C.S. Gromme, 1985. 'The Steens Mountain (Oregon) geomagnetic polarity transition. 2. Field intensity variations and the discussion of reversal models'. J. Geophys. Res., 90, 10417-10488.

Theyer, F., E. Herrero bervera, V. Hsu, and S.R. Hammond, 1985. 'The zonal harmonic model of polarity transitions: a test using successive reversals'. J. Geophys. Res., 90, 1963-1982.

Valet, J.-P. and C. Laj, 1981. 'Paleomagnetic record of two successive Miocene geomagnetic reversals in Western Crete'. Earth Planet Sci. Lett., 54, 53-63.

Valet, J.-P. and C. Laj, 1983. 'Two different R-N geomagnetic reversals with identical VGP paths recorded at the same site'. Nature, 304, 330-332.

Valet, J.-P. and C. Laj, 1984. 'Invariant and changing transitional field configurations in a sequence of geomagnetic reversals', Nature, 311, 552-555.

Valet, J.-P., C. Laj and P. Tucholka, 1986. 'High resolution sedimentary record of a geomagnetic reversal', Nature, 322, 27-32.

Valet, J.-P., C. Laj and C.G. Langereis, 1988. 'Sequential geomagnetic reversals recorded in upper Tortonian marine clays in western Crete', J. Geophys. Res., 93, 1131-1151.

Van Zijl, J.S., K.W.T. Graham and A.L. Hales, 1962. 'The paleomagnetism of the Stormberg lavas, II. The behaviour of the magnetic field during a reversal', Geophys. J., 7, 169.

Weeks, R.J., 1988. 'Paleomagnetic records of the geomagnetic field:
    Reversal records and reversal stratigraphy', Ph.D. thesis,
    University of California Santa Barbara.
Williams, I.S. and M. Fuller, 1981. 'Zonal harmonic models of
    reversal transition fields', J. Geophys. Res., **86**,
    11,657–11,665.
Williams, I.S., R.J. Weeks and M. Fuller, 1988. 'A model for
    transition fields during geomagnetic reversals', Nature, 322,
    719–720.

A PHENOMENOLOGICAL MODEL FOR REVERSALS OF THE GEOMAGNETIC FIELD

A. Mazaud, C. Laj and E. Bard
Centre des Faibles Radioactivités
Laboratoire mixte C.N.R.S. – C.E.A.
Avenue de la Terrasse
91198 Gif-sur-Yvette cedex
France.

ABSTRACT. We consider a system made by N point interacting dipoles, each of them having a finite lifetime. A renewal process for the dipoles gives the time evolution of the system. When the number N of dipoles is low, and for a wide range of the coupling between the dipoles, we observe sudden transitions or excursions between two polarised states. Thus our system provides artificial polarity time-scales which can be compared with the geomagnetic one. It is also possible, by locating the dipoles within a core and calculating the magnetic field produced, to obtain a simple phenomenological modelling of the geomagnetic field. In this paper we discuss first the characteristics and the time evolution of the generator, then we briefly compare in a qualitative way the modelled field with the geomagnetic one.

1. PRESENTATION OF THE MODEL

The simple model developed here is a generator of a roughly dipolar magnetic field, exhibiting spontaneous and sudden collapses and reverses of its magnetization.

As it is a purely phenomenological model, it does not give at all a realistic view of the dynamo process which acts inside the core of the Earth. It simply creates a magnetic field in any point of the space, allowing comparisons between the generated field and the geomagnetic one from both spatial and dynamical points of view.

The field generator, somewhat similar to some models used in solid state physics, is made of an assembly of N interdependent point dipoles. More precisely, a given dipole is only sensitive to the mean magnetic moment of the other ones (mean field approximation). Each elementary dipole presents a unit moment either parallel or antiparallel to a given axis, so that the total moment of the system takes integer values included between $-N$ and $+N$.

The dynamical evolution is computed following a Monte Carlo method acting as follows.

F. J. Lowes et al. (eds.), Geomagnetism and Palaeomagnetism, 205–214.
© 1989 by Kluwer Academic Publishers.

2. THE DYNAMICAL EVOLUTION OF THE SYSTEM

2.1 The dipoles replacement mechanism

The time dependent properties of this system arise from the assumption of a finite lifetime for each dipole and from a renewal process: given any random initial status, at regular intervals T a dipole is suppressed and a new one immediately generated. For simplicity we have assumed a unique lifetime for all the dipoles, equal to N times the interval T, so that a particular dipole is unchanged until the replacement of all the other dipoles is completed.

Changes in the total moment of the system may occur because the polarity of a new dipole is statistically determined by its coupling with the other ones, for which we assume, as a first approach, a mean field interaction.

Thus its polarity is determined by a two levels Boltzmann's statistics with the total moment of the N-1 other dipoles as a field parameter. More precisely, the probabilities for an "up" and "down" new dipole are respectively

$$P(+1) = \frac{\exp(CM_o)}{\exp(CM_o) + \exp(-CM_o)}$$

$$P(-1) = \frac{\exp(-CM_o)}{\exp(CM_o) + \exp(-CM_o)}$$

where M_o is the total magnetic moment of the N-1 other dipoles present when the new dipole is generated. It is clear that the model is driven by only two parameters, N and C, and that the generation of a new dipole parallel to the total magnetization of the N-1 other ones is favoured, especially when the coupling constant is large.

2.2 Existence of self-reversals of the magnetization

Three time evolutions of the magnetic moment can be distinguished.

The first is characterized by large values of the coupling constant, so that a marked difference exists between $P(+1)$ and $P(-1)$, even for low values of M_o. In this case, for any initial status, a spontaneous magnetization occurs immediately, then most of the new dipoles appear with a moment parallel to this magnetization. Consequently, the total moment is always very close to either +N or -N.

In contrast, for very small values of C the interaction is negligible and new "normal" or "reverse" dipoles appear with about equal probabilities. The total moment fluctuates around zero and no stable polarization occurs.

The most interesting behaviour is exhibited by systems

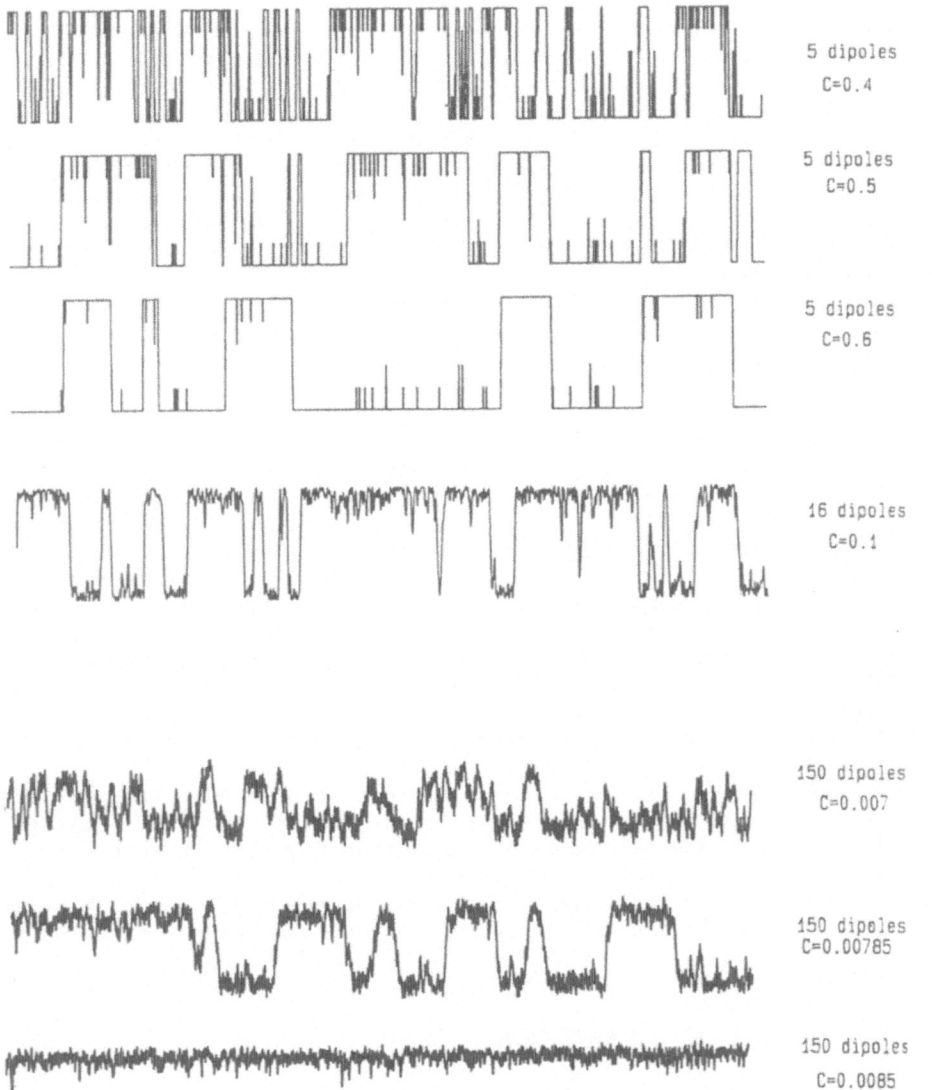

Figure 1. Evolution of the magnetization of models made of N mean field coupled dipoles. The total displayed length of time is equal to 1080xN elementary steps.

(a) N=5 and C=0.4, 0.5 and 0.6. The frequency of the reversals is greater for low values of the coupling constant C.

(b) N=16 and C=0.1. The signal has been smoothed using a 32 width square window.

(c) N=150 and C=0.007, 0.00785, 0.0085. Well defined reversals are observed for C=0.00785.

characterized by intermediate values of the coupling constants (Figure 1). In this case, for any initial random condition a polarized state readily appears, so that the total moment is close to either +N or -N. However, because of the statistical fluctuations, this moment sometimes undergoes sudden decreases and is then restored to either the initial or the opposite polarity. Thus the model exhibits both excursions and reversals. This behaviour results from two competing factors: on one side the mean field interaction favours the creation of dipoles with identical polarity. On the other, sequences dominated by improbable events (i.e. sequences of dipoles all antiparallel to the initial polarity) are possible. When such an improbable sequence begins, at each elementary step the magnetization of the system is lowered so that at each step the difference between $P(+1)$ and $P(-1)$ decreases, in turn enhancing the probability for additional "reverse" dipoles to be created. This positive feedback is particularly efficient for small values of N, because in this case the occurrence of a sequence, even short, of reverse dipoles at each step significantly decreases the total moment of the system. For this reason, for systems with $N < \sim 50$ when a reversal occurs its duration is never significantly longer than N elementary steps, which is of course the minimum possible length. For large values of N, on the contrary, only very long, extremely improbable, sequences of reverse dipoles can significantly lower the total magnetization and thus significantly change the Boltzmann statistics and make this cooperative effect possible. Consequently, reversals and sudden excursions are more easily observed in systems of less than ~ 50 dipoles. In these systems, the frequency of the reversals is directly related to the value of the mean field constant.

Finally, we can ask about the influence of a unique lifetime (NT) for the dipoles on the reversals triggering statistics. It certainly introduces a very short term memory (of about NT), but it is not a crucial characteristic of the generator. Indeed, similar results are obtained when random lifetimes are chosen for the dipoles, so that one may consider that the polarity reverses of our generator are randomly triggered.

3. THE FREQUENCY OF THE REVERSALS

Using first a stationary generator ($N = 16$, $C = 0.1$), we have produced different series made of 200 000 successive replacements of dipoles, so that each series exhibits about 180 reversals of magnetization (i.e. approximately the number of geomagnetic reversals in the last 100 My). We then calculated the frequency of the reversals using a 8000 T width sliding window, strictly equivalent to the 4 My window we already used (Mazaud et al., 1983) for the geomagnetic data (Labrecque et al., 1977). The signal obtained is thus the convolution of the data with the window, which acts strictly as simple low-pass filters for signals with time constant larger than twice the window width.

Most of the frequency curves present peaks (Figure 2a and Figure 2b) similar to those obtained with the geomagnetic data (Figure 3,

Mazaud et al., 1983). Few of them however (<20% of the cases) do
exhibit a periodic oscillating component (Figure 2b) similar to the
geomagnetic one (Figure 3, Mazaud et al., 1983).

   An important point is that a stationary generator never produces
monotonic trends in the frequency of the reversals similar to the
geomagnetic one (Lowrie and Kent, 1983; Mazaud et al., 1983). Such a
trend can by contrast easily be obtained using a non-stationary model,

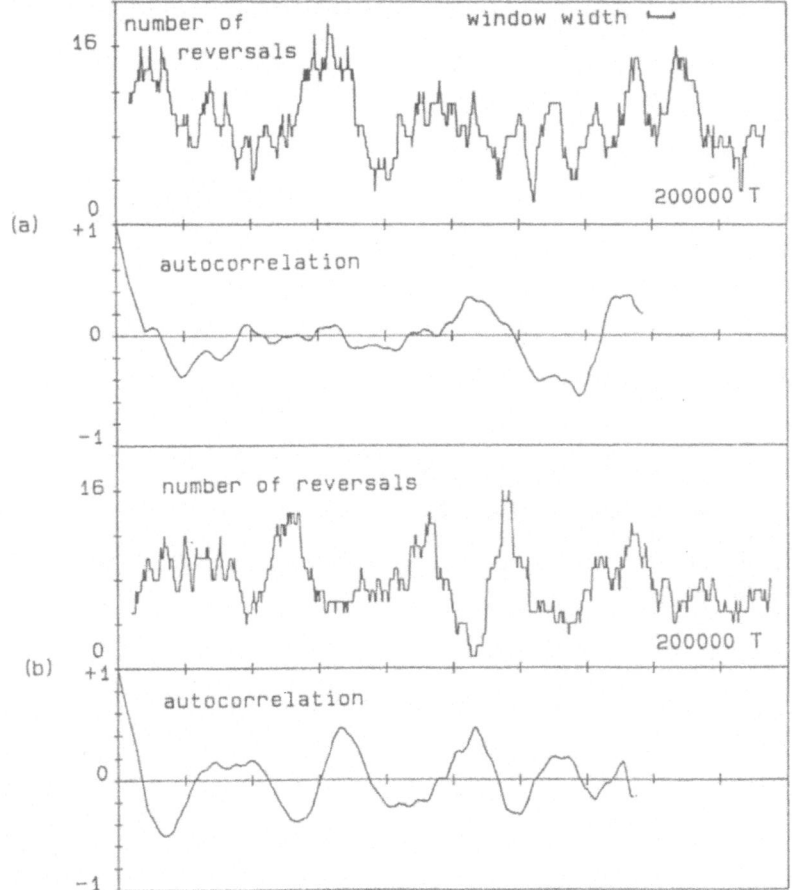

Figure 2. Typical reversal frequencies obtained with our generator
(16 dipoles, C = 0.1, 200 000 successive replacements of the dipoles).
The frequencies are obtained using a 8000 T width overlapping sliding
window, equivalent to the 4 My used for the geomagnetic data. Mean
values have been subtracted before calculating the autocorrelations.
a) In most of the cases, the frequency fluctuations are not periodic.
b) A few of the cases (about 20%) present a periodicity similar to the
   geomagnetic one.

210

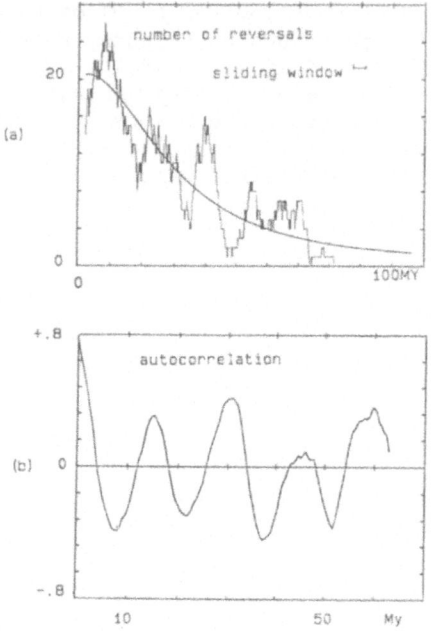

Figure 3. (from Mazaud et al. 1983). (a) Frequency of the
geomagnetic reversals of the Labrecque, Kent and Cande polarity time
scale. The width of the sliding window is 4 My and the mean trend has
been modelled by a fitted Lorentzian curve (full line).
(b) Autocorrelation of the fluctuating part of the signal; a 15 My
periodicity appears.

Figure 4. Frequency of the reversals of a non-stationary generator.
C varies linearly from 0.07 to 0.13 during 100 000 successive
replacements of the dipoles. A striking similarity is observed
between this frequency signal and the geomagnetic one of Figure 3(a).

characterized by a time variation of C or N. As an example, the
records obtained applying our filtering technique on both the
geomagnetic data and on a sequence of reversals issued from a
non-stationary model, characterized by a linear time variation of the
coupling constant are shown in Figures 3a and 4. The similarity of
the synthetic record (Figure 4) with the geomagnetic one (Figure 3a)
is quite striking.

4. THE GENERATED MAGNETIC FIELD

If we randomly locate the dipoles in the space delimited by two
concentric spheres, we can calculate the magnetic field produced
outside, and in particular on an external sphere on which two opposite
poles have been defined by the axis of the system (Figure 5). In a
first approach, we have defined N possible sites for the dipoles and
randomly assigned one of these sites for each appearing dipole.
    Figure 6 shows the field variations calculated at a mid latitude
point (40°) on the external sphere, using a 16 dipoles stationary
generator (C=0.1). Both magnetic moment of the generator and field
characteristics (declination, inclination, intensity and VGP

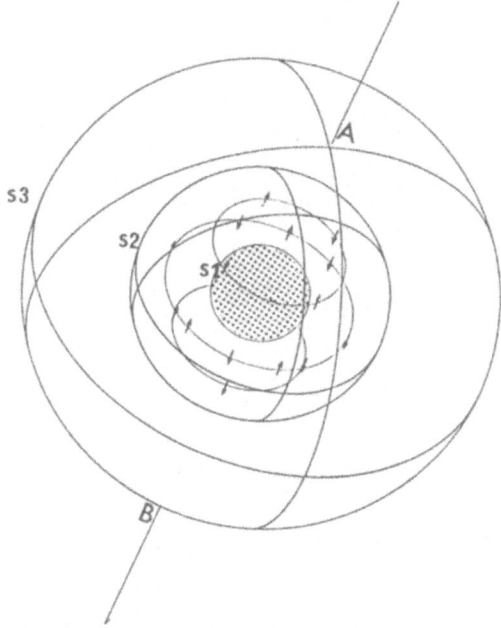

Figure 5. The geometry of the magnetic source. The dipoles are
located between S1 and S2, the field is calculated on S3. A and B are
the poles defined by the two allowed directions for the dipoles.

212

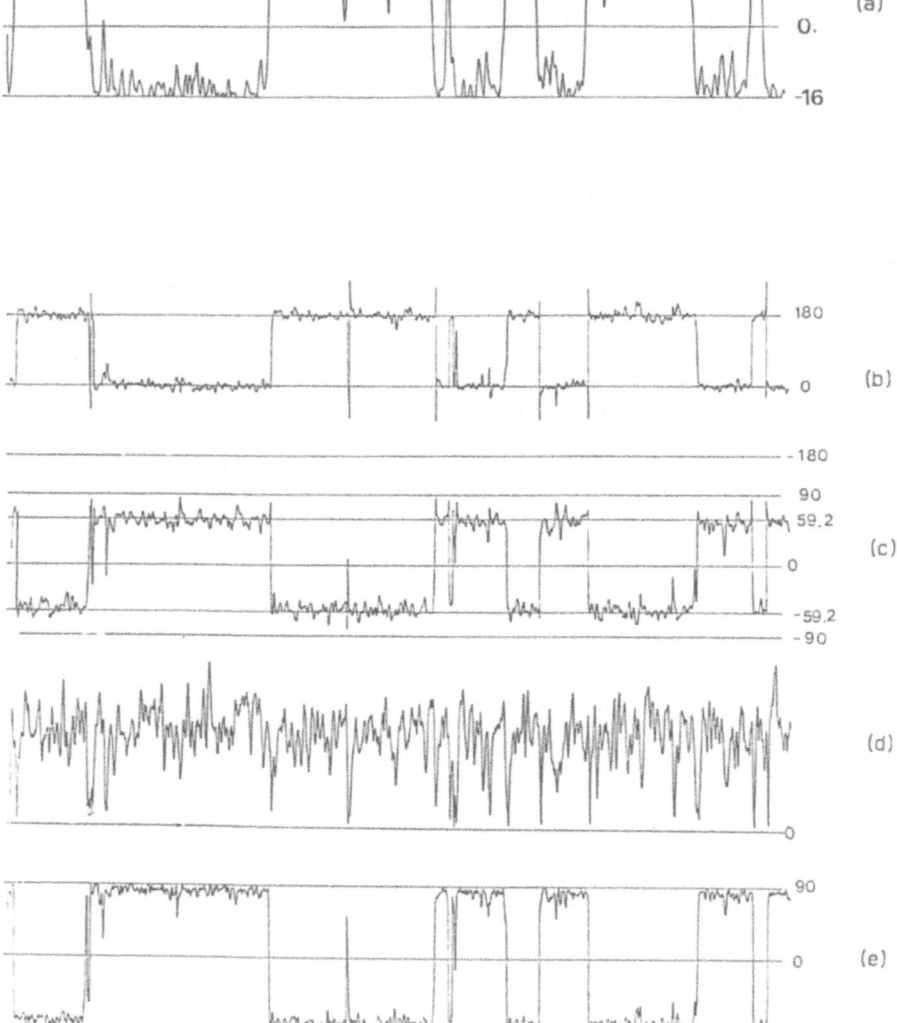

Figure 6. Field variations calculated for a mid latitude (40°) site on the external surface S3 for a 16 dipoles stationary generator (C=0.1). A slight smoothing has been applied (32 T).
(a) Total dipole moment.
(b) Declination, (c) inclination, (d) intensity, at 40° site.
(e) Latitude of corresponding VGP.

latitudes) are displayed on Figure 6, while the corresponding VGP path is shown on Figure 7. The declination and the inclination are stable during chrons and defined polarity and agree quite well with a dipolar field, the dipoles being almost parallel inside the core. The intensity fluctuates strongly, but very marked decreases are always associated with reversals of polarity or strong excursions. The VGP path wanders randomly around the poles during periods of stable polarity, while excursions and reversals are in most of the cases, but not always, characterized by trajectories constrained in longitude. This occurs because only directions parallel or antiparallel to the axis are allowed for the dipoles. Transitional and excursional fields are clearly not dipolar, because they correspond to magnetic sources made of dipoles of opposite polarities in about equal number.

The modeled reversal trajectories are qualitatively similar to some of the ones obtained in sediments (Valet and Laj, 1984), but, because of its random aspect, our model can not easily generate symetrical successive reversals (Valet and Laj, 1984). At least, and because of the discrete mechanism of replacement of the dipoles, some trajectories are in relative good agreement with the existence of successive impulses and standing fields observed in some lava or sediments (Laj et al., 1987; Prévot et al., 1985). Finally, we can notice that the appearance of successive dipoles with identical or closely grouped locations may produce local excursions on the external surface, above that part of the source.

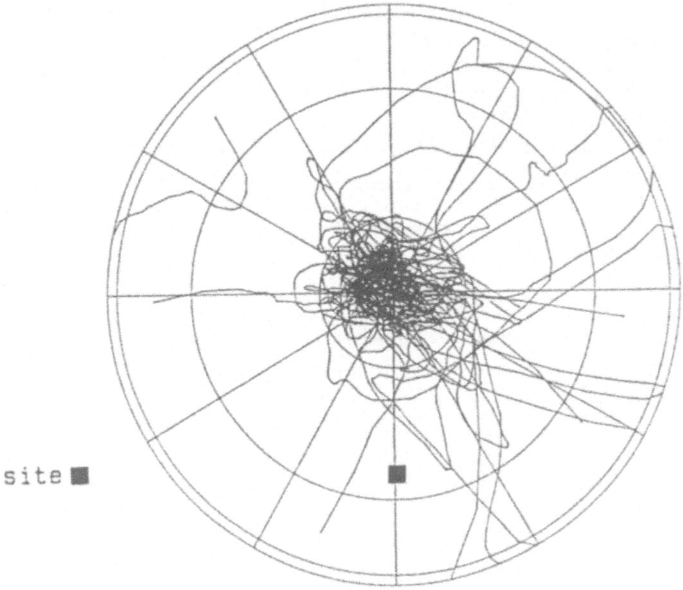

site ■

Figure 7. VGP path corresponding to the magnetic field displayed in Figure 6.

5. DISCUSSION

Our model does not describe in any way the physical processes which act as a dynamo within the Earth's core. Nevertheless, driven by only two parameters, it provides a qualitatively good modelling of the geomagnetic field.

Comparisons between model generated time scale and the geomagnetic polarity time scale show that:

(1) The geodynamo is not, at least during the last 100 My, a stationary generator of reversals; we need to use time-changing parameters to simulate the observed increase in the frequency of the geomagnetic reversals.

(2) Periodic fluctuations similar to the observed one in the geomagnetic data (15 My) are sometimes present (in about 20% of the cases) in generated reversals sets. Consequently, and because of the uniqueness of the geomagnetic data, it is not possible to determine unambiguously whether the detected geomagnetic periodicity arises from a random mechanism or not.

CFR contribution n° 965

6. REFERENCES

Labrecque, J., Kent, D.V. and Cande, S.C., 1977. 'Revised magnetic polarity time scale for the late Cretaceous and Cenozoique time', Geologie, 5, 330–335.

Laj, C., Guitton, S. and Kissel, C., 1987. 'Rapid changes and near stationarity of the geomagnetic field during a polarity reversal', Nature, 330, 145–148.

Lowrie, W. and Kent, D., 1983. 'Geomagnetic reversal frequency since the late Cretaceous', Earth Planet. Sci. Lett., 62, 305–313.

Mazaud, A., Laj, C., de Sèze, L. and Verosub, K.L., 1983. '15-My periodicity in the frequency of the geomagnetic reversals', Nature, 304, 328–330.

Prévot, M., Mankinen, E., Grommé, C. and Coe, R., 1985. 'How the geomagnetic field reverses polarity', Nature, 316, 230–234.

Valet, J.P. and Laj, C., 1984. 'Invariant and changing transitional field configurations in a sequence of geomagnetic reversals', Nature, 311, 552–555.

DETERMINISTIC CHAOS, GEOMAGNETIC REVERSALS, AND THE SPHERICAL PENDULUM

D.J. Tritton,
Department of Physics,
University of Newcastle upon Tyne
NE1 7RU, U.K.

ABSTRACT. This article is an introduction to modern ideas about chaos, as background to the question of whether the geomagnetic reversal sequence is a chaotic one. The chaotic motion of a forced spherical pendulum is used to exemplify these ideas. Reasons are given why geomagnetic reversals might be interpreted as chaotic and the implications discussed. Some statistical data on the pendulum motion are presented in a form that may allow comparison with geomagnetic data.

1. INTRODUCTION

The geomagnetic reversal sequence may be a chaotic one. This article is primarily a brief introduction to the meaning of that statement, together with some discussion of its implications.

Let us be clear in the first place what it does not mean: one is not just saying that the sequence is complicated and "messy-looking", with, for example, no immediately apparent periodicities. As many readers will know, the word "chaos" has taken on a specific scientific meaning in the last couple of decades, a meaning often made more specific by talking of "deterministic chaos". What we are considering is whether it is informative to view geomagnetic reversals within this context.

The topic of geomagnetic reversals may not be the only one through which geophysicists may meet chaotic systems. It is probably not an overstatement to say that modern ideas about chaos may find application in any branch of science (with the word "science" being used in a very broad sense). Perhaps the other applications that readers of this book are most likely to meet are to solar system dynamics: the probable chaotic tumbling of Saturn's satellite Hyperion, and the varied consequences of resonances in asteroid orbits (Wisdom 1987). But arguably the most important reason for an awareness of the ideas about chaos is the possibility of applications that have not yet been identified.

This article firstly (§2) gives a brief introduction to the basic

215

F. J. Lowes et al. (eds.), Geomagnetism and Palaeomagnetism, 215–226.

ideas of deterministic chaos. The ideas can seem vague without
specific examples, and so the chaotic motion of a forced spherical
pendulum is briefly described to lend them substance. The choice of
this example is in part just an author's prerogative to publicise his
own interests, but is appropriate also because the pendulum's motion
involves "reversals" giving a closer analogy with the geomagnetic case
than many other possible examples. In §3 we consider the reasons for
thinking that geomagnetic reversals may be occurring chaotically and
the implications of this inference. It may be useful, in such
discussion, to build up a body of data on the statistical features of
typical chaotic systems, and §4 presents some preliminary results of
this type for the forced spherical pendulum.

2. WHAT IS DETERMINISTIC CHAOS?

Many readers will be already familiar with the basic concepts.
However, some may not, so a brief introduction seems appropriate. In
the space it can only be a very inadequate explanation of a topic that
has now generated a vast literature. Suggestions for readers wanting
a fuller explanation are Bergé, Pomeau & Vidal (1986), Moon (1987),
and Thompson & Stewart (1986).
    As already indicated, chaotic behaviour has been identified in
diverse systems. They do, however, all have one thing in common: they
are described by non-linear equations. Chaotic behaviour can never be
generated by linear equations. With this qualification, systems
exhibiting such behaviour can be remarkably simple. (It is surely a
surprise that simple classical dynamical systems, such as the pendulum
described below, can behave in ways quite unsuspected until recent
years.) Such systems are also, in general, deterministic; that is to
say, they are governed by equations, representing fully deterministic
laws such as the laws of classical mechanics, that have the property
that exact specification of appropriate initial conditions determines
in principle all subsequent behaviour. Solutions of the equations,
and the observed behaviour of corresponding physical systems, may
involve irregular, never exactly repeating, fluctuations. In
addition, the detailed pattern of these fluctuations may be
essentially unpredictable in the long run. One then says that the
behaviour is chaotic.
    The properties of determinism and unpredictability may seem
contradictory, but they are not. The reason is the feature often
known as sensitivity to initial conditions. Two solutions starting
from only marginally different initial conditions ultimately become
quite uncorrelated in the way they fluctuate; the smaller the initial
difference the longer the divergence takes, but it always ultimately
occurs. Since one can never specify the state of any real system
absolutely exactly this implies ultimate unpredictability. That
remark is just as true for numerical solutions of the equations as it
is for physical systems; there are always small imprecisions in any
numerical procedure. In practice, the perturbations responsible for
long-term unpredictability may be introduced continuously (continuous

small disturbances in an experiment, numerical integration in small discrete steps, and so forth) rather than just initially, but that does not alter the point of principle.

Systems exhibiting chaotic behaviour almost always exhibit ordered behaviour for some other values of the governing parameters. The structure of the changes is usually complex; there may, for example, be "windows" of ordered behaviour in a parameter range for which the behaviour is predominantly chaotic - or vice versa. The topic of "routes to chaos" - what happens as a system changes from ordered to chaotic behaviour as a parameter is varied - is an important one. We do not have space for a discussion of these matters here, but just note that there is a remarkable mixture of "universality" and "diversity": that is to say, totally different systems (mathematically and/or physically) may show strikingly similar features, even in quantitative detail; on the other hand, a single system may exhibit quite different routes for different transitions.

Beyond these generalities, the nature of chaotic behaviour is best understood through examples. We consider the forced spherical pendulum. A pendulum, free to swing equally in any direction, is forced by oscillating the point of suspension in a straight line, with an amplitude small compared with the pendulum length and at a frequency close to the small amplitude natural frequency. Resonance will evidently cause the oscillations to become large in the direction parallel to the drive motion. Little else about what the pendulum does is evident. It may develop a component of motion perpendicular to the drive and thus go into an orbit; such an orbit may be either ordered or chaotic. A complex structure to the changes, like that outlined in the preceding paragraph is found. The non-linearity central to all this arises from the swing amplitude being large enough for the gravitational restoring torque not to be directly proportional to the angular displacement.

Our knowledge of the pendulum's behaviour comes primarily from the theoretical and numerical analysis by Miles (1984a). (See also Miles (1984b) for a more elementary introduction.) This theory motivated the construction of a demonstration apparatus (Tritton 1986a). Although the latter relies on Miles's theory to give confidence that it is exhibiting chaos in the strict sense, it does provide a very effective demonstration of chaotic motion. Recently, we have started carrying out computations similar to Miles's original ones but presenting the results in forms that allow direct comparison with the physical pendulum (which, for reasons connected with the form of the equations, is not immediately possible for Miles's own results). The following summary of the pendulum's behaviour is synthesised from observations of our physical and numerical pendulums, although much of it could be obtained from either alone. The reader is referred to Miles's and Tritton's papers for quantitative specification of observations that are presented qualitatively, and with some oversimplification, below. (Miles's theory involves two parameters: α, a non-dimensional damping coefficient, and ν, a non-dimensional quantity related to the difference between the driving frequency and the small amplitude natural frequency. Remarks below

about varying the parameters refer to these. Values are quoted in
captions for readers who wish to relate results to Miles's and
Tritton's papers.)
    When the pendulum's motion is chaotic, its behaviour is typically
as follows. The orbit during a single period of the drive is nearly a
closed ellipse. However, successive orbits are slightly different, so
that, over about 10 periods, the pattern of motion changes markedly.
There is a continuous evolution, involving changes in the size,
ellipticity, and orientation of the orbit. It is this evolution that
exhibits the features of chaos, as outlined above. One way of
illustrating such evolution is to show a sequence of orbits with some
suitably chosen interval between them. Examples of such sequences
have been shown in Tritton (1986b) and (1988, §24.3). Here we use a
different form of illustration, in terms of the variation of M, a
quantity proportional to the angular momentum of the pendulum. (M is
defined precisely in Miles (1984a).) Positive M corresponds to
anticlockwise circulation of the pendulum bob and negative to
clockwise. The size of M relates in part to the size of the orbit,
but it should be noted that other quantities need to be specified for
a complete specification of the orbit; e.g. a particular value of M
can be given either by a smaller nearly circular orbit or a larger
narrowly elliptical one. Figure 1 shows two computed examples, for
different values of the governing parameters, of the variation of M
with an appropriate non-dimensional form of time. They are evidently
very different in their structure, but both are examples of chaotic
fluctuations. In Fig. 1(a), one manifestation of this is the varying
number of peaks on one side before M switches to fluctuating
predominantly on the other side; this number varies in a way that
never repeats in detail. The case in Fig.1(b) does not have this
feature; major changes in the sign of M are occurring all the time.
(It should be noted that the time-scale of these fluctuations is long
compared with the drive period, so this description is consistent with
the earlier remark that successive orbits are only slightly
different.)
    When the pendulum is in chaotic motion, the long-term evolution
of the patterns is unpredictable. Rather precise specification of the
way that it is moving would enable one to forecast the subsequent
motion for a while, but the actual motion will always deviate sooner
or later from any attemptable forecast. This is because of
"sensitivity to initial conditions", or, as we remarked earlier to be
equivalent, sensitivity to the continuous presence of small
perturbations. This property is illustrated by Fig. 2, showing the
computed evolution of the same case from identical initial conditions
with three different values of the tolerance (the maximum uncertainty
in any one step of the integration). It can be seen that the
different cases remain almost identical for a while, then slight
differences become apparent, and these lead remarkably rapidly to
quite different evolution. (This implies, of course, that the details
of Figs.1 (a) and (b) depend on the tolerance, but the broad
appearance of such figures does not.)

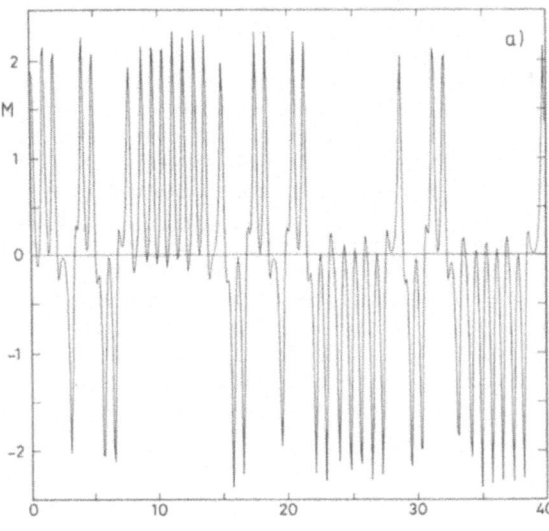

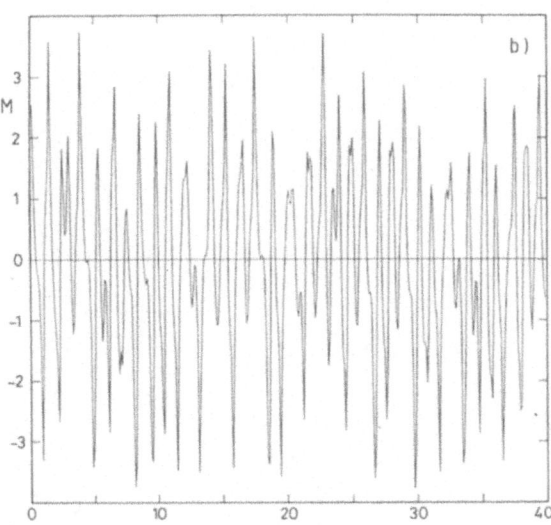

Figure 1. Two examples of variation of angular momentum with time as given by numerical integration of Miles's equations. (a) $\alpha = 0.25$, $\nu = -0.19$; (b) $\alpha = 0.1$, $\nu = -0.5$. Time is in units of τ (Miles 1984a).

220

Chaotic motion is not the only type of motion that can result
from the pendulum developing a component of motion perpendicular to
the drive (through instability of purely planar motion). The pendulum
may go into a completely regular motion; every orbit is identical.
Another possibility is that the orbits evolve in a complicated way –
i.e. observed behaviour over a smallish number of orbits would not
look evidently different from chaotic motion – but long-term
observation shows the presence of a complex periodicity – e.g. changes
in the sign of M occur at equally spaced intervals. We should note
also that there are modes of chaotic motion without changes in the
sign of M; once the pendulum is established in, say, clockwise motion,
it always retains that sense, but there are still chaotic fluctuations
in the orbits. The type of motion that occurs is determined primarily
by the parameter values (e.g. the drive frequency), but there may also
be differences for different initial conditions. The overall
structure of the "regime diagram" is remarkably complex (Miles 1962,
1984a).

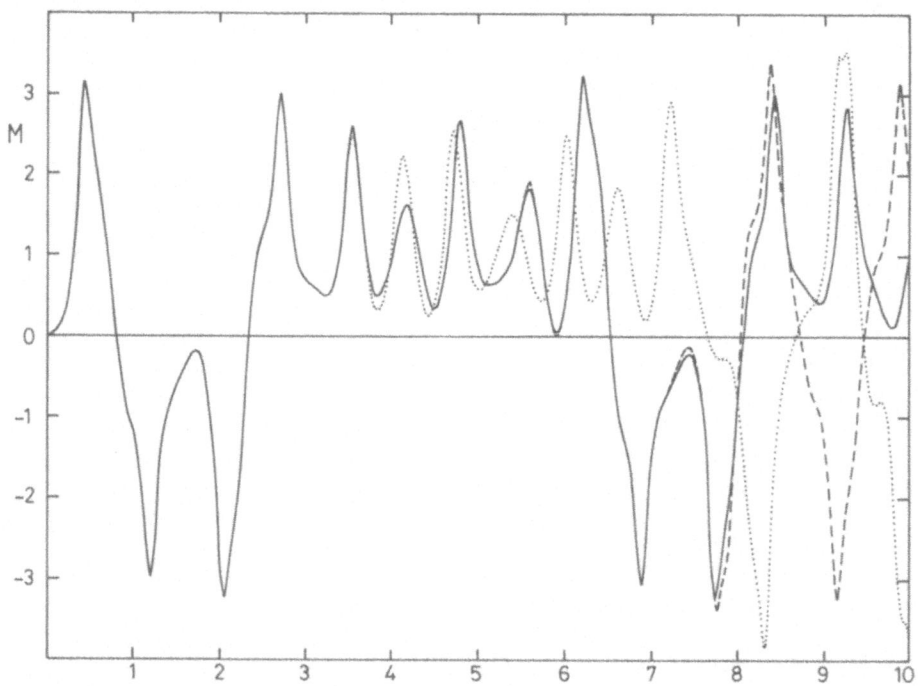

Figure 2. Three integrations of Miles's equations from the same
initial conditions with tolerances of 10^{-7} (solid line), 10^{-5} (dashed
line) and 10^{-3} (dotted line). $\alpha = 0.1$, $\nu = 0$.

3. GEOMAGNETIC REVERSALS

On the basis of the geomagnetic record alone, there is little prospect of a definitive answer to the question: is the reversal sequence a chaotic one? The data are open to different interpretations, and there is little prospect of refining or extending them sufficiently to change this. The evidence for making the suggestion is indirect. In particular, it is known that self-exciting dynamos can fluctuate chaotically. Thus, if one shares the majority view that the Earth's magnetic field derives from dynamo action, the question above automatically poses itself (but it does not follow that the question would not arise on any alternative view).

The most conclusive evidence for chaotic behaviour of self-exciting dynamos has come from numerical analysis of simple mathematical models, such as the Rikitake two-disc dynamo model. This comprises four simultaneous, non-linear, first order, ordinary differential equations. These represent the angular velocities of two driven rotating discs and the currents produced by dynamo action in each of them; the magnetic field for each dynamo is produced by the current generated in the other one. Chaotic fluctuations of this system have been studied by Ito (1980), Shimizu & Honkura (1985), and Kono (1987). Also, a model representing a one disc dynamo with a shunt can be reduced to three equations (Robbins 1977), which may be transformed by change of variable to the Lorenz equations – the best known set of differential equations with chaotic solutions.

Chaotic behaviour has probably been observed with a laboratory self-exciting dynamo, one of the succession developed at the University of Newcastle upon Tyne (Wilkinson 1984). Fig. 3 shows two oscillograms from as yet unpublished work by Dr. D. Kerridge; they represent the magnetic field at a point outside a dynamo with four independently driven rotors. The contrast between the periodic variations in the first case and the irregular ones in the second is apparent. One cannot demonstrate conclusively that the latter is an example of chaotic behaviour (the dynamo cannot be run for long enough without overheating to give data to which specific tests could be applied). However, the nature of the transition from periodic behaviour as conditions are varied suggests this interpretation.

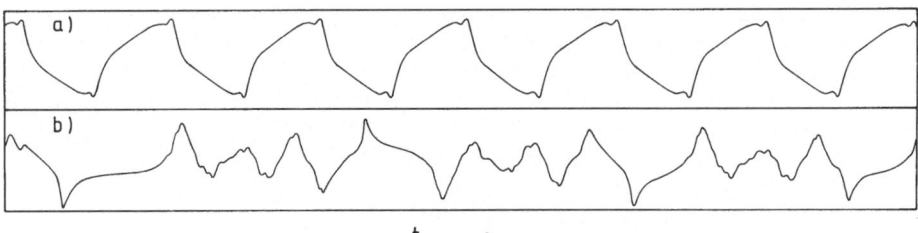

Figure 3. Two examples of the variation of magnetic field with time outside a four rotor laboratory dynamo (Kerridge 1983).

It is, of course, a major extrapolation from highly simplified mathematical models or from a comparatively simple laboratory model to the Earth's core. But as indicators of the possibility that the geo-dynamo reverses chaotically, these models are highly significant. It is not, of course, the only possibility; we have seen that systems that exhibit chaotic behaviour usually exhibit non-chaotic behaviour for different parameter values. Were we, for example, to find (through some at present totally unimagined source of evidence) that Jupiter's magnetic field does not undergo reversals or reverses periodically, that would not imply a different type of origin for the field. In the Earth's case, however, a chaotic reversal sequence is probably the best available interpretation of the data, and it is worth enquiring what are the implications of this interpretation.

The main implication is, in a sense, a negative one, though none the less important for that as it suggests that certain lines of research may be largely a waste of time and effort. It is that individual reversals do not require an "explanation". Once one has said that the system is prone to reverse from time to time, little more needs to be said. The converse of the lack of predictability associated with sensitivity to initial conditions is that one should not and cannot seek to account for every observed fluctuation. If the reversals occurred periodically, we would not look for a separate cause of each reversal; when they occur chaotically, there is no more reason to do so.

Since this is a concept that seems to meet some resistance, it may be useful to develop an analogy (with the usual cautions about the dangers of pressing it too far). Meteorologists seek to understand rain, in that they study the physical processes that produce rain and they consider why it rains more frequently in one place than another or in one season than another. They do not, on the other hand, look for an explanation of every individual rainstorm. The fact that, say, it rained on Monday but not on Tuesday is just the way the fluctuations of the atmosphere happened to go; it does not imply that there was something special about Monday, requiring investigation. This does not mean that it is worthless to say "It is going to rain" (either as an official short-term weather forecast or by a picnicker looking at the sky). It does mean that we are unlikely ever to be able to forecast whether, for example, it will rain on this day next year; more importantly, in the present context, it means that the propensity for rain to occur now and then is an adequate explanation for any particular rainstorm.

Correspondingly, it may be meaningful to speculate whether, for example, the current decline in the strength of the Earth's magnetic field is a precursor of the next reversal. On the other hand, it is probably a fruitless activity either to attempt to predict the future reversal sequence (though we shall not live to prove the point!) or to look for correlations with other events that are to be seen as causes of past reversals.

One can, of course, still ask why there is a propensity for reversals to occur at all. A full understanding obviously remains one of the major objectives of studies of the magnetohydrodynamics of the

Earth's core. However, a partial understanding is provided by the symmetry of the magnetohydrodynamic equations: if a self-exciting dynamo can generate a particular magnetic field pattern, it can also generate the reverse pattern. Thus it only requires the system to be able to find its way from one pattern to another for a reversal to occur.

One can also ask whether changes in the statistical properties of the reversal sequence, such as the average time between reversals, imply causal changes. (One would probably look for a cause of a climatic change involving a systematic difference in the average rainfall.) Discussion of this requires information about typical statistical properties of chaotic fluctuations; we will return to this point at the end of the next section.

4. STATISTICS OF PENDULUM REVERSALS

We return to considering the forced spherical pendulum described in §2. It seems useful to build up a body of information on the statistical properties of the chaotic fluctuations of various systems, looking in particular for common features that may help with the diagnosis of new topics. One cannot say how successful such an approach may be until several cases have been examined, but it seems worth making a start. We have done so with the pendulum, studying both the physical apparatus and the numerical solutions, although we have only very preliminary results so far.

It turns out to be convenient to quantify the pendulum's motion in terms of changes between clockwise and anticlockwise motion – which one rather naturally calls "reversals". The decision to look at this feature was not originally motivated by any possible analogy with geomagnetic reversals, but it is perhaps a useful coincidence as we shall discuss briefly at the end of this section. The decision was actually dictated by the fact that, with an apparatus originally intended purely as a demonstration, the quantifiable features are limited. When one watches the pendulum – or, more conveniently, a video-tape of it – the changes in sense are a conspicuous feature and the times at which they occur may be determined quite precisely. Measurements of the intervals between successive reversals provide the most convincing evidence that the motion is chaotic (Tritton 1986a). (All this, of course, applies only when reversals do occur; we have noted in §2 that there are parameter ranges for which they do not.) In the numerical studies, there is, of course, a much wider choice of determinable quantities. However, since one purpose is to compare the physical and numerical pendulums, we have focussed attention on intervals between reversals there also; i.e. we have determined the intervals between successive times at which M = 0 in numerical runs such as that in Fig. 1(b).

At present we can do no more than give preliminary results with the proviso that we do not yet know how typical these will prove to be when we have more extensive data. (In addition to wider variation of the parameters, we need to investigate more fully whether changing the tolerance in the numerical integrations alters the significant statistical features; first indications are that it does not.) The results are given in Fig. 4. in the form of histograms, similar in their formulation to ones often used to analyse the data on geomagnetic reversals; see, e.g., the article by W. Lowrie in this volume, Fig. 8. Each histogram plots the number of times during the run that the interval betwen successive changes of sense lay within the various ranges. (Clockwise intervals and anticlockwise ones are lumped together.) Each part of the figure shows both a laboratory run

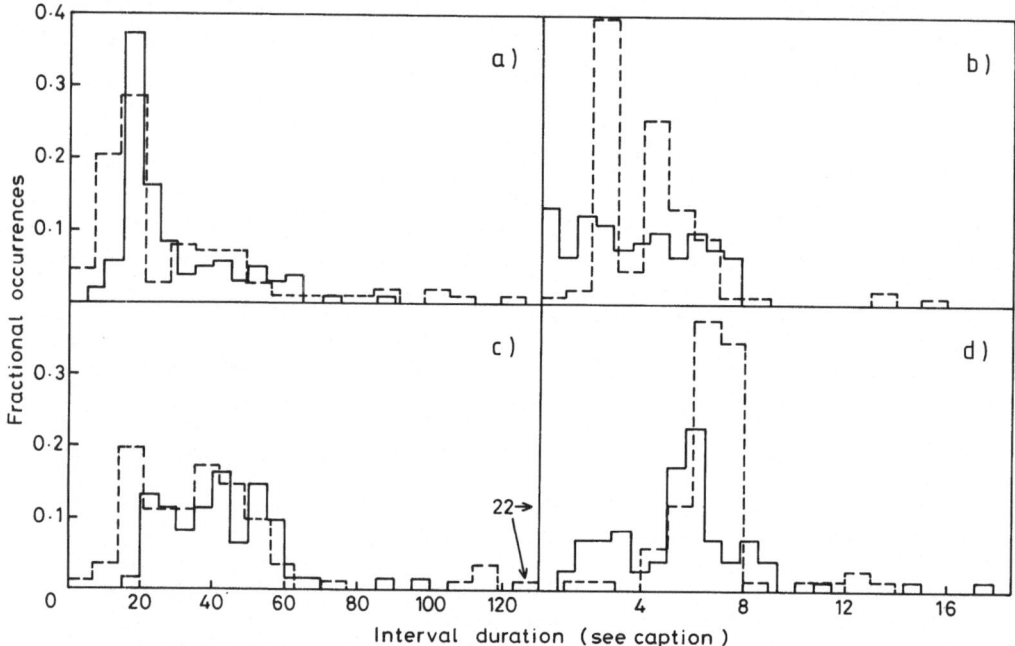

Figure 4. Histograms showing the fraction of the intervals between successive reversals of pendulum motion lying in various ranges. Each diagram shows an experimental histogram (solid line) and computational one (dashed line) for approximately comparable conditions. Left hand abscissa is in units of pendulum period (used in experimental data analysis) and right hand one is in units of τ (Miles 1984) (used in computational analysis), but both scales apply to all four diagrams. Experimental conditions were as in Tritton (1986a).
(a) Experiment: f/f_0 = 0.979 (total number of intervals = 105);
    Computation : α = 0.1, ν = -0.91 (total number = 112).

(b) f/f_0 = 0.984 (91); α = 0.1, ν = -0.70 (106).

(c) f/f_0 = 0.988 (61); α = 0.1, ν = -0.52 (81).

(d) f/f_0 = 0.996 (70); α = 0.1, ν = -0.17 (68).

and a numerical run. The parameter values for superimposed runs are matched as closely as possible, but there is a problem, connected with the damping of the real pendulum being non-linear (Tritton 1986a), in knowing exactly what conditions should be compared. Until we have done further work to find out how much a small change in the damping parameter would alter the results we do not know just how significant any comparison is between the physical and numerical experiments. Tentatively, however, there are signs of similar trends in the two; e.g. we think it is significant that, if the computed histograms in Figs. 4(a) and 4(d) were interchanged, their correspondence with the experimental ones would clearly be much less close. (The use of differently sized time intervals arises from the methods of analysis: in the laboratory experiment, it is natural to use the pendulum period as the basic time period, whilst the theory introduces a scaled time. The two are matched appropriately on the histogram abscissas; this is less problematic as the matching does not involve the damping coefficient. The different scales do not hinder the comparison of theory and experiment, as one is, of course, looking only at the statistical features. An exact match of two histograms would indicate that one did not have a chaotic phenomenon at all!).

The most obvious inference from these preliminary results is that there is no obvious inference! The histograms are not of an obvious simple form, such as a Poisson distribution (which would show as an exponential decrease in the number of occurrences as a function of the time interval). Chaotic fluctuations are not completely random ones. It is also apparent that changing the parameter values significantly changes the average duration of an interval and the way in which the intervals are distributed around that average. Diversity is more evident than any systematic pattern in these few results. This does not mean, however, that no pattern will emerge from a fuller study or from comparison with results from other chaotic systems. As noted in §2, such systems show a remarkable mixture of universality and diversity, and this may (or may not) prove to be the case in this context.

Because of the tendency for different systems to show common features, it is worth considering the possible relevance of these results for geomagnetism. The analogy between pendulum "reversals" and geomagnetic reversals is obviously a very loose one. Nevertheless the pendulum observations have some implications for the discussion in §3 – and may have more when they are extended. In particular, the histograms in Fig. 4 are a clear indication that the statistical structure of chaotic fluctuations may (usually will?) be non-obvious. Hence, if the geomagnetic data show unexpected structure this does not necessarily render the discussion in §3 irrelevant.

In all parts of Fig. 4, there are occasionally very long intervals between reversals. This was noticed in the original observations with the demonstration apparatus (Tritton 1986a, Fig. 3), and we now see that the numerical pendulum shows the same feature. Kono (1987, Fig. 4b) noted a similar feature in numerical solutions of the Rikitake dynamo equations and commented on possible relevance for the geomagnetic record. This poses the question of how long a period might occur without a geomagnetic reversal, as a natural fluctuation rather than as something requiring a special explanation. A related question concerns the extent to which trends in the geomagnetic data, variations in the average reversal frequency for example, reflect major changes in the condition of the core, or are produced by only minor changes to which the statistics are sensitive, or indeed might even occur without any causal change at all. We need more data before attempting answers to such questions. They do, however, strongly suggest the value of pursuing the lines of thought introduced in this article.

ACKNOWLEDGEMENTS

The experimental observations and analysis in Fig. 4 were carried out by Miss P.V. Terzoudi. The computer programs giving Figs. 1 and 2 and the numerical data in Fig. 4 were developed and run by Mr. J.H.E. Cartwright. I am most grateful to both for their skill and interest.

REFERENCES

Bergé, P., Pomeau, Y. and Vidal, C., 1986. 'Order within Chaos', Wiley, New York. (Translation of 'L'ordre dans le chaos', Hermann 1984.)
Ito, K., 1980. Earth Planet. Sci. Lett., 51, 451.
Kerridge, D.J., 1983. Ph.D. thesis, University of Newcastle upon Tyne
Kono, M., 1987. Geophys. Res. Lett., 14, 21.
Miles, J., 1962. Quart. Appl. Math., 20, 21.
Miles, J., 1984a. Physica, D11, 309.
Miles, J., 1984b. Adv. Appl. Mech., 24, 189.
Moon, F.C., 1987. 'Chaotic Vibrations', Wiley, New York.
Robbins, K.A., 1977. Math. Proc. Camb. Phil. Soc., 82, 309.
Shimizu, M. and Honkura, Y., 1985. J. Geomag. Geoelectr., 37, 455.
Thompson, J.M.T. and Stewart, H.B., 1986. 'Nonlinear Dynamics and Chaos', Wiley, Chichester.
Tritton, D.J., 1986a. Eur. J. Phys., 7, 162.
Tritton, D.J., 1986b. New Scientist, 111, 1518 (July 24), 37.
Tritton, D.J., 1988. 'Physical Fluid Dynamics', 2nd Edition, O.U.P., Oxford.
Wilkinson, I., 1984. Geophys. Surv., 7, 107.
Wisdom, J., 1987. Proc. Roy. Soc., A413, 109.

GEOMAGNETIC POLARITY REVERSALS: CAN METEOR IMPACTS CAUSE SPALL DISRUPTION INTO THE OUTER CORE

Alan Rice
Department of Physics
University of Colorado Denver
1200 Larimar Street
Denver, Co. 80204
U.S.A.

K.M. Creer
Department of Geophysics
University of Edinburgh
James Clerk Maxwell Building
Mayfield Road
Edinburgh EH9 3JZ

ABSTRACT. The conditions for spalling to occur are discussed and the possibility that meteor impacts on the Earth's surface might cause spallation at the mantle/core boundary when the shock enters the outer core or at the inner/outer core boundary when the shock wave passes back into the outer core, having crossed the inner core, are evaluated. It is concluded that it is most unlikely that spalling will occur in the first instance and rather unlikely in the second instance. Thus, if geomagnetic polarity reversals are caused by meteor impacts, spalling is unlikely to be the mechanism by which energy of impact is transferred into the core.

1. METEOR IMPACTS AND GEOMAGNETIC FIELD REVERSALS

It has been argued that some bolide impacts on the Earth's surface coincided with polarity reversals of the geomagnetic field (Glass, personal communication; Muller and Morris, 1986).

Three crucial questions arise. First, how is the surface impact energy transmitted into the outer core? Second, how might the pattern of fluid motions in the outer core be disrupted as a consequence of this? Third, how could geomagnetic polarity reversals be thus induced? Here we consider one aspect of the first question only, noting that geomagnetic dynamo theory has not yet been developed to the state where future behaviour of the field can be predicted: in particular core processes associated with polarity reversals are not understood.

2. SPALLATION DYNAMICS

We start by examining shock spallation or 'slabbing' of the mantle/core boundary or of the inner/core boundary as a possible mechanism by which the pattern of fluid motions in the outer core might be perturbed. Spalling is a common phenomenon accompanying impact on the scale of normal human experience. It is well known for

227

F. J. Lowes et al. (eds.), Geomagnetism and Palaeomagnetism, 227–230.
© 1989 by Kluwer Academic Publishers.

instance that a projectile which strikes armour plate may cause a
considerable amount of metal on the inside surface to be thrown off at
high velocity, although little damage is evident at the impact point,
and the surrounding metal is largely undamaged. Spalling occurs when
the difference between the tensile stress associated with the
reflected shock wave and the compressive stress associated with the
incident wave exceeds the tensile strength of the armour material.
The material then fails or 'slabs' at this point, the slab following
the original direction of the shock with a 'throw velocity' of about
twice the particle velocity of the shock (Hino, 1959).

Thus an impact generated shock wave, as it passes through the
target material, tends to be concentrated along the continuation of
the path of the projectile, in contrast to the spherical spreading
associated with shocks from underground explosions.   This 'focussed'
property is evident from observations that the area of the spalled
target material is about equal to that of the impact crater, which
itself is no more than twice the cross sectional area of the
projectile. For example, it is well known that the diameter of a
bullet hole is of the order of that of the bullet itself, provided
that the bullet does not break into fragments that take their own
individual paths. Maximum displacement occurs in the forward
direction in a relatively confined zone called the 'process zone'
(Olsen et al., 1983), and minimum displacement occurs normal to the
projectile path (Chandra and Flaherty, 1983). Most of the shock
energy is carried off in the spall: relatively little is absorbed by
the rest of the system. The process zone is separated from the bulk
of the target material by 'adiabatic shear bands' along which most of
the displacement occurs. This deformation is apparently of a slip
nature similar to that generated by viscous dissipation (Rice, 1977).

As shock waves are uncoupled to the lattice phonons, even at high
temperature (e.g. Karo et al., 1980) they do not show much spreading.
A simple estimate of impact pressure P_I can be obtained from

$$P_I = \frac{F}{A} = \frac{F\Delta T}{A\Delta T} = \frac{I}{A\Delta T} = \frac{m \quad v}{(\pi d^2/4)(d/\tfrac{1}{2}v)} = \frac{2mv^2}{\pi d^3}$$

where the force F is acting over area A, giving an impulse I, as the
projectile is stopped in a time ΔT; m, v and d are the projectile
mass, speed and diameter. Even on the assumption of spherical
symmetry of the shock, i.e.

$$P = \left(\frac{d}{r}\right)^2 P_I \text{ at distance r,}$$

which for reasons given above must lead to underestimates, the impact
of a meteor of 10 km diameter and density 10^4 kg m^{-3}, with an approach
velocity of 70 km s^{-1}, brought to a stop along a path-length of the
order of its diameter, will produce a pressure at least of the order
of a kilobar at the mantle/core interface. The focussed properties of
an impact-induced shock wave should result in stresses several times
greater than this.

3. IMPLICATIONS FOR THE EARTH'S CORE

The pre-impact ambient stress field at the mantle/core boundary should have sufficient time to adjust itself almost to the hydrostatic state, so that vertical compression will be compensated by lateral squeeze. The physical property which determines whether a shock superimposed on the ambient stress field will induce spalling in a material is its tensile strength. The tensile strengths of the base of the mantle and of the inner core are unknown. Crustal material of the Earth has tensile strength of the order of one bar. This suggests that in the geophysical context, spalling could, at least in principle, provide an effective mechanism for transferring energy of surface impact into energy of fluid motions in the outer core.

But are conditions in the Earth favourable for spalling to occur? For spalling it is necessary that a _tensional_ wave should reflect backwards into the mantle when the (compressional) shock wave strikes the mantle/core boundary. The acoustic impedances (product of density and wave speed) on either side of the boundary determine the conditions of reflection and transmission; for a wave travelling from medium 1 to medium 2 the stress reflection coefficient is

$$(\rho_2 c_2 - \rho_1 c_1)/(\rho_2 c_2 + \rho_1 c_1).$$

The acoustic impedance is about $7.4 \times 10^7 \text{kg m}^{-2}\text{s}^{-1}$ for the bottom of the mantle, and about $8.3 \times 10^7 \text{kg m}^{-2}\text{s}^{-1}$ for the outer core. Therefore a (weak) _compressional_ wave rather than a _tensional_ wave will be reflected from the mantle/core interface back into the mantle, and hence spalling of mantle material into the outer core should not occur.

Since the reflection coefficient of the mantle/core interface is only about 6%, most of the incident shock will be transmitted through the outer core towards the inner core surface, where we estimate that about 30% of the incident shock will be reflected back through the outer core towards the mantle. It will be reflected as a tensional wave since the acoustic impedance of the inner core is approximately $15 \times 10^7 \text{kg m}^{-2}\text{s}^{-1}$, and no spalling is expected at this surface. Neglecting attenuation and spreading, the stress associated with the wave transmitted through the inner core will be attenuated to about 60% of the stress incident on the surface of the inner core.

Our knowledge of the metallurgy of the inner core is rather vague, but it is thought to be composed of a nickel-iron alloy, possibly with a low atomic number element as impurity. Thus laboratory experiments on the tensile strengths of Fe-Ni-Si alloys may be of relevance. These suggest, on the foregoing assumption, that the tensile strength of the inner core should exceed the tensile stress of the shock wave reflected from the far side of the inner core by a factor of about 10. This should be so particularly if the slow cooling environment of the inner core allowed grain growth free of defects. However, it has been suggested that the inner/outer core boundary may be 'mushy' (Loper and Roberts, 1978), and it is uncertain what this may mean in terms of mechanical strength. It is just possible that a porous, dendritic structure might possess tensile strength in the kilobar range and hence resist spalling. The possibility that material might be shaken off the outer surface of the inner core as the shock wave undergoes repeated internal reflections should thus be considered finite but small.

Our earlier conclusion that spalling should not be expected to occur at the mantle/core boundary is much more secure, however.

4. REFERENCES

Chandra, J. and Flaherty, J.E., eds., 1983. Computational Aspects of Penetration Mechanics, Springer-Verlag, New York.

Hino, K., 1959. The Theory and Practice of Blasting. Yamaguchi-ken, Nippon Kayaku, Japan.

Karo, A.M., Walker, F.E., Cunningham, W.G. and Hardy, J.R., 1983. 'The study of shock-induced signals and coherent effects in solids by molecular dynamics', in Solids by Molecular Dynamics in Shock Waves, Explosions and Detonations, eds. Bowen, J.R., Manson, N., Opennheim, A.K. and Soloukin, R.I., American Institute of Aeronautics and Astronautics Inc., New York.

Loper, D.E. and Roberts, P.H., 1978. 'On the motion of an iron-alloy core containing a slurry'. Geophys. Astrophys. Fluid Dyn. 9, 289-321.

Muller, R.A. and Morris, D.E., 1986. 'Geomagnetic reversals from impact on the Earth'. Geophys. Res. Lett. 13, 117-1180.

Olsen, G.B., Mescall, J.F. and Azrin, M., 1981. 'Adiabatic deformation and strain localization', in Shock Waves and High-Strain Rate Phenomena in Metals: Concepts and Applications, eds., Meyers, M.A. and Murr, L.E., Plenum Press, New York, 220-247.

Rice, A., 1977. Doctoral Dissertation, Columbia University.

MAGNETIC BACTERIA IN LAKE SEDIMENTS

Nikolai Petersen[1], Dieter G. Weiss[2], Hojatollah Vali[3]

[1] Institut für Geophysik,
Ludwig-Maximilians-Universität München,
Theresienstr. 41,
D-8000 Munich 2, FRG

[2] Institut für Zoologie,
TU-München,
Lichtenbergstr. 4,
D-8046 Garching, FRG

[3] Institut für Angewandte Geologie und Mineralogie,
TU-München
Lichtenbergstr. 4,
D-8046 Garching, FRG

ABSTRACT. The occurrence of magnetic bacteria in several lakes and ponds of the Alpine foreland of Southern Germany has been studied. The habitat of the bacteria comprises the uppermost 2 cm of the sediment, with a maximum population density of 10^7/ml occurring at a depth of two to five millimetres below the water/sediment interface. A variety of different forms of magnetic bacteria was found, spirilli, cocci, vibrios and rod-shaped bacteria.

Most of the magnetic bacteria are highly mobile. It is therefore very difficult to study one and the same bacterium over a significant length of time under the optical microscope. To overcome this difficulty we developed a special configuration of rotating magnetic fields that keeps a certain bacterium within the field of view. This arrangement provides a convenient method to measure the swimming speed of an individual bacterium and also to determine its magnetic moment.

1. INTRODUCTION

To a geophysicist the interaction of the Earth's magnetic field with magnetic minerals and rocks is a familiar phenomenon which is dealt with in a special field of research, in palaeomagnetism. A geophysicist, however, is normally not so aware of the fact that the Earth's magnetic field may interact with living matter, with organisms such as certain birds, fish and bees (for review see Kirschvink et al. 1985).

231

F. J. Lowes et al. (eds.), Geomagnetism and Palaeomagnetism, 231–241.
© 1989 by Kluwer Academic Publishers.

A real surprise in this context came in 1975 with the discovery of magnetic bacteria by the American biologist Richard Blakemore. While still a graduate student in E.Canale-Parola's laboratory at the University of New Hampshire in Durham, Blakemore found in the mud of a coastal marine marsh pond located near Woods Hole, Massachussetts, a new kind of bacterium that always swam along the field lines of the ambient magnetic field. He found that the cause for this unexpected behaviour were tiny, submicroscopic, magnetite particles inside the bacterium - Blakemore called them magnetosomes - that gave the bacterium a permanent magnetic moment. The bacterium is passively oriented in the ambient magnetic field and moves along a field line by self-propulsion.

Rather than being a rarity of nature, magnetic bacteria are widespread in both marine and freshwater environments (Moench and Konetzka 1978, Towe and Moench 1981, Blakemore 1982, Spormann and Wolfe 1984, Sparks et al. 1986, Oberhack and Süssmuth 1987, Vali et al. 1987). They exist in both the northern and southern hemispheres and also near the geomagnetic equator (Kirschvink 1980, Blakemore et al. 1980, Frankel et al. 1981).

The magnetic bacteria vary considerably in size and shape. Best described is Aquaspirillum Magnetotacticum (Blakemore et al. 1979); most frequently observed, however, are cocci varying in size from smaller than one to a few micrometres in diameter. Other forms are vibrios, rod-shaped bacteria and spirilli, the latter extending in length up to a few tens of micrometres. They have been found primarily in the poorly oxygenated zone near the mud/water interface (Blakemore 1975 and 1982).

Common to all magnetic bacteria is the occurrence of intracellular single-domain magnetite particles (the above mentioned magnetosomes) normally forming single or multiple chains. These chains occur along the long axis of the bacterium in the case of spirilli and rod-shaped bacteria. However an irregular distribution of magnetosomes has also been observed (Vali et al. 1987).

2. MAGNETIC BACTERIA IN LAKE SEDIMENTS

Our research aim was to study the occurrence of magnetic bacteria in lake sediments, to define the environment and measure the natural population density, to determine their magnetic moment and pursue the question of whether they provide a noteworthy contribution to the magnetization of a sediment.

We are still a long way from finding satisfactory answers to these questions. The results presented here are only our first approach to the problem.

So far we have studied several lakes in the Alpine foreland of Southern Germany. Here we mainly describe results from Lake Chiemsee, approximately 80 km SE of Munich, and from a pond just outside our palaeomagnetism Laboratory near Landshut, 75 km NE of Munich. The results from other lakes and ponds in this area are similar.

Lake Chiemsee covers an area of 80 km² with a maximum water depth
of 74 m. The water is moderately high in oxygen at the lake floor,
while O₂ approaches essentially zero within the first few millimeters
in the detritus-rich sediment (personal communication Dr. O. Siebeck,
Limnological Station at Seon/Lake Chiemsee, University of Munich).
The Landshut pond has a maximum water depth of only 1.5 m.

As opposed to the Landshut pond, the Lake Chiemsee magnetic
bacteria are observed only in water depths greater than 5 m. The
habitat of the bacteria comprises the uppermost 2 cm of the sediment,
with a maximum population density of approximately 10⁷/ml occurring at
a depth of two to five millimetres below the water/sediment interface.

A variety of different forms of magnetic bacteria was found,
spirilli, cocci, vibrious and rod-shaped bacteria. The spirilli show
a particularly wide range of sizes from less than a few to about
100 μm in length. Some of the forms seem to be identical to those
recovered by Blakemore from Little Styx River, South Island, New
Zealand and Ceda Swamp, near Woods Hole, Massachusetts (Blakemore
1982).

The magnetosomes usually form single or multiple chains, which in
the case of spirilli are oriented along the long axis of the bacterium
(Fig. 1). In cocci the magnetosomes are aligned along the diameter of
the bacterium with the bundle of flagella perpendicular to the chain
of magnetosomes (Fig. 2a and b).

Under natural conditions the magnetosomes in magnetic bacteria
almost always form single or multiple chains. However after several
months in an aquarium a large number of the bacteria show an irregular
distribution of the magnetite particles (Fig. 3).

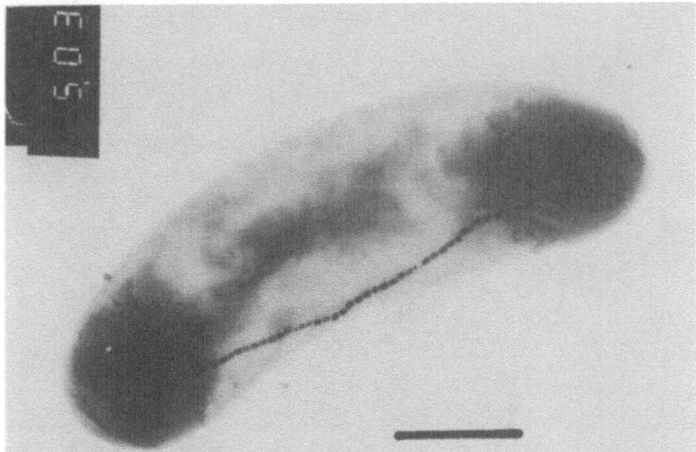

Figure 1. Magnetic spirillum recovered from Lake Chiemsee with single
chain of prismatic magnetosomes aligned along the axis of the
bacterium. The bacterium was obtained from the uppermost 2 cm of the
sediment at a water depth of 20 m. For TEM observation the bacteria
were fixed with 2.5% glutaraldehyde in a 0.1 M cacodylate buffer
(pH 7.2), rinsed with water and stained with a 2% uranylacetate
solution. The bar represents 2 μm.

234

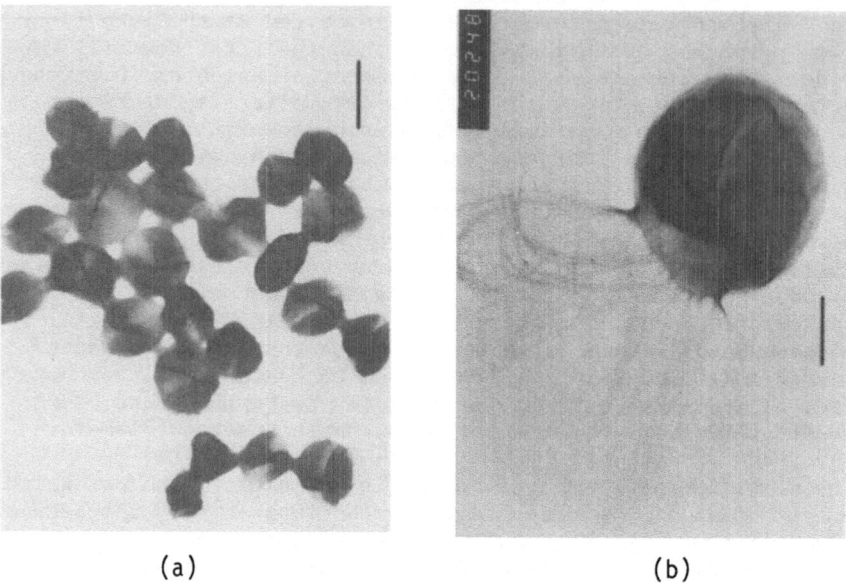

(a)   (b)

Figure 2. Magnetic cocci recovered from a pond near Landshut, 75 km
NE of Munich, Southern Germany. In (a) the magnetosomes are aligned
along the diameter of the bacterium (the bar = 2 μm), in (b) the
bundle of flagella is perpendicular to the chain of magnetosomes
(bar = 0.5 μm).

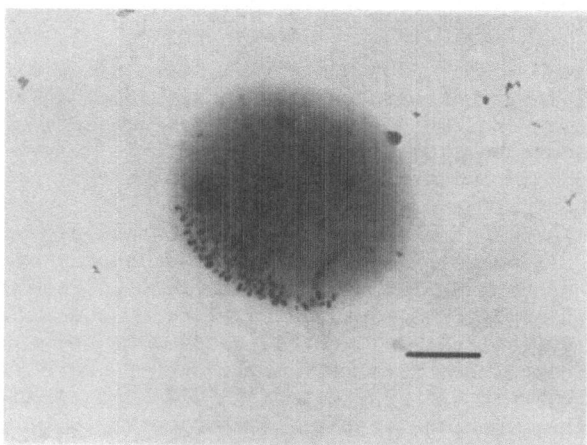

Figure 3. Magnetic coccus recovered from sediment from the Landshut
pond. Before harvesting the bacterium, the sediment was kept 2 months
in an aquarium in the laboratory. Irregular distribution of the
magnetosomes. (Bar = 1 μm)

3. SHAPE AND SIZE OF MAGNETOSOMES

The shape of the magnetosomes of the Lake Chiemsee and Landshut pond magnetic bacteria is essentially the same as already described by Blakemore (1982), Man et al. (1987), von Dobeneck et al. (1987) and Vali et al. (1987). We found: (1) chains of octahedra (Fig. 4); (2) parallelepipeds with varying ratios of width to length (Fig. 1); and (3) tear-drop shaped cones (Fig. 5).

When plotted in the diagram for single-domain magnetite parallelepipeds, as calculated by Butler and Banerjee (1975), the tear-drop shaped cones fell well within the single-domain region. The parallelepipeds and the octahedral magnetosomes, however, fell partly outside the single-domain region and plot in the two-domain region (Fig. 6).

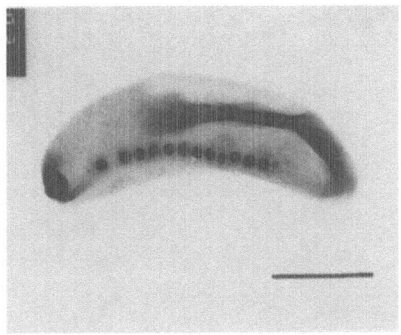

Figure 4. Magnetic spirillum recovered from the Landshut pond. The magnetosomes form a chain of octahedra. (Bar = 0.5 μm)

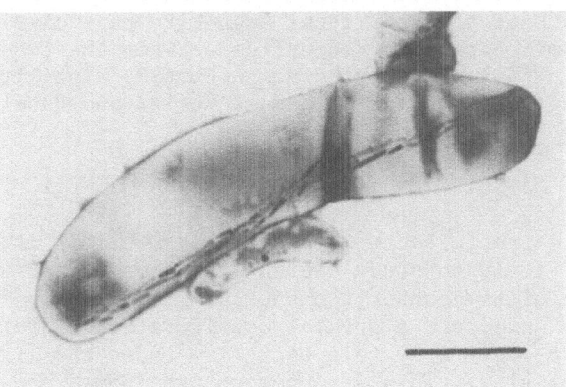

Figure 5. Magnetic spirillum recovered from Lake Chiemsee at a water depth of 20m. The magnetosomes form a double chain of tear-drop shaped cones. Attached is a smaller bacterium with octahedral magnetosomes. (Bar = 1 μm)

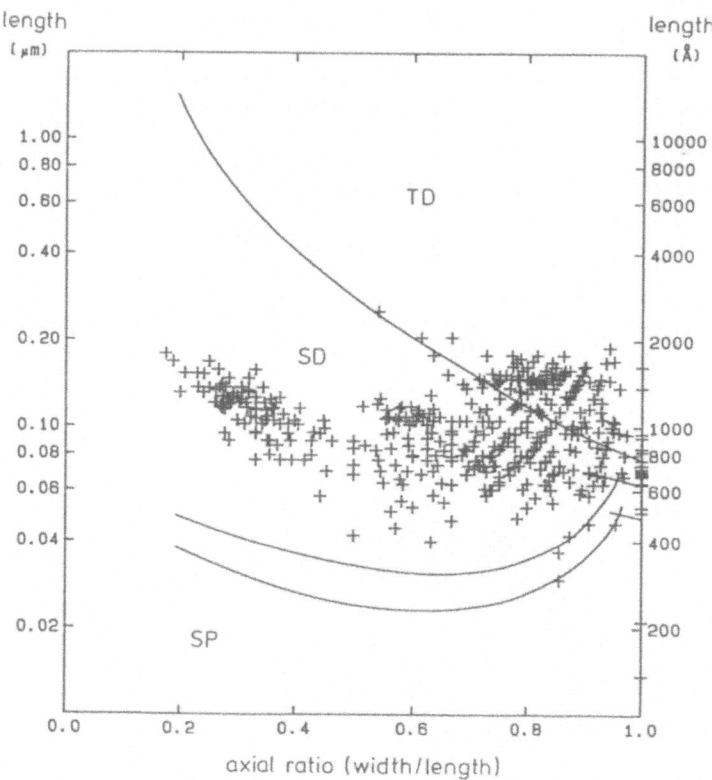

Figure 6. Grain size distribution of the magnetosomes from different types of magnetic bacteria from Lake Chiemsee and the Landshut pond. The length and width of the individual magnetite particles were measured from TEM micrographs. Boundaries between the two-domain (TD), single-domain (SD) and superparamagnetic(SP) stability regions are drawn according to the calculations of Butler and Banerjee (1975).

4. OBSERVATION OF MAGNETIC BACTERIA IN ROTATING MAGNETIC FIELDS

The Lake Chiemsee magnetic bacteria are highly mobile. It is therefore very difficult to study one and the same bacterium over a significant length of time under the optical microscope. To overcome this difficulty we developed a special configuration of rotating magnetic field that keeps a certain bacterium under the microscope within the field of view. This arrangement provides a convenient method to measure the swimming speed of an individual bacterium, and also to determine its magnetic moment (Petersen et al. 1988).

A Leitz Laborlux D microscope, stripped of all magnetic parts, is placed inside a set of large Helmholtz-coils (2.5 m diameter) compensating for the Earth's magnetic field. A second set of two pairs of much smaller Helmholtz-coils (40 cm diameter) is oriented perpendicular to each other in the horizontal plane, and attached to the microscope. The centre of the coils coincides with the field of view of the microscope.

Alternating magnetic fields are generated in the small Helmholtz-coils, with the field in the x-direction being

$$H_x = H_o \sin \omega t,$$

and in the y-direction

$$H_y = H_o \sin(\omega t + \pi/2).$$

The resulting magnetic field is then a homogeneous field that rotates in the horizontal plane with a frequency $\omega = 2\pi/T$, where T is the period of one cycle. T can be varied between ∞ (steady field) and 0.5 sec, the field intensity between 0 and 1.6 Oe (0.16 mT). This arrangement was given the name "bacteriodrome", as the rotating magnetic field causes a circular swimming pattern of the bacteria.

The swimming speed v of a single bacterium can be determined from the diameter D of the circular path (Fig. 7a and b) according to

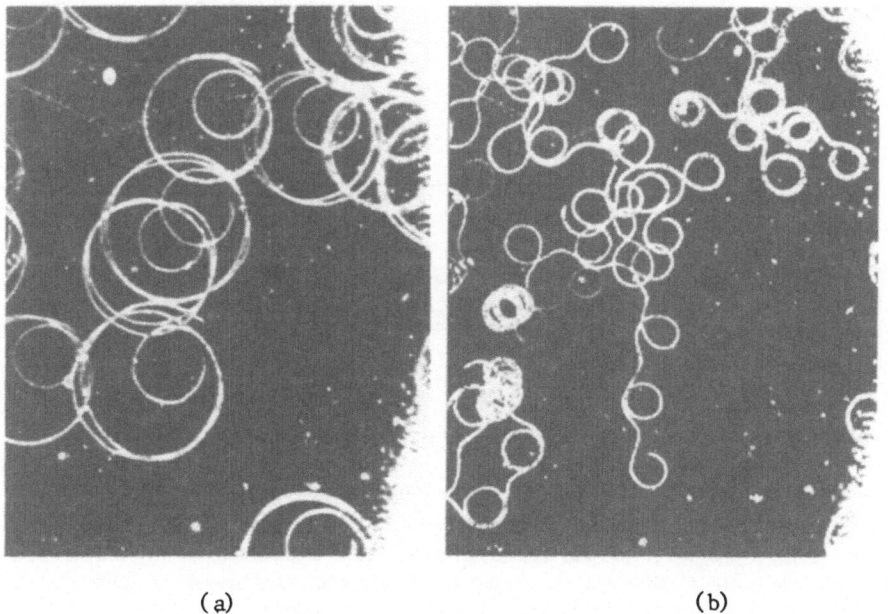

(a)   (b)

Figure 7. Swimming path of magnetic bacteria in a rotating magnetic field of 1.6 Oe (Lake Chiemsee spirillum of the type shown in Fig. 1). (a) Rotation period T = 10 sec (larger circle) and T = 5 sec (smaller circle). (b) Rotation period T = 2.5 sec. Now the magnetic torque is not sufficient, and the bacteria break out of their circular path. The bar represents 100 μm.

$$v = \tfrac{1}{2} D \omega = \tfrac{1}{2} D \; 2\pi/T = \pi D/T.$$

For the type of Lake Chiemsee spirillum shown in Fig. 1, we measured the circular diameter of the swimming path for different periods of field rotations. The swimming speed v thus determined is 36 µm/s. The linear relationship between diameter D and period T shows that for a given field strength the swimming speed is independent of the frequency of the rotating magnetic field (Fig. 8)

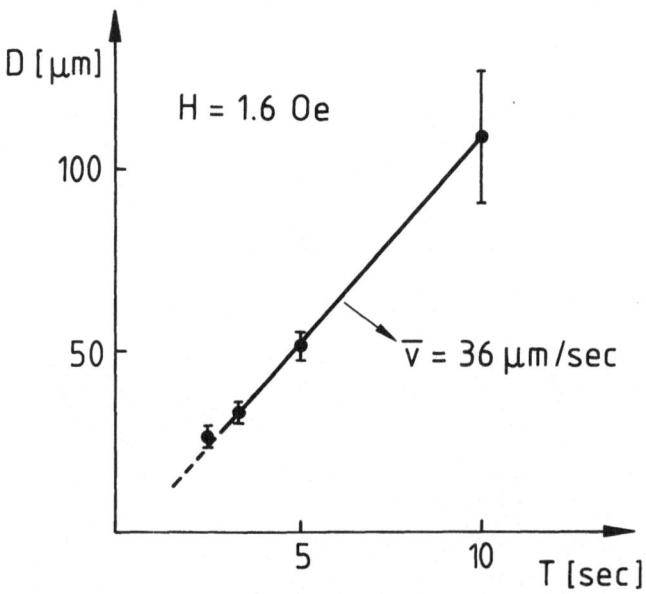

Figure 8. Diameter D of circular swimming path of the magnetic bacteria of Fig. 7 plotted against the period T of the rotating magnetic field. A swimming speed of 36 µm sec is determined from the slope of the straight line.

5. DETERMINATION OF THE MAGNETIC MOMENT OF A SINGLE BACTERIUM

Assuming that all magnetosomes within a bacterium are single-domain magnetite particles and that the magnetic moments are aligned parallel to the long axis of the particle chain, it is possible to calculate the total magnetic moment of a bacterium from the size of the magnetosomes as determined by electron microscopy.

For the bacterium shown in Fig. 1, we measured an average magnetosome volume of $6.62 \times 10^{-16} \mathrm{cm}$. With a spontaneous magnetization of $M_S = 450$ G (1 G $= 10^3$ A m^{-1}) for magnetite we calculate a magnetic moment of 2.98×10^{-13} G cm^3 (1 G cm$^3 = 10^{-3}$ A m^2) per magnetosome. With 50 magnetosomes in the bacterium we obtain a total magnetic moment of 1.49×10^{-11} G cm^3.

The bacterial magnetic moment can also be determined from the analysis of the movement of the bacterium in rotating magnetic fields. This method is similar to the "U-turn" method described by Frankel (1984).

The constant forward speed of the bacterium is given by the balance between its propulsive force and the opposing viscous drag. Similarly, the rate of rotation of the bacterium, needed to keep it directed along the circular path, is given by the balance between a magnetic torque and an opposing viscous torque, the latter increasing with rotational speed, i.e. with frequency. The instantaneous direction of the magnetic moment **m** of the bacterium is not parallel to the applied (rotating) magnetic field H, the direction of the moment lagging behind that of the field by an angle θ so as to give a torque **m** × **H**, of magnitude $mH\sin\theta$, which tries to rotate the bacterium to be more nearly parallel to **H**. For a bacterium of length ℓ, rotating at angular velocity ω in water of viscosity η, the opposing viscous drag is $c\eta\omega\ell^3$, where c is a shape factor. Thus as frequency is increased the lag angle θ increases to maintain equilibrium, but this is only possible up to $\theta = 90°$; for higher frequencies the circular path can not be maintained, and the bacterium breaks out of its circular path as shown in Fig. 7b. The corresponding frequency can be used to give another estimate of the magnetic moment m.

It is difficult to determine the shape factor c, but approximating it by 8π gives

$$m \simeq 16\eta\pi^2 \ell^3/HT$$

for the critical value of T. For H = 1.6 Oe, T = 2.5 s, $\ell \simeq 3 \times 10^{-4}$ cm, we obtain

$$m \simeq 4 \times 10^{-12} \text{ G cm}^3,$$

which is slightly smaller than the magnetic moment calculated from the TEM micrograph.

6. CONTRIBUTION OF BACTERIAL MAGNETITE PARTICLES TO THE MAGNETIZATION OF LAKE SEDIMENTS

We can try to set an approximate upper limit of sediment magnetization carried by biogenic magnetite by assuming that the above described measurements on the Lake Chiemsee sediments are a good representation. A bacterial population density of 10^7/ml with an individual magnetic moment 10^{-12} G cm^3 yields a maximum magnetization of 10^{-5} G, a magnetization intensity that can conveniently be measured with conventional magnetometers in use in palaeomagnetism. Kirschvink and Lowenstam (1979) assumed a natural population density of bacterial magnetite of 10^5/ml and calculated a corresponding sediment magnetization of $(2-12) \times 10^{-7}$ G.

There are however good reasons to assume that the actual remanent magnetization underneath the uppermost 2 cm of sediment containing

living bacteria will be smaller than the value calculated here:
(a) Only a fraction of the bacteria will die off and thus provide
magnetite particles as carrier of remanent magnetization. The rest of
the bacteria will migrate upwards with the growing sediment column.
(b) The chemical stability of the small magnetite particles will
strongly depend on the local environment. Whereas the slightly
alkaline marine environment seems to preserve bacterial magnetite very
well (Petersen et al. 1986), this may be different in lake sediments.
Vali et al.(1987) have shown that magnetosomes in recent deposits in
Lake Ammersee (Southern Germany) show distinct signs of corrosion that
may eventually lead to a dissolution of the magnetite particles.
(c) The above estimate of magnetization assumes perfect alignment of
all the magnetosomes in the Earth's magnetic field.

7. ACKNOWLEDGEMENTS

We wish to thank J. Pohl and E. Schmidbauer for stimulating
discussions. Chris Wold critically read the manuscript. Financial
support has been given by the Deutsche Forschungsgemeinschaft.

8. REFERENCES

Blakemore, R.P., 1975. 'Magnetotactic bacteria'. Science, 190,
    377-379.
Blakemore, R.P., 1982. 'Magnetotactic bacteria'. Ann. Rev.
    Microbiol., 36, 217-238, 1982.
Blakemore, R.P., Frankel, R.B. and Kalmijn, A.J., 1980. 'Southseeking
    magnetotactic bacteria in the southern hemisphere'. Nature, 286,
    384-385.
Blakemore, R.P., Maratea, D. and Wolfe, R.S., 1979. 'Isolation and
    pure culture of a freshwater magnetic spirillum in chemically
    defined medium'. J. Bacteriology, 140, 720-729.
Butler, R.F. and Banerjee, S.K., 1975. 'Theoretical single-domain
    grain size range in magnetite and titanomagnetite'. J. Geophys.
    Res., 80, 252-259.
Dobeneck, von, T., Petersen, N. and Vali, H., 1987. 'Bakterielle
    Magnetofossilien'. Geowiss. in unserer Zeit, 37, 27-35.
Frankel, R.B., 1984. 'Magnetic guidance of organisms'. Ann. Rev.
    Biophys. Bioeng., 13, 85-103.
Frankel, R.B., Blakemore, R.P., Torres de Araujo, F.F., Esquivel,
    D.M.S. and Danon, J., 1981. 'Magnetotactic bacteria at the
    geomagnetic equator'. Science, 212, 1269-1270.
Kirschvink, J.L., 1980. 'South-seeking magnetic bacteria'. J. Exp.
    Biol., 86, 345-347.
Kirschvink, J.L. and Lowenstam, H.A., 1979. 'Mineralization and
    magnetization of chiton teeth: palaeomagnetic, sedimentologic and
    biologic implications of organic matnetite'. Earth Planet. Sci.
    Lett., 44, 193-204.

Kirschvink, J.L., Jones, D.S. and McFadden. B.J., 1985. 'Magnetite Biomineralization and Magnetoreception in Organisms.' A New Biomagnetism., Plenum Press, New York.

Mann, S., Sparks, N.H.C. and Blakemore, R.P., 1987. 'Structure, morphology and crystal growth of anisotropic magnetite crystals in magnetotactic bacteria'. Proc. R. Soc. London, B231, 477-487.

Moench, T.T. and Konetzka, W.A., 1978. 'A novel method for the isolation and study of a magnetotactic bacterium'. Arch. Microbiol., 119, 203-212.

Oberhack, M. and Süssmuth, R., 1987. 'Magnetotactic bacteria from freshwater'. Z. Naturforsch., 42c, 300-306.

Petersen, N., Dobenenck von, T. and Vali, H., 1986. 'Fossil bacterial magnetite in deep-sea sediments from the South Atlantic Ocean'. Nature, 320, 611-615.

Petersen, N., Weiss, D.G. and Vali, H., 1988. 'Magnetic bacteria in rotating magnetic fields: motile behaviour and magnetic moment'. Submitted to Nature.

Sparks, N.H.C., Courtaux, L., Mann, S. and Board, R.G., 1986. 'Magnetotactic bacteria are widely distributed in sediments in the U.K.'. FEMS Microbiology Letters, 37, 305-308.

Spormann, A.M. and Wolfe, R.S., 1984. 'Chemotactic, magnetotactic and tactile behaviour in a magnetic spirillum'. FEMS Microbiology Letters, 22, 171-177.

Towe, K.M. and Moench, T.T., 1981. 'Electron-optical characterization of bacterial magnetite'. Earth Planet. Sci. Lett., 52, 213-220.

Vali, H., Forster, O., Amarantidis, G. and Petersen, N., 1987. 'Magnetotactic bacteria and their magnetofossils in sediments'. Earth Planet. Science Lett., 86, 389-400.

MAGNETIZATION OF SEDIMENTS AND DEPOSITIONAL ENVIRONMENT

Reidar Løvlie
Institute of Geophysics
Allegt. 70
N-5007 Bergen, Norway

ABSTRACT. Most records of claimed palaeomagnetic excursions are not accepted as manifestations of geomagnetic field variations, due to poor chronostratigraphic correlation. Post-depositional acquisition of detrital remanent magnetization may attenuate and modify, but not erase, high amplitude palaeomagnetic directions. It is proposed that the complete absence of excursions in continuously accumulated sediments, may be due to remagnetization by transient environmental processes. Results of an **in situ** bioturbation experiment, performed at 12 metres water depth during 1 year, are presented and demonstrate remagnetization of the upper 2 cm of a marine sand sediment. The results may, however, also be interpreted to reflect the influence of waves affecting the sediment surface of shallow-water sediments.

1. INTRODUCTION

Sediments have been highly successful in retrieving directional information about the past behaviour of the geomagnetic field; a palaeomagnetic reversal stratigraphy has been established back to the Cretaceous, continuous palaeosecular variation records now extend beyond 30 ka, and high resolution records of transitional paths of the geomagnetic field during reversals are reported in increasing numbers. Palaeomagnetic records of inferred short duration, high amplitude excursions of the geomagnetic field, were first reported two decades ago in sediment cores raised from the Caribbean and N-Atlantic Ocean (Smith and Foster 1968). Comparable directional features have since then been encountered in magmatic rocks as well as in sediments, but ambiguity still exists regarding the reality of most reported palaeomagnetic excursions in terms of genuine records of geomagnetic field behaviour (Verosub, 1982).

Palaeomagnetic excursions are encountered in a variety of relatively rapidly deposited sedimentary facies, and it is likely that some reflect either depositional processes influencing the fidelity of the remanent magnetization carried by detrital grains, or disturbances of the recording medium in question, rather than ambient field

243

F. J. Lowes et al. (eds.), Geomagnetism and Palaeomagnetism, 243–252.
© *1989 by Kluwer Academic Publishers.*

directions.

A more fundamental problem, however, is the absence of palaeomagnetic excursions in sediments deposited during times the geomagnetic field performed an excursion, as inferred from synchronous palaeomagnetic records at nearby localities (Verosub 1977, Verosub et al. 1980). Although this disagreement may readily be accounted for by undetected depositional hiati and/or poor age control, a mounting number of apparently high quality palaeomagnetic records from inferred continuously deposited lake sediments, covering times of reported excursions, do not show any evidence for large departures of palaeomagnetic directions (Turner et al. 1982, Smith and Creer 1986).

It is proposed that the absence of excursions may reflect transient processes acting after deposition and which effectively may cause realignement of magnetic grains otherwise physically inhibited from reorientation. Biological mixing of surface sediments has been considered a plausible process for modifying or erasing palaeomagnetic reversal zones in deep-sea sediments (Watkins 1968). Experimental investigations of magnetic remanence acquisition due to bioturbation, have been performed on tidal flat sediments (Graham 1974, Suttill 1980, Ellwood 1984), which is not a very representative environment for retrieving palaeomagnetic information. A bioturbation experiment in situ has therefore been performed at 12 metres water depth during 1 year, demonstrating partial remagnetization of the top layer of a marine sediment. Due to some non-ideal experimental conditions, the results can also be interpreted to represent effects of wave action affecting the top layer of the sediment.

2. DETRITAL REMANENT MAGNETIZATION

2.1 Depositional DRM – dDRM

Sediments acquire a detrital remanent magnetization (DRM) by preferred physical orientation of magnetic mineral grains. Magnetic grains settling in still water become perfectly aligned with the ambient field direction in a matter of seconds only. At the sediment surface, however, this alignement may be distorted by gravitational and/or hydrodynamic forces, giving rise to directional errors of this depositional DRM (dDRM). Hence, depending on grain size distributions and depositional environment, a dDRM does not necessarily record the true ambient field direction at the time of deposition.

2.2 Post-depositional DRM – pDRM

Post-depositional rotation of detrital grains may give rise to a remanent magnetization perfectly aligned with the ambient magnetic field (Irving and Major 1964, Kent 1973). A post-depositional DRM (pDRM) is retained when magnetic grains become physically inhibited from orientation after deposition by the decrease of interstitial voids during consolidation (Løvlie 1974). This is a gradual process during which larger grains will become locked before smaller grains.

Accordingly, a pDRM will at any level represent an integrated record of ambient field variations. The physical reality of this model awaits confirmation, but progressive alternating field (af) or thermal demagnetization results of redeposited and natural sediments suggest that deviating components can be resolved and related to different blocking temperature/coercivity spectra, and hence grain-size distributions (Payne and Verosub 1982, Valet et al. 1988).

2.3 Reactivated pDRM

DRM is carried by magnetic mineral grains with a preferred, spatial orientation. Any process distorting this arrangement is also likely to remobilize already locked-in grains, causing complete or partial remagnetization of a sediment.
     Biological mixing of the top layer of sediments is common in deep-sea environments. The observation that deep-sea sediments record true ambient field inclinations (Opdyke and Henry 1969), implies that

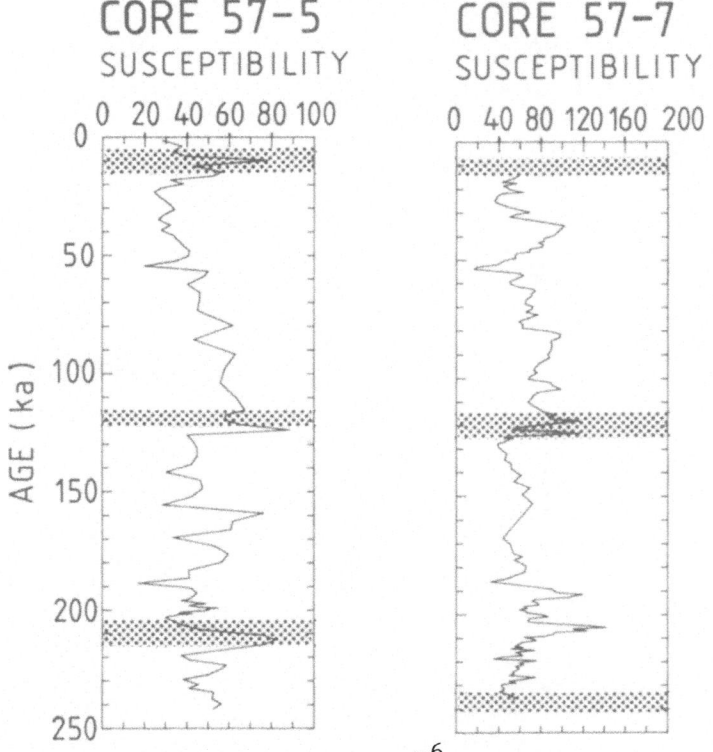

Figure 1. Bulk susceptibilities (units 10^{-6} cgs emu) along two deep-sea cores raised from 1902 and 1620 m water depth ca 80 km apart from the Norwegian Sea (68°N, 13°E). Age estimates derived from linear interpolation between oxygen isotope stage boundaries. Extent of volcanic ash layers represent half-width of maximum particle counts (Sjøhlm 1987).

bioturbation does not induce distortions of any pDRM. Although
bioturbation certainly represents a remagnetizing process, the actual
process is poorly known and it is still unclear whether grain
remobilization of this reactivated pDRM is due to changes in the
volume of interstitial water or due to a mechanism similar to a shear
remanent magnetization (Games 1977).

The significance of biological mixing are demonstrated by the
poor stratigraphic correlation between variations in bulk
susceptibility along two deep-sea cores raised only some 80 km apart
(Figure 1). The influence of bioturbation upon the palaeomagnetic
directions in these cores is illustrated in Figure 2. Three
well-defined zones of anomalous inclinations of characteristic
remanent directions (ChRM) are present in core 57-5, none of which can
readily be recognized in core 57-7.

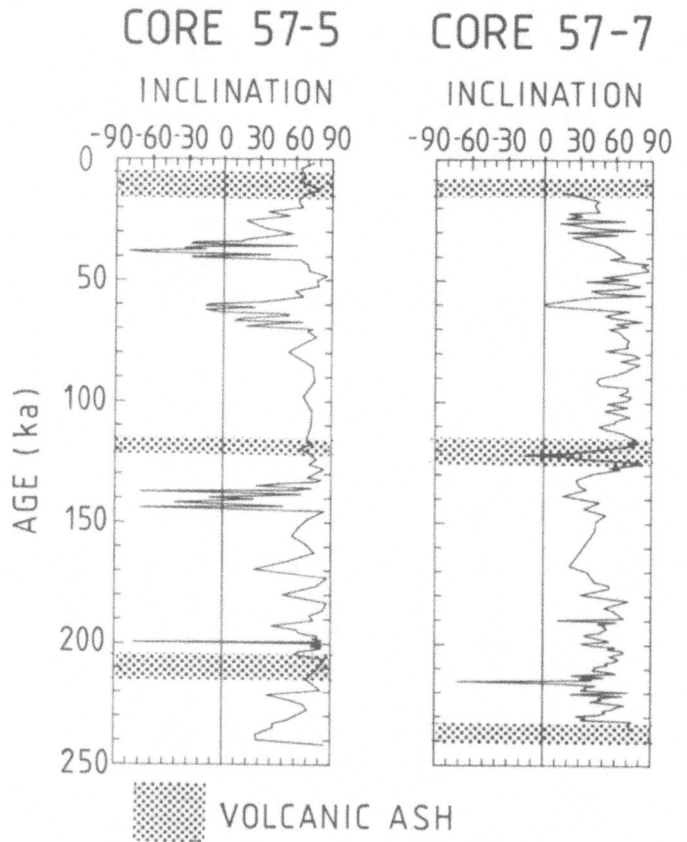

Figure 2. Inclinations of characteristic remanent magnetization in
the same two cores as in Figure 1. Low to negative inclinations are
interpreted to represent palaeomagnetic excursions. Note poor
directional correlation between the cores.

3. BIOTUBATION EXPERIMENTS

3.1 Previous Results

The effect of biological mixing upon magnetic remanence acquisition in the top layer of sediments has previously been investigated by in situ experiments in a biological active tidal flat sediment (Ellwood 1984). Ambient magnetic field directions developed some months after depositing a magnetite layer. Remobilization of the magnetite grains was attributed to bioturbation. However, tidal flat sediments experience two diurnal transgressions of sea water, associated with interstitial water movements which is likely to cause reorientation of magnetic grains. It is thus likely that the experiment does not adequately demonstrate effects due to biological mixing only, an assertion supported by palaeomagnetic results from recent tidal flat sediments, suggesting a time lag in the acquisition of a stable pDRM of the order of 100 years (Suttill 1980).

3.2 Shallow Marine in situ Experiment

In order to avoid the diurnal soaking of sediments on tidal flats, eight 11/50 cm (diameter/length) oriented cores were collected by

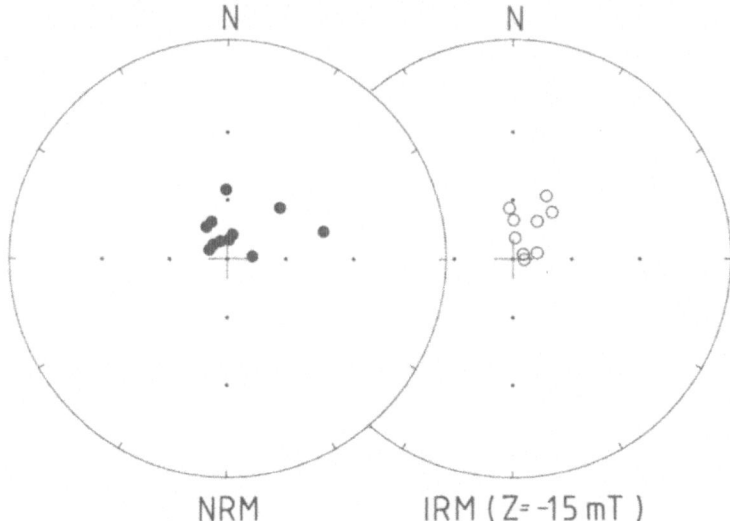

Figure 3. Stereographic plot of palaeomagnetic directions in two oriented cores collected from 12 m depth, frozen to -20°C for 7 days. A vertical, upward directed isothermal remanent magnetization was induced in one core (15 mT) and both cores were subsequently thawed in the earth's magnetic field. Mean NRM direction (Dec=355°, Inc=74°, α_{95}=10°) is not significantly different from the geomagnetic field direction at the locality. Vertical IRM directions retained after thawing.

divers from 12 meters water depth in a small bay south of Bergen
(Kviturdsvikpollen). The sediment consists of sand with shells.
Thermomagnetic and remanent (Dankers, 1981) coercive force analyses
suggest that magnetite is the only magnetic mineral present.

After sterilization by cooling to −20°C for one week, an
isothermal remanent magnetization (IRM) was subsequently induced in a
vertical (reversed) 15 mT field in all but one of the still frozen
cores. No significant effect was observed during thawing of two cores
(one not magnetized) in the earth's magnetic field (Figure 3).

Six cores, two of which were coated with a netting (1 mm) to
prevent intrusion of larger organisms, were then set out at the
original locality. After one year **in situ,** two cores (one coated with
netting) were retrieved.

3.3 Palaeomagnetic Results

The two cores were sub-sampled by cutting 2.2 cm cubes. All
measurements were performed within a week. The cores have retained
strong and vertically directed remanent magnetizations from the bottom
and up to 2 cm below the sediment surface, implying that neither
thawing nor biological mixing have destroyed the induced IRM in this
section of the cores.

In the top 2 cm of the sediment, 8 samples revealed both ambient
field directions as well as intermediate normal to reversed polarity
directions (Figure 4). No significant differences were observed

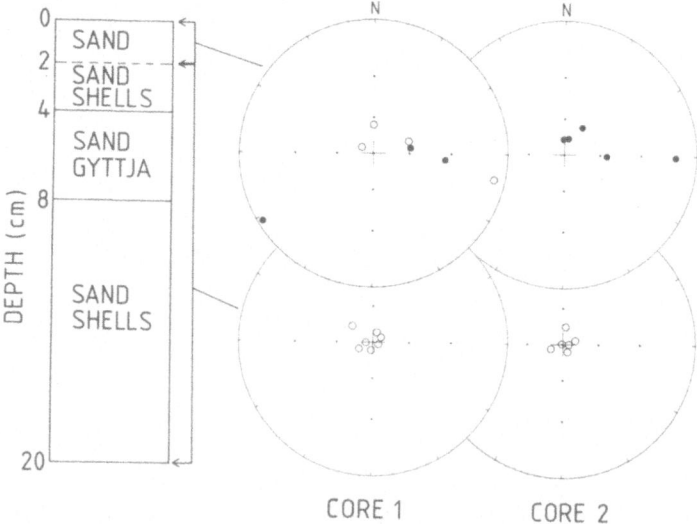

Figure 4. Stereographic plot of remanence directions in two cores
retrieved after 1 year at 12 m water depth. Top 2 cm carry both
normal and reversed directions. The pre-thawing induced IRM is
retained below 2 cm in both cores.

between open and netting-coated cores. Progressive alternating field (af) demagnetization removed shallow to steep reversed components, but the present field direction was encountered in only a few samples (Figure 5). The top 2 cm exhibit large lateral inhomogeneities of palaeomagnetic directions as well as bulk susceptibilities. The latter may indicate either a major break in the supply of detrital material, or reflect transient biogenic magnetite production.

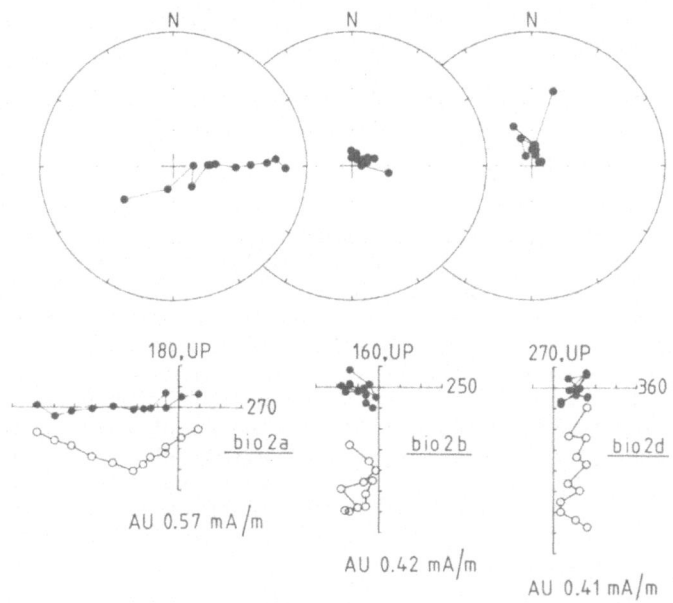

Figure 5. Directional results for progressive af demagnetization of samples from the top 2 cm of core 2. Stereographic (top) and optimum orthogonal vector projection (bottom). AU is the magnetization corresponding to one unit on the axes. Note two-component magnetization: the low-coercivity component represent the induced IRM while ambient field directions reside in high-coercivity grains.

4. DISCUSSION

4.1 Effect of Bioturbation

The palaeomagnetic results are readily explained by realignement of detrital magnetite grains parallel with the ambient geomagnetic field due to reworking caused by biological mixing. The intermediate directions indicate that one year of exposure to biological activity caused only partial realignement of all magnetic grains. The sediment cores were exposed to negligible consolidation and detrital deposition during one year **in situ**. The remagnetization demonstrates the mixing efficiency of small organism (< 1 mm) entering the netting coated tube.

However, photographs of the cores **in situ** revealed that the core tubes had not been buried in level with the sediment surface (due to poor visibility). Hence, reworking organisms must have entered all core tubes through the water column. Although such organisms appear to cause physical reworking capable of reorienting magnetite grains, the undesired experimental conditions suggest an alternative interpretation by considering possible influence of surface waves on this shallow marine sediments.

4.2 Wave Influence

The trajectories of water particles within a travelling surface wave is described by circular motions which vanishes below a wave-length dependent depth (Drake et al. 1978). When a surface wave, originating at deep waters, reach shallower depths, the wave length will decrease when the the wave starts to "see" the bottom. Under these conditions the circular motion of water will deform to ellipsoids above the bottom. Although the water/sediment boundary is ideally treated as an inpenetrable physical boundary, the vertical pressure component of the elliptic water motion is likely to induce movement of the interstitial water in the top layer of the sediment. It is tentatively proposed that this effect may give rise to remobilization of magnetic grains.
    The results obtained from the presented **in situ** experiment is tentatively suggested to reflect the influence of waves travelling across the coring locality. Waves with wavelength of the order of 40 m will start to "see" the bottom at 12 metres depth. In the fjord outside the coring bay, such waves are not uncommon during westerly high winds.

5. CONCLUSIONS

Palaeomagnetic excursions may reflect errors in the acquisition process of dDRM, rather than records of geomagnetic field variations. A pDRM, however, represents a time integrated record of true ambient field directions. The smoothing effect of a pDRM is not well established, but deconvolution of palaeomagnetic records, assuming a specified moment fixing function, suggest lock-in depths of stable pDRM of less than 10 cm in sediments accumulating at a rate of some 0.8 mm/year (Hyodo 1984). Such smoothing may modify the finer details of geomagnetic field variations, but is not likely to completely attenuate records of geomagnetic excursions. Presently there is no general approach to discriminate between a dDRM and pDRM, or recognize and correct for directional errors in a dDRM.
    In order to account for the absence of palaeomagnetic excursions in sediments deposited during inferred geomagnetic excursions, reactivation of pDRM acquisition by environmental processes is proposed. One such process is bioturbation, which is likely to cause complete remagnetization within the zone of active mixing. While biological mixing may be regarded as a more or less continuous process, the proposed remagnetizing effect of wave action upon the

sediment surface, is a transient process which may be envisioned to affect sediments by varying intensity and depths depending on season and climate. During non-seasonal heavy storms, continental shelf areas may experience waves which may "see" the bottom at depths greater than 100 metres. Interpretation of palaeomagnetic results from sediments should hence take into account facies relationships in addition to bulk lithological properties.

6. ACKNOWLEDGEMENTS

The bioturbation experiment emerged from discussions with T. Brattegard. The work at the "bottom" was performed by H. Botnen, I. Døskland and T.O. Wahl. The experiment has been supported by funds from the Norwegian Research Council for Science and the Humanities.

REFERENCES

Dankers, P., 1981. 'Relationship between median destructive field and remanent coercive forces for dispersed natural magnetite, titanomagnetite and hematite', Geophys. J. R. astr. Soc., 64, 447–461.

Drake, C.L., Imbrie, J., Knauss, J.A. and Turekian, K.K., 1978. Oceanography, Holt, Rinehart & Winston, N.Y., pp. 447.

Ellwood, B.B., 1984. 'Bioturbation: minimal effects on the magnetic fabric of some natural and experimental sediments'. Earth Planet. Sci. Lett., 67, 367–376.

Games, K.P., 1977. 'The magnitude of the palaeomagnetic field: a new non-thermal, non-detrital method using sun-dried bricks'. Geophys. J. R. astr. Soc., 48, 315–329.

Graham, S., 1974. 'Remanent magnetization of modern tidal flat sediments from San Francisco Bay, California', Geology, 2, 223–226.

Hyodo, M., 1984. 'Possibility of reconstruction of the past geomagnetic field from homogeneous sediments', J. Geomag. Geoelectr., 36, 45–62.

Irving, E. and Major, A., 1964. 'Post-depositional detrital remanent magnetization in a synthetic sediment', Sedimentology, 3, 135–143.

Kent, D.V., 1973. 'Post-depositional magnetization in deep sea sediments', Nature, 246, 32–34.

Løvlie, R., 1974. 'Post-depositional remanent magnetization in a re-deposited deep-sea sediment', Earth Planet. Sci. Lett., 21, 315–320.

Opdyke, N.D. and Henry, K.W., 1969. 'A test of the dipole hypothesis', Earth Planet. Sci. Lett., 6, 139–151.

Payne, M.A. and Verosub, K.L., 1982. 'The acquisition of post-
    depositional detrital remanent magnetization in a variety of
    natural sediments'. Geophys. J. R. astr. Soc. 68,
    625-642.
Smith, G. and Creer, K.M., 1986. 'Analysis of geomagnetic secular
    variations 10 000 to 30 000 years bp, Lac du Bouchet, France',
    Phys. Earth Pl. Int., 44, 1-14.
Sjøholm, J.I., 1987. 'Kvartaere askesoner og sedimenter på
    Islandsplåtaet, Geokjemiske og sedimentologiske undersøkelser
    av kjerne 57-7'. Hovedfagsoppgave i maringeologi, Univ. Bergen,
    pp. 74.
Suttill, R.J., 1980. 'Post-depositional remanent magnetization in
    recent tidal-flat sediments'. Earth Planet. Sci. Lett., 49,
    132-140.
Turner, G.M., Evans, M.E. and Hussin, I.B., 1982. 'A geomagnetic
    secular variation study (31 000-19 500 bp) in Western Canada',
    Geophys. J.R. astr. Soc., 71, 159-171.
Valet, J.-P., Tauxe, L. and Clark, D.R., 1988. 'The Matuyama-Brunhes
    transition recorded from Lake Tecopa sediments (California)',
    Earth Pl. Sci. Lett., 87, 463-472.
Verosub, K.L., 1977. 'The absence of the Mono Lake geomagnetic
    excursions from the palaeomagnetic record of Clear Lake,
    California', Earth Planet. Sci. Lett., 36, 219-230.
Verosub, K.L., Davis, J.O. and Valastro, S., jr., 1980. 'A
    palaeomagnetioc record from Pyramid Lake, Nevada, and its
    implications for proposed geomagnetic excursions', Earth
    Planet. Sci. Lett., 49, 141-148.
Verosub, K.L., 1982. 'Geomagnetic excursions; a critical assessment
    of the evidence as recorded in sediments of the Brunhes Epoch',
    Phil. Trans. R. Soc. Lond. A., 306, 161-168.
Watkins, N.D., 1968. 'Short period geomagnetic polarity events in
    deep-sea sedimentary cores', Earth Planet. Sci. Lett., 4,
    341-349.

TIMING BETWEEN A LARGE IMPACT AND A GEOMAGNETIC REVERSAL AND THE DEPTH
OF NRM ACQUISITION IN DEEP-SEA SEDIMENTS

C.A. Burns
Department of Geology
University of Delaware
Newark
DE 19716
U.S.A.

ABSTRACT. The hypothesis that large meteorite impacts on earth may
trigger geomagnetic reversals is partly based on the close
stratigraphic occurrence of the Australasian microtektite layer to the
Brunhes/Matuyama reversal boundary in deep-sea sediments. However,
the precise timing between these two events is unclear because the
original stratigraphic position of the microtektites has been altered
by bioturbation and the position of the reversal boundary has been
changed by post-depositional remanent magnetization. The
palaeomagnetic stratigraphy and microtektite distributions from 20
deep-sea cores have been utilized to solve the problem. After
correction for bioturbation, a model to account for the distance (Q)
separating the microtektites and the reversal boundary includes the
time difference (Δt) between the two events, the sediment accumulation
rate (w), and an exponential term ($n\ e^{-w}$) representing the depth of
NRM acquisition in an expression:

$$Q = \Delta t\ w + n\ e^{-w}$$

The fitted value for Δt is -11.8 ± 5.6 kyr, which means the
microtektites were deposited 11,760 yr before the midpoint of the
Brunhes/Matuyama reversal. Taking into account the uncertainty, and a
reversal transition time of 5000 yrs, the impact either slightly
predates or is essentially simultaneous to the reversal. However,
this information does not prove causality between the two events. The
second term in the model shows that depth of NRM acquisition decreases
exponentially with increasing sedimentation rates. Although this
result was unexpected, it probably reflects the relationship between
sediment grain size and water content of the sediment.

1. INTRODUCTION

The discovery of Australasian microtektites in close stratigraphic
proximity to the Brunhes/Matuyama geomagnetic reversal boundary by
Glass and Heezen (1967) led them to speculate that the reversal may

253

F. J. Lowes et al. (eds.), Geomagnetism and Palaeomagnetism, 253–261.
© *1989 by Kluwer Academic Publishers.*

have been caused by the impact of a large cosmic body. Since that
initial speculation, others (Burek, 1984; Muller and Morris, 1986)
have similarly suggested that large meteorite impacts may trigger
geomagnetic reversals and have presented more detailed explanations of
the triggering mechanisms. However, although it may be theoretically
possible for an impact to cause a reversal, the stratigraphic
relationship between the microtektite layer and the reversal boundary
is not straightforward. Therefore, it has not yet been shown whether
the impact producing the Australasian microtektites predates or
postdates the Brunhes/Matuyama reversal. The relative ages of the
Australasian microtektite layer and the Brunhes/Matuyama reversal
boundary are not clear because the microtektites have been displaced
from their original position by bioturbation and the reversal boundary
has been affected by the process of post-depositional detrital
remanent magnetization (PDRM). Once these effects are corrected for,
the objective of this study is to determine the true age relationship
between a large impact (represented by the microtektites) and a
reversal boundary.

2. METHODS

Deep-sea sediment cores collected over the last twenty years from the
Indian and western Pacific Oceans from the research vessels Robert
Conrad, Eltanin, Vema, and the Glomar Challenger have provided the
samples for this study. Determination of palaeomagnetic stratigraphy
has largely been performed in the palaeomagnetic laboratory of
Lamont-Doherty Geological Observatory. The palaeomagnetic directions
for most cores were measured with a spinner magnetometer but samples
measured more recently were analyzed in a cryogenic magnetometer.
Once the position of the Brunhes/Matuyama reversal boundary was
determined (often to within 5 cm), the cores were sampled for
microtektites. These samples were wet sieved through a 63 m sieve
and then the coarse fraction was searched with a binocular microscope.
Because microtektites are glassy spherules, often pale green or pale
brown in colour, they are easy to distinguish from most other deep-sea
sediment components. Furthermore, because the total number of
microtektites in one sample is often small (10 or less), all
microtektites found in a sample were counted, picked out, and then
stored. Sediment accumulation rates have been calculated from the
depth of the B/M and Jaramillo event reversal boundaries in the cores
versus their respective ages. Over this time interval, the rates are
assumed to represent average uniform sediment accumulation rates.
Data such as core latitude and longitude, water depth from which the
core was obtained, and sediment type in the region of the
microtektite/reversal boundary has been obtained from macroscopic core
logs.
    In order to determine the true age relationship between the
impact (that produced the microtektites) and the Brunhes/Matuyama
geomagnetic reversal, the effects of bioturbation must be accounted
for. Many papers have discussed the problems of bioturbation in

deep-sea sediments and have presented models that attempt to quantify it (e.g. Berger and Heath, 1968; Ruddiman and Glover, 1972; Guinasso and Schink, 1975; Robbins, 1986). Guinasso and Schink (1975) suggested that a simple way to quantify the amount of downward displacement that an impulse tracer (such as a microtketite layer) has undergone due to bioturbation is to find the mean depth of the impulse tracer distribution. Another way to describe this mean is as the centre of mass of the distribution. The underlying assumption to this method of determining the original position of the microtektite layer is that the microtektite layer must not have been displaced by some mechanism prior to reworking by organisms. There is no evidence that this assumption has been violated in any of the cores used in this study, so determination of the mean depth was taken to indicate the depth of original deposition for the microtektites.

3. RESULTS

Australasian microtektites have now been discovered in 39 deep-sea sediment cores (Table I). Many of these cores have been sampled for determination of palaeomagnetic stratigraphy. Out of these 39 cores, 19 cores were deemed unusable for the purposes of this study for one of three reasons: the palaeomagnetic data could not be unambiguously interpreted to pick a depth for the B/M boundary; the microtektites were found scattered in the core and thus no single layer could be defined; or the sediment accumulation rate was not constrained by either at least three reversal boundaries or supported by independent evidence.

Determination of the amount of displacement of the microtektites due to bioturbation was made from computing the mean depth of the microtektite distribution in each core (Table II). Note that there is no systematic relationship between the degree of bioturbation and sediment accumulation rate.

Figure 1 is a plot of the difference in depth between the original position of the microtektites and the depth of the B/M boundary versus sediment accumulation rate. Note that positive y-axis values correspond to cases where the original position of the microtektite layer is stratigraphically above the presently recorded reversal boundary and negative y-axis values correspond to cases where the microtektites are found below the reversal boundary.

4. DISCUSSION

Besides bioturbation, the other process that acts to secondarily alter and confuse the stratigraphic relationship between the microtektites and the reversal boundary is post-depositional remanent magnetization. The existence of a zone of uncompacted sediment near the sediment-water interface in the deep-sea allows for the rotation of the magnetic carriers to a new magnetic field during a geomagnetic reversal. The depth at which this rotation of magnetic grains ceases

TABLE I.   Summary of palaeomagnetic and microtektite core data.

| | Depth to: (cm) | | | | (cm) | (cm/kyr) | |
Core	B/M Bound.	top of Jaram.	bottom Jaram.	Tektite peak	Diff.	Sed. Rate	Accept
RC8-52	398	460	480	386	+12	0.35	YES
RC8-53	184	350	380	scat.	x	0.38	NO
RC9-137	445	x	x	413	+32	0.61	NO
RC9-142	346	380	390	341	+5	0.19	YES
RC9-143	362	468	513	349	+13	0.63	YES
RC12-327	629	729	790	611	+18	0.66	YES
RC12-328	Ambig.	x	x	770	x	x	NO
RC12-331	643	737	803	645	-2	0.65	Yes
RC14-23	164	600	695	144	+20	0.43	YES
RC14-24	280	685	802	246	+34	0.45	YES
RC14-46	972	1322	x	975	-3	1.43	YES
V16-70	155	205	225	154	+1	0.29	YES
V16-75	434	525	555	439	-5	0.51	YES
V16-76	1055	x	x	1043	+12	1.44	NO
V19-153	525	637	667	510	+15	0.60	YES
V19-169	235	275	305	190	+45	0.28	YES
V19-171	355	415	475	340	+15	0.47	YES
V20-138	325	x	x	325	0	0.44	NO
V20-184	725	965	1025	735	-10	1.28	YES
V28-238	1200	x	x	1214	-14	1.64	YES
V28-239	725	878	944	730	.-5	0.91	YES
V29-39	107	167	195	105	+2	0.36	YES
V29-40	346	427	467	341	+5	0.50	YES
V29-43	443	x	x	433	+10	0.37	YES
E35-6	Ambig.	x	x	356	x	x	NO
E35-9	233	x	x	241	-18	0.32	NO
E39-45	281	x	x	scat.	x	0.38	NO
E45-71	666	816	x	692	-26	0.91	YES
E45-74	621	x	x	scat.	x	0.85	NO
E45-89	400	x	x	392	+8	0.55	NO
E48-6	Ambig.	x	x	scat.	x	x	NO
E49-4	151	x	x	150	x	0.21	NO
E49-50	365	x	x	365	x	0.50	NO
E49-51	286	x	x	270	+16	0.39	NO
E50-2	291	x	x	scat.	x	0.40	NO
DSDP292	x	x	x	850	x	x	NO
LSDA156G	x	x	x	4	x	x	NO
LSDH23G	x	x	x	60	x	x	NO
MSN48G	x	x	x	121	x	x	NO

Diff. = The B/M reversal depth - the depth of max. tektite abundance
Accept = Cores accepted or rejected for study
x = no data
Ambig. = Ambiguous data
scat. = scattered microtektites
Depth to reversal boundaries taken at reversal midpoint.
Depth to tektite peak taken at depth of maximum abundance.

TABLE II. Summary of microtektite distributions corrected for bioturbation, core latitude, and sediment type.

Core	Max depth (cm)	Mean depth (cm)	Diff. (cm)	Sed. Rate (cm/kyr)	New Diff. (cm)	Lat.	Sediment type
RC9-142	341	339	2	0.19	7	-42.43	foram lutite
V19-169	190	184	6	0.28	51	-10.13	diatom clay
V16-70	154	152	2	0.29	3	-32.06	foram lutite
RC8-52	386	364	22	0.35	34	-41.06	foram lutite
V29-39	105	101	4	0.36	6	-7.42	rad. clay
V29-43	433	423	10	0.37	20	-12.20	rad. clay ooze
RC14-23	144	142	2	0.43	22	-9.10	rad. clay
RC14-24	246	243	3	0.45	37	-6.37	clay
V19-171	340	330	10	0.47	25	-7.04	mud
V29-40	341	333	8	0.50	13	-10.29	rad. clay
V16-75	439	435	4	0.51	-1	-22.13	silty lutite
V19-153	510	509	1	0.60	16	-8.51	lutite
RC9-143	349	342	7	0.63	20	-41.21	calc. lutite
RC12-331	645	640	5	0.65	3	2.30	foram marl
RC12-327	611	612	-1	0.66	17	1.44	foram lutite
V28-239	730	728	2	0.91	-3	3.15	foram ooze
E45-71	692	688	4	0.91	-22	-48.01	foram ooze
V20-184	735	727	8	1.28	-2	-25.48	lutite
RC14-46	975	979	-4	1.43	-7	-7.49	rad. lutite
V28-238	1214	1208	6	1.64	-8	1.01	foram ooze

Max. depth = depth of maximum tektite abundance
Mean depth = depth of original deposition of tektites
New diff. = B/M reversal depth - original depth of tektite layer
Lat. = latitude;   rad. = radiolarian; calc. = calcareous

is referred to as the depth of acquisition of natural remanent
magnetization (NRM). While it is obvious that any depth of NRM
acquisition greater than zero means the reversal is being recorded in
the sediment at an apparent time rather than the real time of its
occurrence, there have been few measurements of the depth of NRM
acquisition. Estimates have ranged from about 5 cm (Kent, 1973)
through 10-15 cm (Raisbeck and others, 1985) to 30 cm (Raisbeck,
personal communication). While it is assumed that depth of NRM
acquisition depends on such factors as water content, magnetic moment,
and grain size (Kent, 1973), little is known about the specific
relationship between depth of NRM acquisition and any of these
variables. Thus, it becomes necessary to derive information on the
depth of NRM acquisition from the differences in position of the
microtektite layer relative to the reversal boundary in order to
determine their age difference.

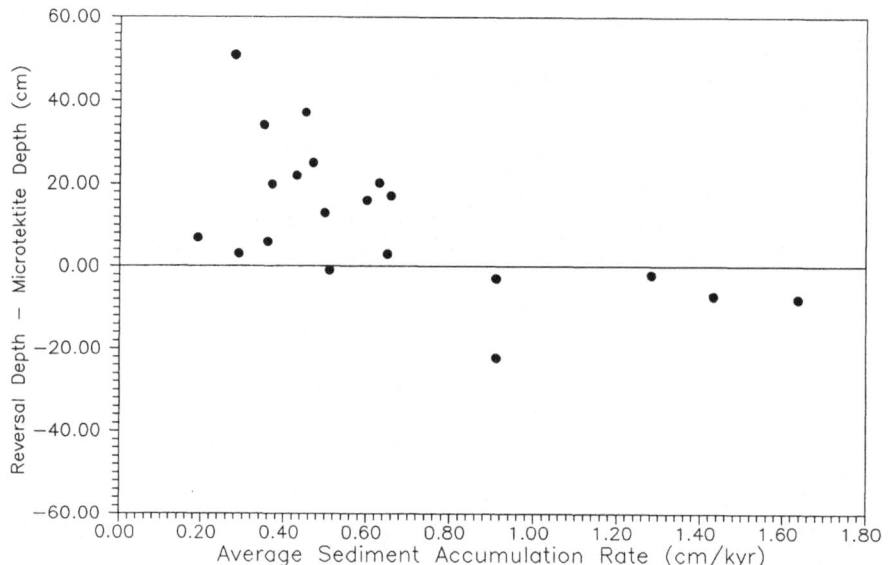

Figure 1. Distance between Australasian microtektite layer and B/M
reversal boundary vs. sediment accumulation rate.

The most simplistic model that would relate the difference in
position of the microtektite layer and that of the reversal boundary
to sediment accumululation rate would be:

$$Q = \Delta t \ w + D$$

where Q represents the spatial difference between the original
position of the microtektites and the midpoint of the B/M reversal, Δt
the time difference between the two events, w the sediment

accumulation rate, and D the depth of NRM acquisition. For this model it is found that $\Delta t = -26.7 \pm 6.2$ kyr and $D = 28.8 \pm 8.2$ cm, both of which are nonzero at the 95% confidence level. The negative value for Δt indicates that the impact would have preceded the reversal by 26,700 yr. However, the r^2 value of this model is only 0.37.

A slightly more sophisticated model is one that still invokes some time difference between the two events but does not treat depth of NRM acquisition as a constant. It is reasonable to suspect that depth of NRM acquisition may vary systematically with sediment accumulation rate. Hence the model:

$$Q = \Delta t \ w + n \ e^{-w}$$

where the variables are the same as above with n representing the intercept, or depth of acquisition when $w = 0$. In this model, $\Delta t = -11.8 \pm 5.6$ kyr and $n = 34.3 \pm 7.3$ cm. Again, both are nonzero at the 95% confidence level. Also as above, the negative Δt value indicates that the microtektites were produced and deposited before the reversal occurred. Depth of NRM acquisition would be defined by the second term in the equation; so $D = n \exp(-w)$. The linear transformation of this model gives an r^2 value of 0.58.

While it is possible that other models may fit the data better, it seems prudent to stick with these models that are simple and physically reasonable. In the first model, depth of NRM acquisition is treated as a constant. The concept that the reversal boundary is displaced by PDRM is confirmed since the intercept (D) is nonzero at the 95% confidence interval when $w = 0$. However, the low r^2 value may indicate that may not be appropriate to treat depth of NRM acquisition as a constant.

In the second model, depth of NRM acquisition is treated as nonconstant and varying inversely with sediment accumulation rate. While it may seem intuitive that the depth of NRM acquisition increases with increasing sediment accumulation rate, that does not fit the data. If water content of the sediment and grain size of the magnetic carriers are controlling factors in depth of NRM acquisition, then an inverse function between sediment accumulation rate and depth of NRM acquisition may not be so surprising. In general, as sediment accumulation rate increases in the deep-sea, the sediment type changes from clays to oozes (Table II). An increase in grain size is related to a decrease in porosity in deep-sea sediments (Sverdrup and others, 1942). Thus, a decrease in the depth of NRM acquisition with increasing sediment accumulation rate may be physically plausible.

While the second model proves more satisfactory than the first model, it does not account for all the variability in the data. One reason for this may be the intervals at which the cores were sampled for magnetics and microtektites. Figure 2 is similar to Figure 1 but shows the best fit line from the model $Q = \Delta t \ w + n \exp(-w)$, and error bars on the data points which indicate the sampling interval for each core. It can be seen that some of the variability in the data may be accounted for by sampling interval. However, some variability is still unexplained. It was suspected that core latitude may play a role in determining depth of NRM acquisition, but models incorporating latitude (Table II) proved insignificant. At this point, no explanation can be given for the unaccounted variability.

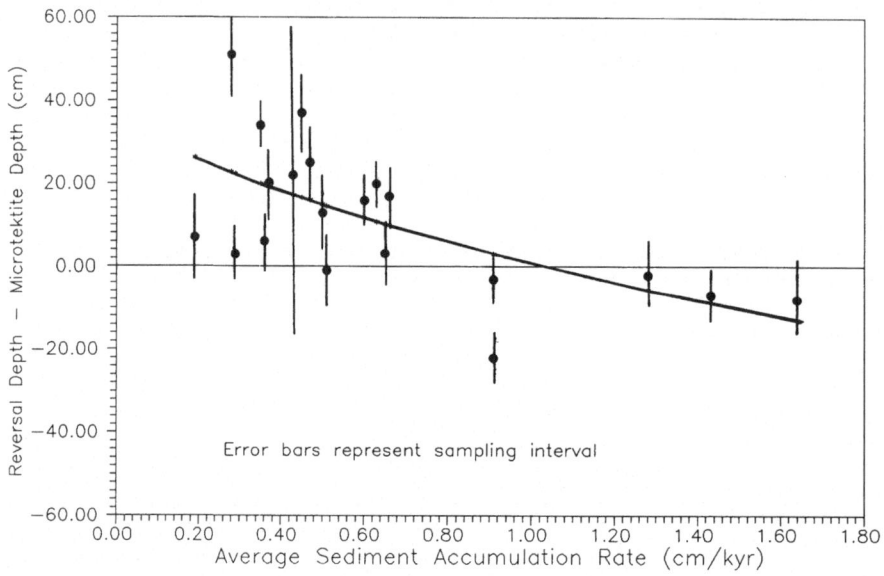

Figure 2. Distance between microtektite layer and reversal boundary vs. sediment accumulation rate. Error bars represent sampling interval for core. Best fit line relates time difference between two events and exponential decrease in depth of NRM acquisition with increasing sedimentation rate.

5. SUMMARY AND CONCLUSIONS

By comparing the difference in original depth of deposition of the Australasian microtektite layer to the presently recorded depth of the Brunhes/Matuyama reversal boundary, two important points of information were learned. First, over the range of sediment accumulation rates observed in these cores, depth of NRM acquisition varies systematically from about 28 to 7 cm as sediment accumulation rates increases from 0.19 to 1.64 cm/kyr respectively. These values seem reasonable in light of previous estimates. Secondly, the present data indicates that the impact producing the microtektites occurred 11.7±5.6 kyr before the measured midpoint of the reversal. Given that the transition length for a reversal may be between 4.9 and 8.5 kyr (Clement and others, 1982); the impact either predates or is essentially simultaneous with the reversal. This does not prove causality between the two events but it does not preclude it. Other close stratigraphic occurrences between microtektite layers and reversal boundaries have been cited (Glass and others, 1979) but these must be similarly investigated in more detail before the hypothesis that impacts can trigger geomagnetic reversals can be treated more seriously.

6. ACKNOWLEDGEMENTS

I would like to thank Dr. B. Glass, Dr. J. Pizzuto, and Mr. J. Smullen for helpful discussions. I would also like to thank Dr. D. Kent and Mr. D. Schneider of Lamont-Doherty Geological Observatory for providing some of the palaeomagnetic data. I am grateful for core samples from L-DGO and DSDP. This work has been supported by NSF grant OCE-8314522.

7. REFERENCES CITED

Berger, W.H. and Heath., 1968. 'Vertical mixing in pelagic sediments', Journal of Marine Research, 26, 134-143.
Burek, P.J., 1984. 'Impacts, magnetic field reversals and tectonic episodes', Abstracts and Papers of the 15th Lunar and Planetary Science Conference, 102-103.
Clement, B.M., Kent, D.V. and Opdyke, N.D., 1982. 'Brunhes-Matuyama polarity transition in three deep-sea sediment cores', Phil. Trans. Roy. Soc., A306, 113-119.
Glass, B.P. and Heezen, B.C., 1967. 'Tektites and geomagnetic reversals', Scientific American, 217, 32-38.
Glass, B.P., Swincki, M.B. and Zwart, P.A., 1979. 'Australasian, Ivory Coast and North American tektite strewnfields: size, mass and correlations with geomagnetic reversals and other earth events', Proceedings of the 10th Lunar and Planetary Sciences Conference, 2535-2545.
Guinasso, N.L. and Schink, D.R., 1975. 'Quantitative estimates of biological mixing rates in abyssal sediments', J. Geophys. Res., 80, 3032-3043.
Kent, D.V., 1973. 'Post-depositional remanent magnetisation in deep-sea sediment', Nature, 246, 32-34.
Muller, R.A. and Morris, D.E., 1986. 'Geomagnetic reversals from impacts on the earth', Geophys. Res. Lett., 13, 1177-1180.
Raisbeck, G.M., Yiou, F., Bourles, $D_{10}$and Kent, D.V., 1985. 'Evidence for an increase in cosmogenic ^{10}Be during a geomagnetic reversal', Nature, 315, 315-317.
Robbins, J.A., 1986. 'A model for particle-selective transport of tracers in sediments with conveyor belt deposit feeders', J. Geophys. Res., 91, 8542-8558.
Ruddiman, W.F. and Glover, L.K., 1972. 'Vertical mixing of ice-rafted volcanic ash in North Atlantic sediments', Geol. Soc. of America Bull., 83, 2817-2836.
Sverdrup, H.U., Johnson, M.W. and Fleming, R.H., 1942. The Oceans, Prentice Hall Inc, Englewood Cliffs, NJ.

TESTS OF MAGNETIC PROPERTIES OF NEW ZEALAND PLIOCENE MARINE MUDSTONES

D.M. McGuire
Research School of Earth Sciences
Victoria University of Wellington
PO Box 600
Wellington
New Zealand

ABSTRACT. Three tests have been used in a magnetostratigraphic
investigation of Pliocene mudstones in the Wanganui Basin of New
Zealand to examine the viscous behaviour and the stability of
secondary overprints, including those acquired after sampling. A
storage test evaluated longer term viscosity. The other tests
measured remanent magnetization acquired during drying and short term
viscosity. The results of these tests have led to the adoption of the
following procedures for collecting and handling palaeomagnetic
samples: 1. Sampling sites are located to provide fresh, unweathered
samples from saturated sediments that have never dried out. 2.
Samples are stored moist and unshielded until sample preparation is
complete. 3. The specimens are then dried in mumetal shields and
stored shielded until measurements are finished. These procedures
permit reliable determination of reversed and normal palaeodirections
from all but the most unstable specimens.

1. INTRODUCTION

I am attempting to define Pliocene magnetostratigraphy in 2 km thick
sequence of massive sandy mudstones deposited in the Wanganui Basin
approximately 4.5 to 2 myr ago and exposed in the Turakina River, New
Zealand. The Turakina River flows N-S through roughly the centre of
the Wanganui Basin, which is on the West coast of the North Island,
South of the Taupo Volcanic Zone. The basin has a relatively simple
structure, with a few N-S fault zones. As a result, the stratigraphy
of the exposed sequence is simple. Regional dip averages 5 degrees,
and the exposure stretches over 25 km. There are three
lithostratigraphic units in the section, varying from mudstones to
very sandy mudstones, thought to represent depositional environments
from outer shelf to shallow inner shelf. While deposition rates do
not appear to have been constant, they have been high throughout. In
the Rangitikei River, less than 30 km to the west, Seward et al.
(1986) calculated sedimentation rates ranging from 1.7 to 1.2 m/kyr
through the same time interval. The high sedimentation rates result

F. J. Lowes et al. (eds.), Geomagnetism and Palaeomagnetism, 263–269.
© 1989 by Kluwer Academic Publishers.

in potentially good time resolution in the sequence, but also very low concentrations of magnetic minerals.

Difficulties in interpreting palaeomagnetic results from some sites led me to carry out additional tests of the rock magnetic properties of the mudstones, to ensure that the collection and storage methods are not introducing errors or systematic bias. Besides being only weakly magnetized, the sediments display magnetic viscosity, and it was the viscosity I sought to test.

2. MATERIALS AND METHODS

The initial investigation of the demagnetization response of the mudstones using thermal and AF methods resulted in the decision to use thermal demagnetization, with each specimen's stability assessed individually.

Specimens are drilled in the field from the freshest permanently water saturated sediments that can be located. They are stored damp, wrapped in plastic film and unshielded until specimen preparation is complete. The specimens are then dried in Mumetal shields and stored shielded until treatment and measurement are finished. At least two specimens from each of three cores are measured for each site. Thermal demagnetization is carried out in four to twelve steps, and bulk susceptibility measurements are used to monitor the thermal alteration of the magnetic minerals.

The first test is a standard storage test (Banerjee 1981). Because the New Zealand Tertiary sediments have been known to alter their properties during drying (Walcott and Mumme 1982, Mumme and Walcott 1985), the storage test was applied to wet specimens that were not allowed to dry out. Two to four specimens were chosen from various lithological units. Sample collection and preparation were as usual, except that after specimen preparation, they were kept wet and stored unshielded. After initial measurement of their natural remanent magnetization (NRM), the specimens were placed in a Mumetal shield and remeasured after intervals of 1, 2, 4, 8, 16 and 29.5 days. The experiment is continuing, with additional measurements planned at 6, 12, and perhaps 18 months.

The second test is designed to examine the changes that take place in specimen remanence as they dry. Trios of specimens from sites in the different lithogical units were chosen and again stored wet and unshielded until their NRM's could be measured. They were then placed in different environments to dry: one in a Mumetal shield, the others in a controlled field of about 50,000 nT, similar to the Earth's field, directed along the Z axis of the specimens, with the specimens aligned opposite to each other. All specimens were measured when partially dry (after 4 days) and fully air dry (8 days), then thermally demagnetized in the normal manner.

The third test was designed to show the short term (24 hour) viscous behaviour of the dried specimens. Sets of five to 10 specimens that had been dried in a shield were placed into the same artificial field described above and were remeasured at 24 hour

intervals. After each measurement, one specimen in turn was inverted
in the field. Remeasurements of all specimens continued until the
last specimen had been inverted for 24 hours.

3. RESULTS

All the material in the section is very weakly magnetized; dry NRM
intensities vary from 0.04 to 0.9 mA/m, with an average value of about
0.2. Thermal magnetic cleaning often reduces specimen intensities to
the limits of reliable measurement at about the point where specimen
susceptibilities begin to increase due to thermal alteration (340–380°
C). Susceptibility increases are accompanied by a marked increase in
viscous behaviour with very short decay times.
    The storage test results show a decrease in total intensity for
all sites (specimens), with a single exception (site 21T, 2 specimens)
where intensity rose for 8 to 16 days, then began to drop (Figure 1).
This is interpreted as the decay of a laboratory acquired viscous
component oppositely directed to the specimens' remanent

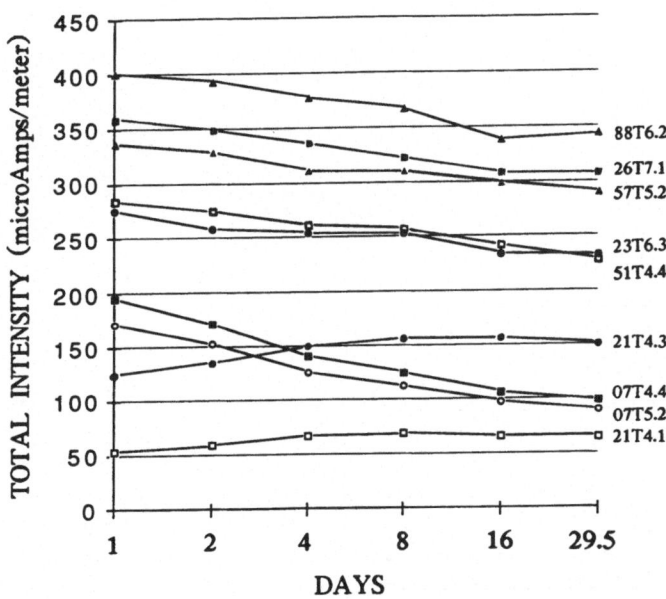

Figure 1. Storage test: plot of total intensity versus log of time in
days. The general trend is one of logarithmic decay for all specimens
except the two from site 21T. Their increase in total intensity over
the first 8–16 days is due to the decay of a VRM acquired during
unshielded storage in the laboratory which is oppositely directed to
the remanence of the specimens.

magnetization. Intensity values have not stabilized over the current
duration of the test, indicating that most sites are likely to carry a
viscous overprint directed towards the present field, and initial data
analysis supports this. Viscous remanent magnetizations (VRM's)
acquired during laboratory storage can be cleaned from specimens by
keeping them wet and shielded for about two weeks.

The test of remanence acquired during drying clearly shows the
growth of a remanence when specimens dry in an external field. The
more intensely magnetized specimens display the effect more clearly
(Figure 2). Drying in a Mumetal shield results in a loss of 30-60% of
NRM intensity. Drying in the external field usually displays a
greater or lesser decrease in intensity, depending upon the relative
directions of the NRM and the external field. Zijderveld and stereo
plots of remanent vectors during drying show a divergence of direction
between the three specimens from a site that is not removed by the
subsequent demagnetization, even after the intensities of the
specimens are roughly equal. This behaviour is repeated for different

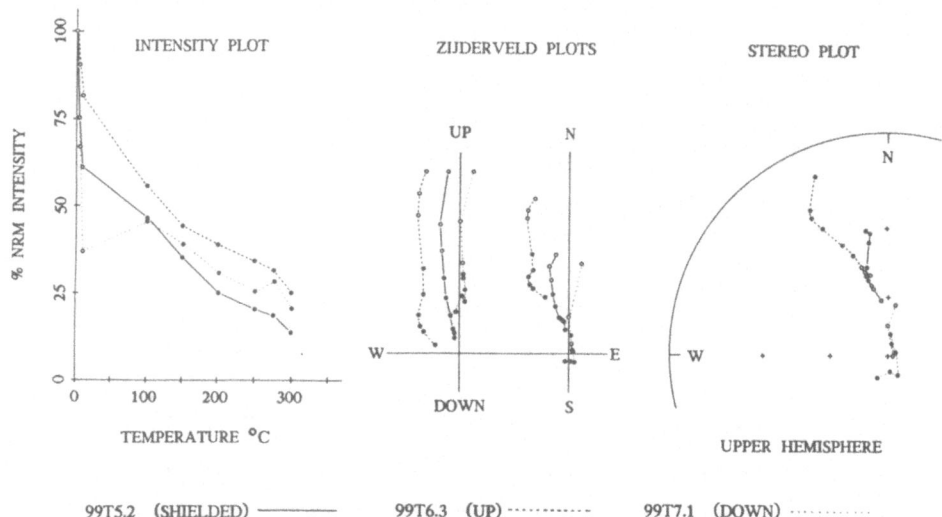

Figure 2. Remanence Acquired During Drying: specimen 5.2 was dried in
a Mumetal shield, while specimens 6.3 and 7.1 were dried in an
approximately 50,000 nT field directed along their Z axis, 6.3 'up',
and 7.1 'down'. The open circles plotted for each specimen follow
their change from wet through semi-dry to dry. The solid circles
follow their behaviour through thermal demagnetization. The
Zijderveld plots are normalized to the maximum 'up' intensities.
Normalized total intensities diverge during drying, as do the
directions. Upon thermal cleaning, intensities assume similar values
while directions do not, and in fact directions continue to diverge
throughout the cleaning process.

lithologies, polarities, and NRM intensities, though somewhat masked
in low intensity specimens by the larger element of randomness between
individual measurements. Variations in porosity from 3% to 35% are
not reflected in the behaviour of the specimens in this test. It must
also be noted that the changes in the remanence Declination and
Inclination that result are not sufficiently large to obscure the true
polarity of the sites in question.

Throughout the test of short term viscosity, the X and Y
components of remanence (perpendicular to the applied field) did not
change significantly, showing negligible zero-field decay over short
time periods (Figure 3). However, all specimens grew a component
along the Z axes, parallel to the applied field. Some ceased to
acquire additional remanence after only 24 hours, while others
continued to slowly increase in intensity for the duration of the
test. The largest portion was acquired in less than 24 (sometimes 48)
hours. Upon inversion in the field, a new oppositely directed
remanence was again largely acquired in less than 24 hours, and
continued to grow for the duration of the test (presumably partly due
to the decay of the initially acquired, now oppositely directed
remanence) (Figure 4).

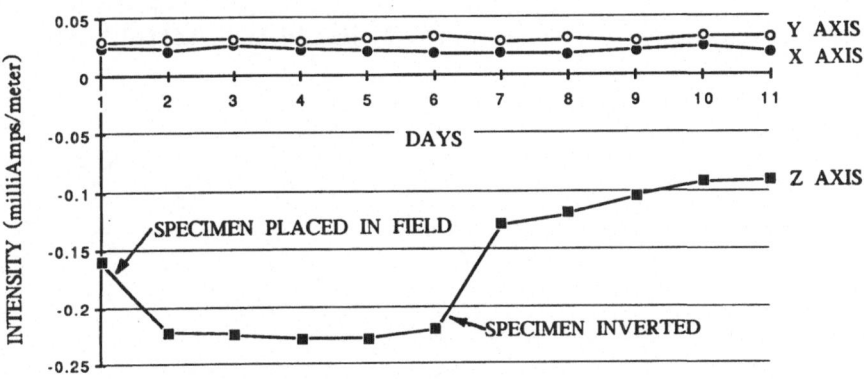

Figure 3. Short Term Viscosity Test: plot of daily values of X, Y,
and Z axis intensities of a specimen in an approximately 50,000 nT
field directed along its Z axis. The large changes in Z axis
intensity occur within 24 hours of introduction into, and inversion in
the field.

4. DISCUSSION

Viscous behaviour has been noted in many of the Tertiary marine
mudstones in New Zealand (Walcott, 1988 personal communication), and
the Pliocene section in the Turakina River is a good example. Wet or
dry, specimens are liable to develop a longer or shorter term VRM
whenever they are exposed to a magnetic field. The errors that result
from this viscous magnetic behaviour can be minimized by proper

handling and storage, but since the behaviour is an inherent property of the magnetic grains, it cannot be avoided, and it limits the amount of useful information that can be obtained. The stability of individual specimens and sites varies.

Changes in remanence of specimens upon drying is potentially a larger source of error than the grain viscosity discussed above, but it can be minimized by drying in zero field. This results in a significant reduction of intensity, but has no effect on remanent direction. A new test using AF cleaning of wet rather than dry specimens has given results that appear to be as good or better than thermal cleaning in determining a stable palaeodirection. This has the benefit of avoiding both the acquisition of stray remanent directions and the loss of NRM intensity. Problems with AF cleaning in the original investigations appear to have been due partly to the available equipment not generating a sufficiently pure AF signal, particularly at higher fields, and partly to using dry instead of wet specimens. Evaluation of the cleaning response of a sediment should fully reflect the advantages of the method employed: AF cleaning should be applied to specimens that have not been allowed to dry out, if they are wet when collected. Thermal cleaning requires that specimens be dry, and that drying should only be allowed to take place in a shielded environment at temperatures below 50–60° C. The method adopted for this study is similar to methods used for much younger sediments with higher water contents (Thompson and Oldfield 1986).

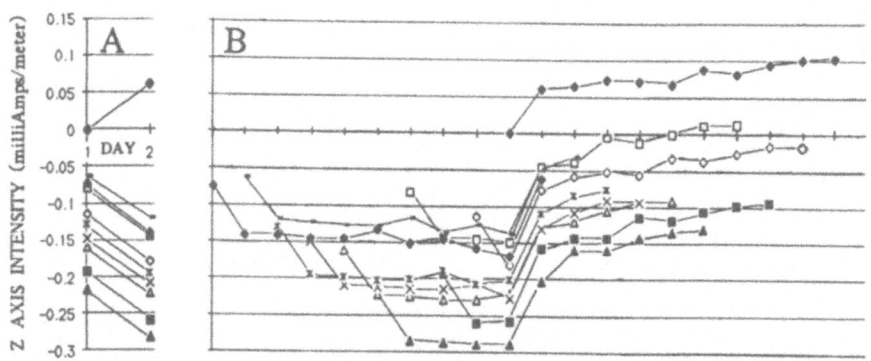

Figure 4. Short Term Viscosity Test. A: plots of change in Z axis intensities of all specimens in one test set, before and after 24 hours in the external magnetic field. The specimens are from various sites within a single lithostratigraphic unit. All specimens show a similar change in intensity (one specimen inverted relative to the others). B: plots of Z axis intensities, same set as A, but with all 11 days' data shown, and each specimen plot adjusted horizontally to align the point at which the specimens were inverted. The displacement of intensity values upon inversion, and intensity behaviour both before and after, is remarkably similar for all specimens.

5. ACKNOWLEDGEMENTS

I wish to gratefully acknowledge the advice and review of this paper by Prof. P. Vella, and Dr. G.M. Turner. Internal Research Grants provided by the Victoria University Internal Research Committee have supported the costs of field and laboratory work associated with this research.

6. REFERENCES

Banerjee, S.K., 1981. 'Experimental Methods of Rock Magnetism and Paleomagnetism'. In Saltzman (ed.) Advances in Geophysics, Academic Press Inc., New York 23, 25-99.

Mumme, T.C. and R.I. Walcott, 1985. 'Paleomagnetic Studies at Geophysics Division 1980-1983', Geophysics Division Report 204, 62 pp. Department of Scientific and Industrial Research, Wellington, New Zealand.

Seward, D., D.A. Christoffel and B. Lienert, 1986. 'Magnetic Polarity Stratigraphy of a Plio-Pleistocene Marine Sequence of North Island, New Zealand', Earth and Planetary Science Letters, 80, 353-360.

Thompson, R. and F. Oldfield, 1986. Environmental Magnetism, Allen & Unwin Ltd., London, p. 70.

Walcott, R.I. and T.C. Mumme, 1982. 'Paleomagnetic Study of the Tertiary Sedimentary Rocks from the East Coast of the North Island, New Zealand', Geophysics Division Report 189, 44 pp. Department of Scientific and Industrial Research, Wellington, New Zealand.

SOME ASPECTS OF THE MEASUREMENT OF MAGNETIC ANISOTROPY

A. Stephenson & D.K. Potter
Department of Physics
University of Newcastle upon Tyne
Newcastle upon Tyne, NE1 7RU, England

ABSTRACT. Anisotropy of magnetic susceptibility (AMS) measurements
are commonly used to assess the degree and type of alignment of ferro-
or ferri-magnetic particles in rocks i.e. they are used by many
workers to decide whether rocks are anisotropic and if so whether the
axes of the constituent particles exhibit a predominantly planar
(foliated) or linear (lineated) distribution. However AMS
measurements as commonly interpreted can give erroneous results both
for the type and degree of alignment in that apparently isotropic
samples can be anisotropic and that apparently foliated rocks can
actually be lineated (and vice-versa). To obtain an unambiguous
indication of the degree and type of alignment, anisotropy of magnetic
remanence (AMR) methods such as those described here must be employed.

1. INTRODUCTION

Anisotropy of low-field magnetic susceptibility (AMS) can be a very
sensitive, useful and rapid measurement using modern instruments. A
commonly used method is to use the Molspin anisotropy delineator which
applies a weak 10 kHz field at right angles to the vertical rotation
axis of a slowly rotating sample. Any horizontal component of induced
10 kHz magnetization produced at right angles to the inducing field by
the intrinsic anisotropy of the sample is then detected by two pick-up
coils with a common horizontal axis normal to both the inducing field
and the rotation axis. By measuring each sample in three different
orientations it is possible to compute the orientation of the
principal axes of an anisotropy ellipsoid (for reviews of AMS and its
uses see Hrouda 1982 and MacDonald and Ellwood 1987). Since the
signal produced by the sample depends only on the susceptibility
differences between the principal axes a further measurement of an
actual susceptibility along a known direction (in this case z) is
necessary before the magnitudes of the principal susceptibilities can
be found. (See e.g. Hrouda et al. 1983).
   There are two aspects of the interpretation of AMS which are
discussed here. The first is the use of AMS to check whether or not

F. J. Lowes et al. (eds.), Geomagnetism and Palaeomagnetism, 271–278.
© *1989 by Kluwer Academic Publishers.*

there is preferential alignment of the ferro- or ferri-magnetic particles within rocks. This is sometimes used as a quick test in palaeomagnetism to check whether anisotropy is likely to be important in deflecting the natural remanent magnetization (NRM) vector away from the local ambient field direction when the NRM was acquired.

A second common use is to determine whether samples are predominantly foliated (non-spherical particles predominantly orientated with their long-axes distributed in a plane) or lineated (particles predominantly along an axis). Foliation may indicate such processes as compaction (e.g. in a sediment) and lineation might indicate flow axes in a lava, water flow in sediments or other similar processes which tend to line up any elongated particles (including ferro- or ferri-magnetic ones).

It is the purpose of this paper to point out that the use of AMS alone to decide whether a rock or similar sample is anisotropic, and if so whether it is (a) foliated or (b) lineated, can in some circumstances give an erroneous answer in that a rock which shows very little susceptibility anisotropy can contain very strongly aligned ferro- or ferri-magnetic particles. Moreover on those occasions when AMS indicates a foliated rock, the ferro- or ferri-magnetic particles within it may actually be lineated and conversely when AMS indicates lineation the particles within it can be foliated. AMS measurements alone therefore do not give an unambiguous answer to questions regarding the degree and type of alignment of ferro- or ferri-magnetic particles. These observations follow from an analysis of anisotropy carried out by Stephenson et al. (1986).

2. SUSCEPTIBILITY ANISOTROPY AND SINGLE DOMAIN/MULTIDOMAIN PARTICLE CHARACTERISTICS

Consider the simple model illustrated in Fig. 1 of a foliated rock (a) and a lineated rock (b). The elongated particles illustrated are imagined to be magnetite with shape anisotropy. If the particles are large (say $\gtrsim 10\mu m$) then they are multidomain (MD) and the susceptibility of each particle will be greatest along the long axis where the demagnetizing factor is small, and smallest normal to this axis where the demagnetising factor is larger. Thus when a field is applied along the z axis in Fig. 1a the susceptibility χ_z will be smaller than if it is applied in the x or y directions ($\chi_x = \chi_y$) and the rock will be correctly identified from AMS measurements as being foliated. Conversely in Fig. 1b χ_z will exceed χ_x or χ_y (which are equal) and the rock will correctly be identified as lineated.

The problem arises when the particles are of the same overall shape but are much smaller (say $\lesssim 1\mu m$). In this case they become essentially single-domain (SD) and their behaviour is then ideally described by the well-known Stoner-Wohlfarth model (1948) which gives zero susceptibility along the long axis but a finite susceptibility normal to that axis. Thus uniaxial SD particles with the type of distribution shown in (a), i.e. foliation, would give a maximum susceptibility along z since the measuring field (along z) is then

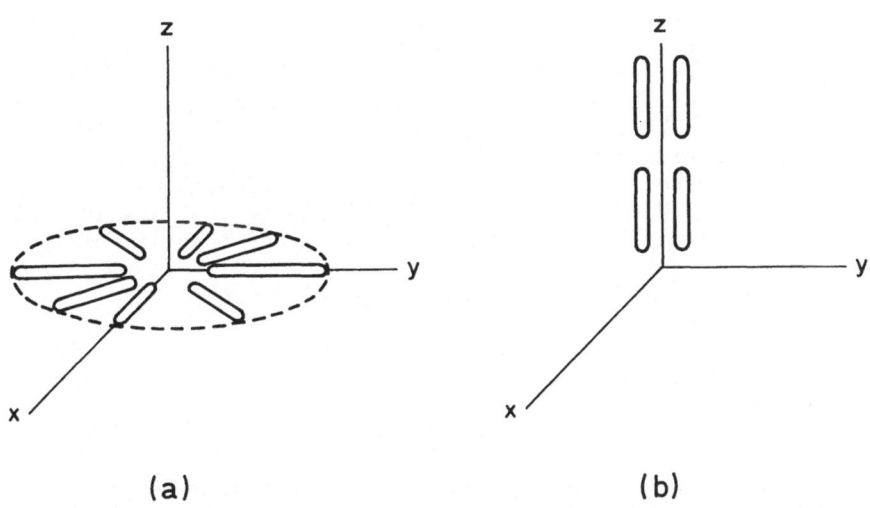

(a)   (b)

Figure 1.   Illustrating (a) a foliated rock containing elongated
particles exhibiting shape anisotropy and (b) a lineated rock.   The
shape of the susceptibility ellipsoid of either rock will be prolate
or oblate depending on whether the particles are multidomain or
uniaxial single-domain.

normal to all the long axes of the particles.   AMS alone therefore
would incorrectly identify distribution (a) as lineated.   Distribution
(b) would give zero susceptibility along z and would from AMS
measurements alone be erroneously identified as foliated.
        Clearly any magnetic anisotropy method which sets out to identify
lineation or foliation must, since particle characteristics are
crucial but generally unknown, be independent of such characteristics.
Thus any single magnetic parameter, if it is to be useful in this
context, must always give a maximum along the long axis of each
particle irrespective of particle size.   Such a parameter is remanence
(it does not matter what type of remanence as will be shown later).
Anisotropy of magnetic remanence (AMR) unlike AMS will thus always
correctly identify a foliation or a lineation irrespective of particle
size.
        A convenient remanence to use is isothermal remanent
magnetization (IRM) given in a field of say 50 mT, a field which is
strong enough to change the magnetization state of most naturally
occurring elongated SD particles.   Although remanence generally
(unlike initial susceptibility) exhibits a non-linear dependence on
field it may to a first approximation be regarded as linear, in which
case the anisotropy method of Stephenson et al. 1986 may be used as
follows.
        The direct field is applied along the x direction of a
demagnetized sample and the resulting remanences M_{1x}, M_{1y}, M_{1z} are
measured after removal of the field. M_{1y} and M_{1z} will of course be
zero if the sample is isotropic or if the x axis marked on the sample

happens to coincide with a principal axis of the ellipsoid describing the particle distribution. If the sample is then demagnetized by tumbling in an alternating field, and the direct field is applied along the y axis of the sample, remanence components M_{2x}, M_{2y}, and M_{2z} will result. The same procedure is once more followed with regard to the z axis, giving remanences M_{3x}, M_{3y}, and M_{3z}. From these 9 remanences a "remanence ellipsoid" can be computed with principal values M'_{xx}, M'_{yy}, M'_{zz}. The shape of the remanence ellipsoid will then indicate whether for instance a rock is predominantly foliated or lineated.

To illustrate the fundamental differences between AMS measurements and AMR measurements it is very useful to be able to compare the shapes of the respective ellipsoids. This can be done by normalising the principal values of the remanence ellipsoid such that

$$r_x = M'_{xx}/(M'_{xx} + M'_{yy} + M'_{zz}) \text{ etc}$$

and comparing the resulting 3 parameters with the normalized parameters obtained from the corresponding susceptibility ellipsoid, i.e,

$$p_x = X'_{xx}/(X'_{xx} + X'_{yy} + X'_{zz}) \text{ etc.}$$

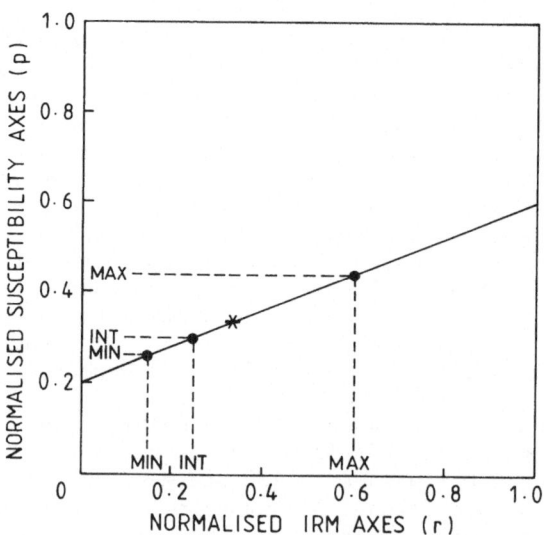

Figure 2. The schematic dependence of the normalised principal susceptibilities on the normalised principal IRM ellipsoid values for an anisotropic sample containing multidomain magnetite. Both sets of ellipsoid axes are coincident, with the two maximum and two minimum axes coinciding with each other. The susceptibility anisotropy is less than the IRM anisotropy. (The star is at 1/3, 1/3).

A plot of the points (r_x, p_x), (r_y, p_y), (r_z, p_z) then compares the shapes. Note that the p and r values lie between 0 and 1 and their sum is unity. The 3 points also lie on a straight line passing through the point (1/3, 1/3). Clearly if both ellipsoids were of identical shape the line would have an intercept of zero (slope of unity). For magnetite particles of any shape or size, susceptibility anisotropy is expected to be smaller than IRM anisotropy (Stephenson et al. 1986). The line shown in fig. 2 is that appropriate to MD magnetite particles and to some metamorphosed granites, slates and schists (Stephenson et al. 1986 – see figs 8 and 9). Note that for these MD particles, the maximum χ axis coresponds to the maximum IRM axis but the difference between the maximum and minimum susceptibility is less than the difference between the maximum and minimum IRM, i.e. AMS indicates a smaller anisotropy than AMR.

Fig. 3, however, illustrates the same plot for a sample containing uniaxial SD particles where the maximum susceptibility axis corresponds to the minimum IRM axis. The result illustrated (intercept 0.45) was obtained for a dilutely dispersed set of elongated γFe_2O_3 'tape' particles made anisotropic by setting them in resin in the presence of a magnetic field (Stephenson et al. 1986). Uniaxial SD particles give a plot with negative slope (intercept 1/3) and there is evidence that as the particle size decreases from MD the slope gradually decreases, reversing sign at an estimated particle size of just under 1 μm (for magnetite).

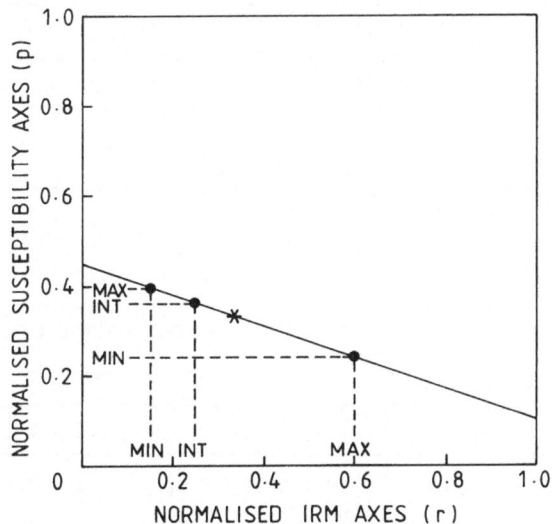

Figure 3. The dependence of the normalised principal susceptibilities on the normalised principal IRM ellipsoid values for an anisotropic sample containing prolate single-domain ferrimagnetic particles. This time the maximum susceptibility axis coincides with the minimum IRM axis and vice-versa. Once again IRM anisotropy exceeds susceptibility anisotropy.

To test whether rocks can exhibit this interchange of maximum and minimum axes (for susceptibility and remanence) a survey of several rocks taken arbitrarily from a rock store was carried out. One of these did indeed exhibit these characteristics (Potter & Stephenson 1988). The result is shown in Table I. Normalised susceptibilities are shown and these are correct to the number of figures quoted. The AMS method has the advantage of being more sensitive than most AMR methods with the possible exception of gyroremanent magnetization (GRM) which is very sensitive for rocks containing fine particles. GRM is in fact the remanence analogue of the AMS measurements obtained by the Molspin anisotropy delineator. The first row in the table shows (from the normalised magnitudes) that the rock has a minimum magnitude (along an axis denoted x') which is some 2.4% lower than those along the other two principal axes which are almost equal. The orientation (declination and inclination) of the principal axes are also shown. (Note that the y' and z' axes are not well-determined because the ellipsoid is almost an ellipsoid of revolution). Susceptibility measurements would thus indicate a weak but definite foliation in the y'z' plane.

When the IRM measurements are carried out, however, it is clear that the interpretation based on AMS measurements alone is false and that the rock is actually strongly lineated, i.e. there is a maximum axis which, to within the errors of measurement, agrees in orientation with that of the minimum susceptibility axis. Moreover the difference between the maximum and minimum magnitudes for the IRM measurements is approaching 20% so not only is the type of alignment deduced from the AMS measurement incorrect but it also seriously underestimates the degree of alignment. This latter effect is also apparent in the results of Daly and Zinsser (1973) who also measured susceptibility and IRM anisotropy.

TABLE I. A comparison of ellipsoid shapes for susceptibility and four different types of remanence. The fifth remanence (GRM) does not yield an ellipsoid shape but is very sensitive in identifying the maximum axis. For further details see Potter & Stephenson (1988).

	DIRECTIONS x' DEC	x' INC	y' DEC	y' INC	z' DEC	z' INC	NORMALISED MAGNITUDES x'	y'	z'	AVERAGE DIMENSION OF ELLIPSOID
χ	3	−11	98	−25	251	−62	MIN 0.328	0.336	0.335	$430 \times 10^{-8}\,m^3\,kg^{-1}$
IRM	14	−18	108	−15	236	−67	MAX 0.37	0.33	0.30	$150 \times 10^{-3}\,Am^2\,kg^{-1}$
ARM$_s$	12	−14	284	8	224	−74	MAX 0.37	0.32	0.31	$780 \times 10^{-6}\,Am^2\,kg^{-1}$
ARM$_R$	12	−20	278	−16	158	−65	MAX 0.37	0.32	0.31	$740 \times 10^{-6}\,Am^2\,kg^{-1}$
RRM	13	−22	276	−17	152	−62	MAX 0.37	0.32	0.31	$2340 \times 10^{-6}\,Am^2\,kg^{-1}$
GRM	13	−10	280	−15	137	−72	MAX	INT	MIN	

The rest of Table I shows the normalised remanence values and axis orientations of four other types of room temperature remanence methods all of which give very similar results to the IRM measurements. ARM_s and ARM_R involve giving an anhysteretic remanent magnetization either to a static sample or to a rotating sample. McCabe et al. (1986) have also used an ARM_s method to measure anisotropy. The last two rows involve the production of a gyroremanent magnetization (GRM) either to a static sample (last row) or to a rotating sample (rotational remanent magnetization (RRM) – penultimate row). These latter methods are discussed in more detail by Potter and Stephenson (1988). It should be noted that any method which attempts to determine anisotropy by the use of a weak direct field in the presence of a strong alternating field (i.e. the conditions for producing ARM) is inherently defective if the possible presence of GRM is ignored. GRM will not always be a problem but if magnetite or titanomagnetite particles in the approximate size range 0.1 to 10µm are present then GRM will inevitably be produced along with the ARM. Only for tumbling ARM – see Stephenson and Collinson (1974) – will GRM be insignificant.

3. CONCLUSION

The results of anisotropy measurements carried out on the rock sample in Table I confirm the theoretical argument that AMS measurements can in some instances seriously underestimate the true magnitude of anisotropy. This can arise either because the particles within the rock are on the SD–MD boundary or could occur if there were a mixture of uniaxial SD and MD particles with the same alignment direction. In addition it is clear that AMS measurements alone can give an erroneous indication of the type of alignment. As expected theoretically, all remanence methods (AMR) give an unambiguous measurement of the type of alignment and give a better indication of the degree of alignment, and, except for the GRM measurement (final row) which cannot determine the ellipsoid shape, all methods for this sample turn out to yield the same shape for the remanence ellipsoid.

Remanence anisotropy is thus more relevant in assessing the importance of anisotropy in palaeomagnetism and in correcting for its effects than any method based on susceptibility anisotropy alone, such as that proposed by Cogné (1987) which is only applicable to samples containing fully aligned MD particles.

REFERENCES

Cogné, J.P., 1987. "TRM deviations in anisotropic assemblages of multidomain magnetite". Geophys. J.R. astr. Soc., 91, 1013–1023.

Daly, L. and Zinsser, H., 1973. "Étude comparative des anisotropies de susceptibilité et d'aimantation rémanente isotherme. Conséquences pour l'analyse structural et le paléomagnétisme". Ann. Geophys. 29, 189–200.

Hrouda, F., 1982. "Magnetic anisotropy of rocks and its application
    in geology and geophysics". Geophys. Surv., 5, 37-82.
Hrouda, F., Stephenson, A. and Woltär, L., 1983. "On the
    standardization of measurements of the anisotropy of magnetic
    susceptibility". Phys. Earth Planet. Int., 32, 203-208,
McCabe, C., Jackson, M. & Ellwood, B.B., 1985. "Magnetic anisotropy
    in the Trenton limestone : results of a new technique, anisotropy
    of anhysteretic susceptibility". Geophys. Res. Lett., 12,
    333-336.
MacDonald, W.D. & Ellwood, B.B., 1987. "Anisotropy of magnetic
    susceptibility : sedimentological, igneous and structural-
    tectonic applications. Rev. Geophys., 25, 905-909.
Potter, D.K. and Stephenson, A., 1988. "Single-domain particles in
    rocks and magnetic fabric analysis". Geophys. Res. Lett.
    (in press).
Stephenson, A. and Collinson, D.W. 1974. "Lunar magnetic field
    palaeointensities determined by an anhysteretic remanent
    magnetization method. Earth & Planet Sci. Lett. 23,
    220-228.
Stephenson, A., Sadikun, S. & Potter, D.K., 1986. "A theoretical and
    experimental comparison of the anisotropies of magnetic
    susceptibility and remanence in rocks and minerals". Geophys.
    J. R. astr. Soc., 84, 185-200.
Stoner, E.C. and Wohlfarth, E.P., 1948. "A mechanism of magnetic
    hysteresis in heterogeneous alloys". Phil. Trans. Roy. Soc. A.,
    240, 599-642.

THE MAGNETISM OF ORDINARY CHONDRITES AND SNC METEORITES: POSSIBLE IMPLICATIONS FOR ANCIENT SOLAR SYSTEM MAGNETIC FIELDS

D.W. Collinson
Department of Physics
The University
Newcastle upon Tyne, NE1 7RU, England.

ABSTRACT. The characteristics of the natural remanent magnetism (NRM) of ordinary chondrites are described and are related to processes of magnetization which the meteorites may have experienced and possible ancient magnetizing fields. The latter include a magnetic field associated with the solar nebula at the time of formation of iron-nickel grains, which carry the NRM in chondrites, and an intrinsic magnetic field arising from internal processes in meteorite parent bodies. The results suggest that chondrules and iron-nickel grains were magnetized in a nebula field and not significantly re-magnetized later, although there are difficulties associated with the occurrence and formation of the ordered iron-nickel mineral, tetrataenite, and with metamorphic heating of chondrites.

SNC meteorites are believed to originate on Mars, and some of them possess a highly stable primary magnetization which appears to have been acquired prior to their ejection from the planet. The evidence is described, together with field intensity estimates and the origin of the remanence as either a thermoremanent or shock remanent magnetization is discussed.

1. INTRODUCTION

The investigation of natural remanent magnetization (NRM) and bulk magnetic properties of meteorites is of interest in two aspects of solar system research, namely the detection of ancient magnetic fields and the elucidation of meteorite history and evolutionary processes. Ancient magnetic fields may be associated with the solar nebula before or during the planetary formation process, or an interplanetary field at a later time, or magnetic fields generated within planetary bodies.

Meteorites, with a few interesting exceptions, were formed at the same time as the planets, about 4.6 Ga ago, and originally formed part of meteorite parent bodies located in the asteroid belt. Unlike the Moon and terrestrial planets, however, meteorite compositions range from that which is believed to be closely similar to primitive pre-accretion planetesimal material (in type 1 carbonaceous

279

F. J. Lowes et al. (eds.), Geomagnetism and Palaeomagnetism, 279–295.
© *1989 by Kluwer Academic Publishers.*

chondrites) to those which are products of melting and differentiation (the achondrites). Between these extremes are the most common type of meteorites, the ordinary chondrites, in which varying amounts of mineralogical alteration have been brought about by moderate heating at some stage in their evolution.

Because they have in general undergone less severe alteration during their history, the chondrites offer the best possibility of providing information about early solar system magnetic fields. If the magnetic carriers in a meteorite acquire a remanent magnetism at or near their time of formation, then under appropriate conditions the remanence can persist with little alteration through the subsequent evolution of the meteorite until the present. If this primary magnetization is partially overprinted with a secondary component acquired at a later time, then this may be evidence for an interplanetary field or one generated within the parent body. The first part of this contribution describes the magnetism of ordinary (non-carbonaceous) chondrites and its interpretation in terms of magnetic fields and meteorite evolutionary processes.

A relatively recent development in meteorite studies has been the emergence of evidence that some unusual differentiated meteorites originated on Mars (Wood and Ashwal 1981, McSween 1985). These meteorites are few in number (13) and are commonly referred to as SNC meteorites, from the initial letters of the type samples Shergotty (India), Nakhla (Egypt) and Chassigny (France). The evidence includes the close similarity of Shergotty and EETA 79001 (an Antarctic shergottite) major element compositions to that of the Martian surface material as measured by the Viking spacecraft, isotopic ratios in trapped noble gases (e.g. $^{40}Ar/^{39}Ar$, $^{129}Xe/^{132}Xe$) and in nitrogen ($^{15}N/^{14}N$) in EETA 79001, comparable with those found in the Martian atmosphere, and the anomalously young ages of ~1.3 Ga or less (McSween 1985). There is still some discussion about the significance of the data, but if the SNC meteorites are indeed of Martian origin then the study of any remanent magnetization they possess may provide evidence for the existence of an ancient Martian magnetic field. Some work on EETA 79001 by the author is described in the second part of this paper, together with magnetic studies of other SNC meteorites.

2. REMANENT MAGNETISM OF ORDINARY CHONDRITES

2.1 Magnetic carriers in chondrites

The dominant magnetic carriers in ordinary chondrites are iron-nickel particles of varying grain size and composition. There is a broad division between kamacite or α-iron (<~7% Ni) and taenite (γ-iron, >~25% Ni), both of which can carry NRM. Recently, an ordered iron-nickel compound has been found in chondrites and other meteorites, with composition typically close to 50% nickel (Clarke and Scott 1980). This material, tetrataenite, has emerged as an important carrier of NRM in ordinary chondrites, because of its very high remanent coercive force. The ordering breaks down at about 580°C, which is the effective Curie point of tetrataenite, and it reverts irreversibly to ordinary taenite of the same composition and Curie point.

Tetrataenite forms in meteorites through very slow cooling from above the ordering temperature of iron-nickel of the appropriate composition. Over long time intervals the ordering temperature is ~320°C. It can only be synthesised in the laboratory by a complicated process involving high magnetic fields and neutron irradiation (Néel et al. 1964).

In some chondrites other minerals carry significant magnetization. Sugiura and Strangway (1981) report cohenite (iron carbide, Fe_3C) as the carrier of much of the NRM of the Abee (E4) chondrite. In Chainpur (LL3) plessite (fine-grained intergrowth of kamacite and taenite) carries a hard component of NRM, as is also the case in Mezo Madaras (L3) (Sugiura and Strangway 1982). In most ordinary chondrites some NRM is carried by kamacite and taenite, and the former has been found to contribute to the NRM of chondrules. If sufficiently fine-grained, both kamacite and taenite can possess an NRM of high coercivity.

2.2 Chondrite history and the acquisition of NRM

A characteristic feature of chondrites is the presence in them of chondrules, more or less well-preserved, small, spherical glassy objects. Although there is uncertainty as to their mode of origin (Wasson 1985), the two favoured theories are that they condensed directly out of the solar nebula as liquid droplets, followed by solidification, or they were melted debris resulting from collision processes among meteorite parent bodies. However, it seems clear that their formation pre-dated the formation of the parent bodies which ultimately provided the chondrite meteorites, and therefore they were accreted with other material to form these bodies.

The broadly accepted outline of chondrite formation and evolution, and possible associated magnetic history, is as follows. At some stage in the contraction and cooling of the solar nebula kamacite grains condensed out, cooled, and if a magnetic field existed, acquired a thermoremanent magnetization (TRM). If chondrules formed also at this time through condensation, they would also acquire a TRM in an ambient field. At some later time the primitive condensates accreted into planetesimals and later into larger objects, the meteorite parent bodies. Metamorphic heating occurred either before or soon after accretion into parent bodies, and was either of external (solar?) or internal origin. Depending on the maximum temperature reached and the presence or absence of a magnetic field, the metal grain and chondrule NRM could be demagnetized, completely re-magnetized or partially re-magnetized. Any existing magnetic field could be a residual interplanetary field or one generated within the parent body. During subsequent collisions, including the one which ultimately perturbed the fragments into Earth-crossing orbits, shock effects could modify the chondrite NRM, through shock demagnetization or magnetization (in an ambient field). Although severe surface heating of meteorites occurs during their passage through the Earth's atmosphere, the heated layer ablates and the low thermal conductivity of silicate-rich meteorites inhibits the penetration of surface heat into the interior. Only a thin (~ 1 mm) fusion crust is commonly observed, below which thermal alteration is usually minimal. The final opportunity for NRM acquisition occurs during the period between meteorite fall and discovery, when a viscous magnetization (VRM) in the Earth's field can be acquired.

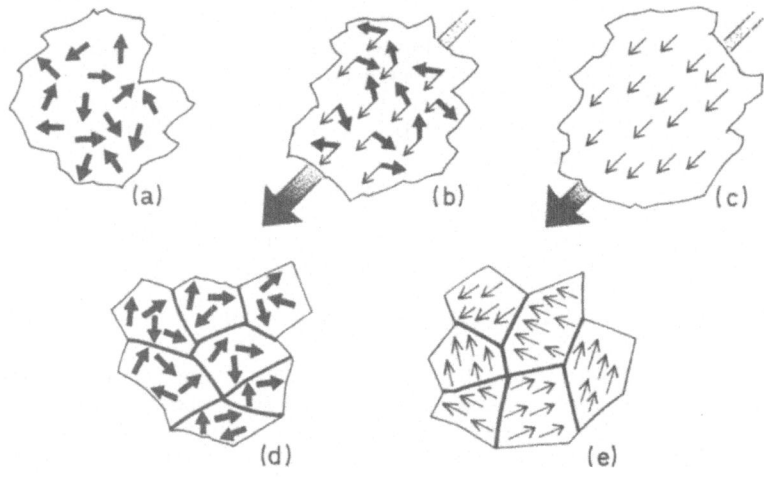

Figure 1. Possible types of NRM contributed by metal particles or chondrules in chondrites. (a) accumulation into parent body with no subsequent alignment. (b) uniform secondary component producing scattered but not random directions. (c) complete overprint in uniform field. (d),(e) brecciated chondrite produced from material with (a) or (c) type NRM respectively.

It is apparent from the above that chondrites may have a complex magnetic history, but also that they offer the possibility of deriving information about solar system magnetic fields and meteorite evolutionary processes. Fig. 1 shows some examples of NRM characteristics which would be observed in chondrites according to their different magnetic histories. Fig. 1(a) shows the expected inhomogeneous magnetization of a chondrite in which magnetized chondrules and metal particles have accreted at low temperature into a parent body. Any magnetic field present is unlikely to be strong enough to align the chondrules or metal during the accretion process, and if there are no subsequent secondary magnetizations imposed, the randomly-scattered NRM directions of chondrules and matrix fragments will be observed at present. Fig. 1(b) and (c) show the effect of partial and complete remagnetization in a uniform ambient field at some stage during the meteorite's history, for example during cooling after the metamorphic heating event. The degree of scatter of the NRM directions where partial remagnetization has occurred is a measure of the relative magnitude of the secondary NRM. In the absence of an early solar nebula magnetic field, where chondrules and metal are accreted unmagnetized, then only uniform, primary magnetization will be observed if the above later magnetizing event occurs, and there are no subsequent randomizing events. In the absence of both a solar nebula magnetic field and one during the later heating event, no NRM from those sources would be expected, which would also be the case if an NRM acquired earlier was demagnetized by sufficiently severe heating followed by cooling in a zero or only very weak ambient field.

Some evidence has emerged recently (Scott et al. 1985) that brecciation may be rather common in ordinary chrondrites. The mode of origin appears to be the break-up of a parent body through collision some time after the heating event and subsequent re-accretion and lithification of small fragments into a secondary body. Possible consequences of brecciation which might be observed in magnetic studies are shown in Figs. 1(d) and (e). In the former, the brecciated body resulted from fragments in which an early, random magnetization was preserved, while in Fig. 1(e), homogeneity of NRM within different fragments indicates the first break-up occurred after the acquisition of a uniform primary or secondary NRM. In the absence of visual evidence enabling the different fragments to be recognized, the situation of Fig. 1(d) would be difficult to distinguish magnetically from a randomly magnetized non-brecciated meteorite.

2.3 Observations on the magnetization of ordinary chondrites

Early systematic studies of the magnetic properties of ordinary chondrites (Brecher and Ranganayaki 1975, Brecher et al. 1977) were generally exploratory in nature, in that NRM characteristics and bulk magnetic properties were measured with a view to correlating them with petrologic history, metamorphic grade and shock effects. Because of the complex magnetic history of many chondrites and lack of knowledge of the iron-nickel minerals responsible for their magnetism, this approach was not always very successful, and later studies tended to concentrate on intensive examination of specific meteorites, rather then extensive, comparative studies of several samples. The development of very sensitive cryogenic magnetometers enabled a greater number of measurements to be made on the NRM of separated chondrules and small matrix fragments.

Several studies have been reported of the NRM and bulk magnetic properties of the matrix, iron-nickel particles, clasts and chondrules in ordinary chondrites. Potentially important evidence has been obtained from NRM measurements on the above meteorite constituents whose mutual orientation in the meteorite has been maintained, enabling the NRM directions between and among the different constituents to be compared. This procedure is analogous to the conglomerate test of stability in terrestrial palaeomagnetic studies.

The Abee (E4) brecciated chondrite, which contains clasts up to ~5 cm across, has been studied by Sugiura and Strangway (1981). It is strongly magnetized (~0.1 $Am^2 kg^{-1}$) and cohenite (Fe_3C) and kamacite–cohenite intergrowths carry a substantial part of the NRM. Although rather complicated behaviour was observed on demagnetization, the stable NRM in each of six clasts was different in direction and different from that of the surrounding matrix. The evidence is that the clast magnetization is of high temperature origin and pre–dates the formation of the meteorite, whereas the matrix NRM is of low temperature origin, but has been partially remagnetized (Rubin and Keil 1983, Sugiura and Strangway 1983). NRM directions within clasts are approximately uniform, and the NRM appears to have been acquired during rapid cooling after excavation from the clast parent body.

Funaki et al. (1981) examined an Antarctic chondrite, ALHA 76009, and found that the initial NRM of bulk samples was essentially uniform in direction but unstable to a.f. demagnetization. Mutually oriented chondrules showed good stability against a.f. demagnetization with their NRM directions almost randomly distributed in the meteorite. There is evidence here that the chondrules were magnetized prior to the accumulation of the meteorite parent body, and were not subsequently remagnetized in a uniform field. However, this conflicts with the petrologic evidence that ALHA 76009, an L6 chondrite, has undergone severe metamorphism, with maximum temperatures in the range 700–900°C (Dodd 1969, Wasson 1974). If cooling has occurred in the presence of a magnetic field, or in zero field, then the chondrules would undergo remagnetization, uniform in direction, or demagnetization respectively, since the above temperature is high enough to unblock the remanence of kamacite and plessite particles carrying the NRM. This problem is discussed further below.

Among the less metamorphosed chondrites, Bjurbole (L4) has been studied by Sugiura and Strangway (1982). Both chondrules and matrix sub–samples show scattered directions of NRM, although there appears to be a weak overprinting secondary component giving the whole sample a poorly defined mean NRM direction. Type 4 chondrites are believed to have been heated to ~600°C during metamorphism (Dodd 1981), again suggesting that Bjurbole might be expected to have been remagnetized or demagnetized. Chainpur (LL3) and Mezo–Madaras (L3) show scattered directions of primary NRM in small (50–100 mg) sub–samples, as do sub–samples of an L3 Antarctic chondrite, Y–74191.

The present author has made an intensive study of the Olivenza chondrite (LL5) (Collinson 1987). On removal of secondary magnetizations (of uncertain origin), the initially uniform NRM directions in mutually oriented sub–samples diverge to a widely scattered distribution (Fig. 2). Breaking each sub–sample to smaller fragments (down to ~1 mm^3) shows that inhomogeneity of NRM (of high stability) is present down to a very small scale, possibly to individual iron–nickel particles (Fig. 3). Similar inhomogeneity of primary NRM is observed in other LL5 chondrites currently being investigated, namely Khanpur, Oberlin and Cherokee Springs. Only a few chondrules in these samples are fresh and well–rounded, most appearing altered and somewhat broken in appearance, and their NRM has not yet been studied.

Nagata (1983) has studied the magnetic properties of St. Severin (LL6) and the Antarctic chondrites Yamato 74160 (LL6) and ALHA 77260

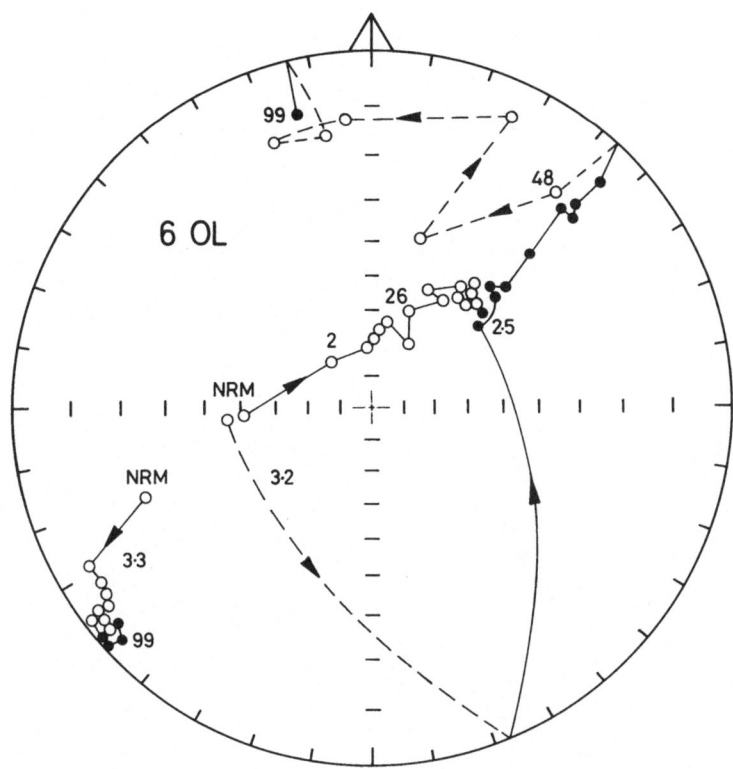

Figure 2. NRM directional changes during a.f. demagnetization of three mutually oriented fragments of Olivenza. Note the widely divergent stable NRM directions.

(L3), with special reference to tetrataenite, which is present in all the samples. All three chondrites contain a highly stable component of NRM carried by tetrataenite, but no information is given on homogeneity or otherwise of the NRM.

2.4 Discussion

Among the aims of studies of the magnetism of ordinary chondrites and other meteorites is the detection of ancient planetary and interplanetary magnetic fields and the obtaining of evidence for meteorite evolutionary processes. In the interpretation of the results described above these two aspects are closely connected in reconciling the widespread occurrence of scattered NRM of chondrules and metal particles in the chondrites with their metamorphic history. As already mentioned, the maximum temperatures which type 4, 5, and 6 ordinary chondrites are believed to have reached during metamorphism

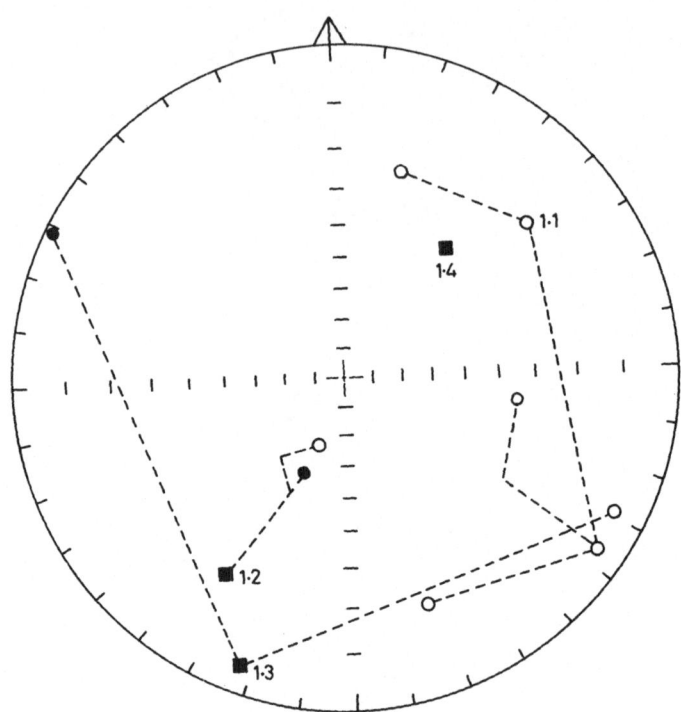

Figure 3. Directions of stable NRM of four mutually oriented sub-samples (numbered) of an Olivenza fragment, and of smaller fragments broken from sub-samples.

~(600°-900°C) are close to, or exceed, the maximum blocking temperatures of the likely magnetic carriers, kamacite, plessite and tetrataenite. Thus demagnetization or uniform remagnetization is expected on cooling, according to the absence or presence of an ambient magnetic field during the cooling period.

Before considering the problem further, it is pertinent to enquire about the time of formation of tetrataenite in chondrites, since it is clear that it is an important carrier of primary NRM. The metal grains condensing out of the nebula would be essentially of kamacite composition (Wood 1967). Later condensates may have had a higher Ni content and migration of nickel may then have formed taenite regions in the grains of appropriate composition for later ordering to tetrataenite. This is believed to have occurred during very slow cooling at depth through the ordering temperature (~320°C) in the meteorite parent body, after post-accumulation heating. If the

tetrataenite carries the presently-observed primary NRM and this NRM
is of nebula magnetic field origin, it appears necessary to postulate
that the NRM was first acquired by original kamacite grains and then
persisted through nickel content changes and ordering to tetrataenite.
There is some doubt about the validity of this process, evidence for
which must await further investigations of the properties of
tetrataenite.

Other possible causes of very scattered NRM directions of metal
particles and/or very small rock fragments are anisotropy and
brecciation. Tetrataenite has a high susceptibility anisotropy, and
it is possible that a thermoremanent magnetization acquired in a
uniform field either in tetrataenite or pre-existing kamacite or
taenite could be pulled out of parallelism with the field towards the
easy axis of particle magnetization. However, this mechanism could
not result in the TRM direction in any particle being more than 90°
away from the field direction, and thus cannot explain the Olivenza
observations of essentially random NRM directions.

Scott et al. (1985) adduce evidence, based on differences in
metamorphic grade among minerals and rock fragments within chondrites,
that many ordinary chondrites may be brecciated. If this is the
explanation of the Olivenza data, the constituent fragments must be
very small, and there are as yet no data of the above nature that
Olivenza is brecciated: it may also be relevant that Olivenza is only
lightly shocked (Ashworth and Barber 1975).

Assuming that Olivenza is not brecciated, it is tentatively
concluded that it and possibly other ordinary chondrites record a
pre-accumulation magnetic field. It is not at present clear how this
can be reconciled with accepted temperatures reached during
metamorphic heating, an intriguing problem which must await further
research for its elucidation.

An associated problem is whether chondrite parent bodies
possessed intrinsic magnetic fields at any stage in their evolution.
The existence of iron and stony-iron meteorites is generally accepted
as evidence for melting and differentiation within some parent bodies,
and the existence in them of a molten iron core, in which in principle
a magnetic field could be generated. Uncertainties about small core
sizes and rotation rates and the limitations of current dynamo theory
make it difficult to assess the likelihood of such fields, but current
opinion is probably against their existence.

Other possible magnetic fields post-dating parent body
accumulation could be of external origin. Depending on the time since
parent body formation, the solar magnetic field could still have been
of significant strength, or there could be weak field amplification
associated with collision processes.

Any acquisition of NRM in these fields by chondrite material most
likely occurred during cooling after metamorphic heating or during a
shock event. If the metamorphism occurred in a partially
differentiated body in which a core-generated magnetic field existed,
a TRM could be acquired, providing the field existed at the same time
as the material cooled and for long enough so that cooling in the
field took place through a reasonable range of blocking temperatures.

Some theories of chondrite formation envisage metamorphism as a result of external heating of small planetesimals prior to their accumulation into a parent body (Wasson 1972). If this occurred in the presence of an ambient magnetic field (of probable solar origin), a TRM could be acquired during cooling of the planetesimal. However, rotation (or tumbling) and rapid orbital motion of the planetesimal in the field would result in a reduced, or even zero, TRM acquired over an extended period.

There is no firm evidence concerning the importance or otherwise of shock effects in chondrite magnetism. It is known (Cisowski et al. 1975) that in the presence of a magnetic field shock of sufficient magnitude can impart a shock remanent magnetization (SRM) to magnetic minerals: in the absence of a field, any NRM already present can be demagnetized. Another possible acquisition process involves impact heating of surrounding material and TRM acquisition on cooling in a magnetic field. With a sufficiently high shock level (which is difficult to specify) a prior NRM can be completely overprinted by the new SRM.

If any of the above magnetizing processes occur they are expected to impart, over a length scale of up to tens of metres, a uniformly-directed NRM to the parent body or planetesimal. Where uniform NRM is observed in chondrite fragments, or scattered but not random NRM directions (suggesting the presence of a random primary magnetization with a superimposed, weaker, uniform secondary NRM), its most likely origin appears to be a form of TRM or SRM acquired in an external field.

Although it is of considerable interest to determine the strength of meteorite magnetizing fields (palaeointensities), it is clear from the foregoing discussions that there are difficulties where ordinary chondrites are concerned. These arise from the possible existence of inhomogeneous primary NRM, and of the tetrataenite which may carry it, for which the origin and magnetization process is far from clear. Testing for NRM homogeneity would appear to be essential prior to any palaeointensity determination, particularly in L and LL chondrites. Those in which some or all of the NRM is carried by tetrataenite should be avoided. It should be noted that complications are also possible both in thermal magnetization processes involving kamacite, taenite and plessite (Wasilewski 1974) and when these minerals are re-heated during Thellier palaeointensity investigations. For all these reasons, the present author feels that Thellier palaeointensity results from ordinary chondrites reported by various workers (Nagata and Sugiura 1977, Nagata 1979, Nagata and Funaki 1981, 1982, Sugiura and Strangway 1982 and Westphal and Whitechurch 1983) should be treated with extreme caution pending greater understanding of chondrite history and magnetization processes. Reported palaeointensities from ordinary chondrites are typically in the range 10–100 μT (see Cisowski 1987 for useful review), but without improved knowledge of the magnetization history of the specific chondrites, it is not generally possible to assign the values to the very early nebula field or a later field present during metamorphism or shock events.

3. MAGNETISM OF SNC METEORITES

3.1 Magnetic properties

One of the most interesting recent developments in meteorite studies
has been the investigation of shergottite, nakhlite and chassignite
(SNC) achondrite (basaltic) meteoritics and the emergence of evidence
that their parent body may be the planet Mars (Wood and Ashwal 1981,
Bogard et al. 1984, McSween 1985). There is still some controversy
about the significance of the data, but if these meteorites are of
Martian origin, their petrological and physical (including magnetic)
properties offer the opportunity of deriving evidence for Martian
history and evolutionary processes. Any remanent magnetism possessed
by the SNC meteorites may be evidence of the existence of a Martian
magnetic field at the time when they experienced a magnetizing event,
e.g. through heating or shock. The present Martian magnetic field is
very weak and not well characterized (Russell 1979) and any evidence
of a past Martian field is of considerable interest because of its
possible origin in a molten, electrically conducting core.

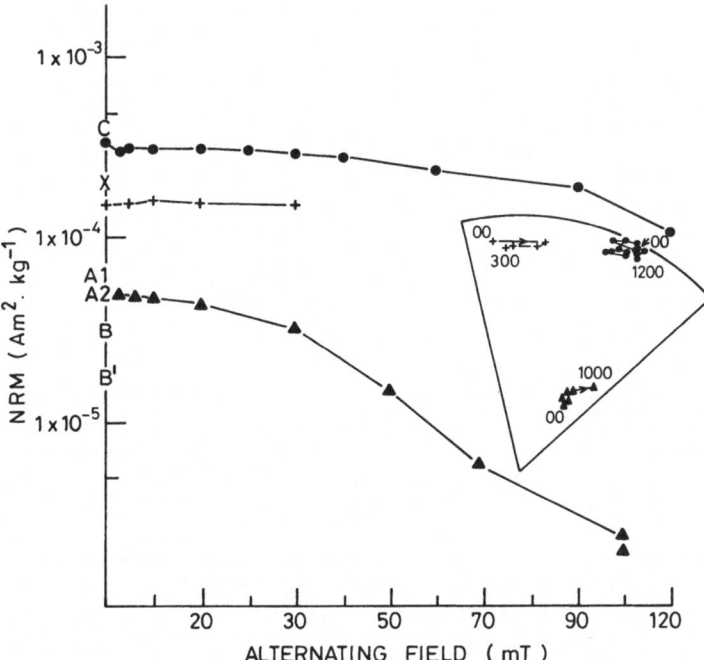

Figure 4. Alternating field demagnetisation of Shergotty sub-samples.
(Reprinted with permission from Geochimica et Cosmochimica Acta. 50,
S.M. Cisowski, 'Magnetic studies on Shergotty and other SNC
meteorites', copyright 1986, Pergamon Press plc.)

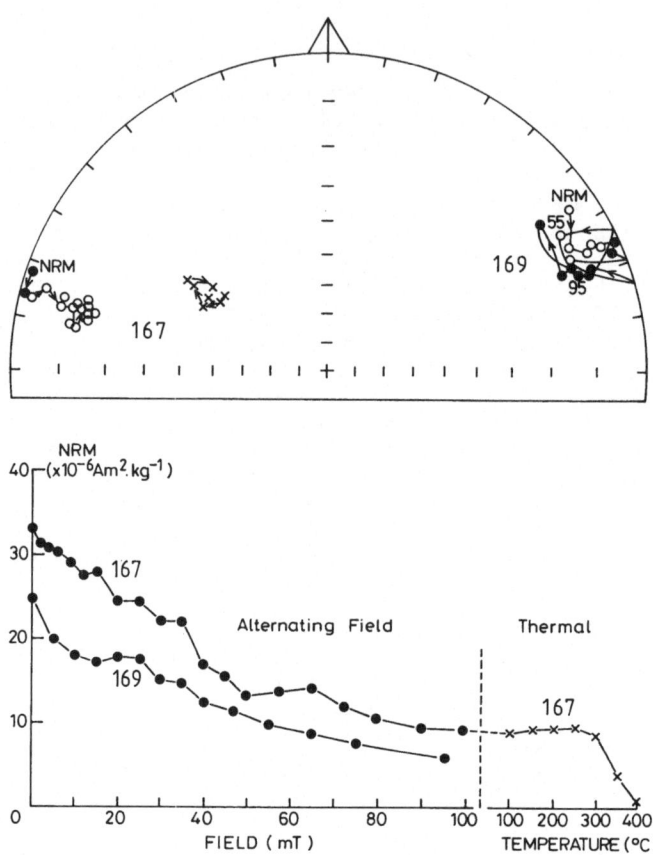

Figure 5. Alternating field (● o) and thermal (x) demagnetization of two samples of EETA 79001. Fragment 167 was thermally demagnetized after a.f. demagnetization to 99 mT. The change in direction of 167 prior to thermal demagnetization was caused by difficulty in accurate orientation of the sample in a different holder.

At the present time there are 8 documented SNC meteorites, 4 shergottites, 3 nakhlites and 1 chassignite, named after the type samples from Shergotty (India), Nakhla (Egypt) and Chassigny (France). Their dominant magnetic minerals, with the exception of the Antarctic shergottite ALHA 77005, are single and multidomain titanomagnetites containing varying proportions of ulvospinel, with Curie points between 580°C (magnetite) and 150°C (65% Fe_2TiO_4) (Cisowski 1986). In ALHA 77005, chromite appears to be the dominant magnetic phase (McSween et al. 1979) and superparamagnetic particles appear to be present (Nagata 1980, Collinson 1986).

The NRM of SNC meteorites lies in the range (1-35) x $10^{-6}Am^2kg^{-1}$. Their demagnetization characteristics are somewhat variable, both within and between members (Cisowski 1986). However directional stability is generally good particularly in Shergotty (Fig. 4) (Cisowski 1986) and in two fragments of the Antarctic shergottite EETA 79001 (Fig. 5) (Collinson 1986). ALHA 77005 shows only a poorly stable NRM, with some movement of the vector during demagnetization. The dominant NRM in this meteorite is likely to be a viscous magnetization (VRM), acquired in the geomagnetic field since it reached the Earth's surface and during storage.

Thermal demagnetization of NRM of a Shergotty sub-sample showed a decay curve in which 80% of the NRM was blocked in below 135°C, similar to an earlier result from another Shergottite, Zagami (Cisowski 1986). A fragment of EETA 79001, after alternating field demagnetization to 30% of its initial intensity in 100 mT, showed blocking temperatures lying between 300°C and 400°C, with no directional changes (Fig. 5) (Collinson 1986).

Although not universally present in SNC meteorites, there is evidence from a.f. demagnetization of a stable, probably primary NRM in EETA 79001, Nakhla, Governador Valadares (a nakhlite) and some fragments of Shergotty. The few results of thermal demagnetization are more ambiguous, EETA 79001 indicating relatively high stability and Shergotty and Zagami chips showing a low maximum blocking temperature. It is possible that some of the observed NRM is of viscous origin but in EETA 79001 at least it cannot explain the observed intensity (based on the laboratory-determined viscosity coefficient and estimated residence time on Earth of 10^4-10^5 years) or the high a.f. stability. VRM probably dominates in the samples of ALHA 77005 so far studied.

It is tentatively concluded that at least part of the NRM observed in SNC meteorites was acquired in an ambient magnetic field on the parent body, either by a thermoremanent process or as a result of shock.

3.2 Discussion

There is as yet no reliable evidence concerning the strength of the SNC magnetizing field. Cisowski (1986) reports an attempt to carry out a Thellier-Thellier palaeointensity determination on a Shergotty sub-sample, giving poorly-defined palaeointensities of 2000 nT in the range 20°-134°C and 250-600 nT in the range 338°-590°C. Some thermal alteration of magnetic carriers contributes to the difficulty of interpreting the result. During an attempted Thellier determination by the present author on a fragment of EETA 79001 a problem arose in that a PTRM acquired in cooling from a certain temperature was apparently not completely removed on re-heating to the next higher temperature. Nagata (1980) used an ARM method (Stephenson and Collinson 1974) on ALHA 77005, but his result of 1000 nT must be considered of doubtful validity because of the uncertainty of the origin of the NRM and the use of the ARM method in rocks in which titanomagnetites carry the remanence, for which the ARM technique has not been proven. The above results, together with other evidence based on the ratio of NRM to saturated IRM, the latter being used as a normalizing factor for magnetic mineral content, suggests that the SNC magnetizing field was of 1000 nT in order of magnitude.

The igneous crystallisation ages of the SNC meteorites are ~1.3 Ga or less, and their characteristics are not consistent with an impact melt origin (McSween 1985). Asteroidal bodies could not have undergone differentiation so late in Solar System history, indicating that the SNC parent body was planet-sized. There is evidence in Shergotty and EETA 79001 of a severe (~300 kbar) shock event at ~180 Ma, and SNC magnetization could have occurred during either the igneous cooling period or the shock event, presumably when the meteorites were ejected from Mars. It is also possible that an earlier TRM was partially or completely overprinted by a shock magnetization acquired during the 180 Ma event. Cisowski (1986) is of the opinion that it is unlikely that any prior magnetization would survive the shock event, and that the weak, high temperature field derived from the Thellier determination on Shergotty reflects the ambient magnetic field at that time.

The present Martian magnetic field is very weak ($<$~50 nT) (Russell 1980). The directional stability of the NRM of Shergotty and EETA 79001 suggests a single magnetizing event, but, at present, firm evidence is lacking as to the time of acquisition. However, the evidence that a significant field existed on Mars at some period during the last billion years is of considerable interest through its possible implication of past dynamo action in a Martian core. It may be stated with some conviction that the observed NRM of Shergotty, EETA 79001 and Nakhla could not have been acquired in space or during residence on the Earth's surface. Hopefully, further SNC meteorites will be found (or even fall) in the future so that further evidence can be accumulated concerning both their Martian origin and a Martian magnetic field.

Another source of Martian rocks and dust will result from a Mars sample return mission, now being actively discussed. Our present knowledge of the magnetic properties of the Martian surface material comes from experiments carried out by the Viking landers, which showed the presence of a strongly magnetic mineral, probably maghemite (γ-Fe_2O_3) (Hargraves et al. 1977, 1979). This is most probably an oxidation product of magnetite-bearing basalt, a process for which there is evidence from Earth-based spectral reflectance studies of Mars (Adams and McCord 1969). Although it may be possible to do some NRM studies on maghemite particles, or maghemite-coated particles, clearly the original basalt is more likely to carry a primary NRM acquired in a Martian field, although a comparison between maghemite NRM and basalt NRM would be of some interest. It is to be hoped that any sample return will include some basaltic pebbles or fragments, although it may be significant that acquisition of such material up to ~1.2 cm across for inorganic analysis during the Viking programme was unsuccessful (Moore et al. 1977), possibly suggesting that in this size range pebble-like objects are either absent or highly weathered and friable.

4. REFERENCES

Adams, J.B. and McCord, T.B., 1969. 'Mars: interpretation of spectral reflectivity of light and dark regions'. J. Geophys. Res., 74, 4851-4856.

Ashworth, J.R. and Barber, D.J., 1975. 'Electron petrography of shock-deformed olivine in stony meteorites'. Earth Planet. Sci. Lett., 27, 43-50.

Bogard, D.D., Nyquist, L.E. and Johnson, P., 1984. 'Noble gas contents of shergottites and implications for the Martian origin of SNC meteorites'. Geochim. Cosmochim. Acta., 48, 1723-1729.

Brecher, A. and Ranganayaki, R.P., 1975. 'Palaeomagnetic systematics of ordinary chondrites'. Earth Plan. Sci. Lett., 25, 57-67.

Brecher, A., Stein, J. and Fuhrman, M., 1977. 'The magnetic effects of brecciation and shock in meteorites: 1. The LL chondrites'. The Moon, 17, 205-216.

Cisowski, S.M., 1986. 'Magnetic studies on Shergotty and other SNC meteorites'. Geochim. Cosmochim. Acta., 50, 1043-1048.

Cisowski, S.M., 1987. 'Magnetism of meteorites'. In Geomagnetism Vol. 2, Ed. by J.A. Jacobs, Academic Press, New York.

Cisowski, S.M., Fuller, M.D., Wu, Y.M., Rose, M.F. and Wasilewski, P.J., 1975. 'Magnetic effects of shock and their implications for magnetism of lunar samples'. Proc. 6th Lun. Sci. Conf., 3, 3123-3142.

Clarke, R.S. and Scott, E.R.D., 1980. 'Tetrataenite-ordered Fe-Ni, a new mineral in meteorites'. Am. Min., 65, 624-630.

Collinson, D.W., 1986. 'Magnetic properties of Antarctic shergottite meteorites EETA 79001 and ALHA 77005: possible relevance to a Martian magnetic field'. Earth Planet. Sci. Lett., 77 159-164.

Collinson, D.W., 1987. 'Magnetic properties of the Olivenza meteorite - possible implications for its evolution and an early Solar System magnetic field'. Earth Planet. Sci. Lett., 84, 369-380.

Dodd, R.T., 1969. 'Metamorphism of the ordinary chondrites: a review'. Geochim. Cosmochim. Acta., 33, 161-203.

Dodd, R.T., 1981. 'Meteorites, a petrologic-chemical synthesis', Cambridge University Press, 368 pp.

Funaki, M., Nagata, T. and Momose, K., 1981. 'Natural remanent magnetization of chondrules, metallic grains and matrix of an Antarctic chondrite ALHA 76009'. Proc. 6th Symp. Antarctic Meteorites, Nat. Inst. Polar Res., 300-315.

Hargraves, R.B., Collinson, D.W., Arvidson, R.E. and Spitzer, C.R., 1977. 'The Viking magnetic properties experiment: primary mission results'. J. Geophys. Res., 82, 4547-4558.

Hargraves, R.B., Collinson, D.W., Arvidson, R.E. and Cates, P.M., 1979. 'Viking magnetic properties experiment: extended mission results'. J. Geophys. Res., 84, 8379-8384.

McSween, H.Y., Taylor, L.A. and Stolper, E.M., 1979. 'Allan Hills 77005: a new meteorite type found in Antarctica'. Science, 204, 1201-1203.

McSween, H.Y., 1985. 'SNC meteorites: clues to Martian petrologic evolution?'. Rev. Geophys., 23 391-416.

Moore, H.J., Hutton, R.E., Scott, R.F., Spitzer, C.R. and Shorthill, R.W., 1977. 'Surface materials of the Viking landing sites'. J. Geophys. Res., 82, 4497–4523.

Nagata, T., 1979. 'Meteorite magnetism and the early Solar System magnetic field'. Phys. Earth. Planet. Inter., 20, 324–341.

Nagata, T., 1980. 'Palaeomagnetism of Antarctic achondrites'. Mem. Nat. Inst. Polar Res., Special Issue 17, 233–242.

Nagata, T., 1983. 'High magnetic coercivity of meteorites containing the ordered FeNi (tetrataenite) as the major ferromagnetic constituent'. Proc. 13th Lun. Planet. Sci. Conf., J. Geophys. Res., 88, A779A784.

Nagata, T. and Sugiura, N., 1977. 'Palaeomagnetic field intensity derived from meteorite magnetization'. Phys. Earth Planet. Inter., 13, 373–393.

Nagata, T. and Funaki, M., 1981. 'The comparison of an Antarctic chondrite, ALHA 76009 (L6)'. Proc. 12th Lun. Planet. Sci. Conf., 2, 1229–1241.

Nagata, T. and Funaki, M., 1982. 'Magnetic properties of tetrataenite-rich stony meteorites'. Mem. Nat. Inst. Polar. Res., Special Issue 25, 222–250.

Néel, L., Paulevé, J., Pauthenet, R., Laugier, J. and Dautreppe, D., 1964. 'Magnetic properties of an iron-nickel single crystal ordered by neutron bombardment'. J. Appl. Phys., 35, 873–876.

Rubin, A.E. and Keil, K., 1983. 'Mineralogy and petrology of the Abee enstatite chondrite breccia and its dark inclusions'. Earth Planet. Sci. Lett., 62, 118–131.

Russel, C.T., 1979. 'Planetary magnetism'. Rev. Geophys. Space Phys., 17, 295–301.

Scott, E.R.D., Lusby, D. and Keil, K., 1985. 'Ubiquitous brecciation after metamorphism in equilibrated ordinary chondrites'. Proc. 16th Lun. Planet. Sci. Conf., J. Geophys. Res., 90, D137–D148.

Stephenson, A. and Collinson, D.W., 1974. 'Lunar magnetic field palaeointensities determined by an anhysteretic remanent magnetization method'. Earth Planet. Sci. Lett., 23, 220–228.

Sugiura, N. and Strangway, D.W., 1981. 'The magnetic properties of the Abee meteorite: evidence for a strong magnetic field in the early Solar System'. Proc. 12th Lunar. Planet. Sci. Conf., 12, 1243–1256.

Sugiura, N. and Strangway, D.W., 1982. 'Magnetic properties of low-petrologic grade non-carbonaceous chondrites'. Proc. 7th Symp. Antarctic meteorites, 260–280.

Sugiura, N. and Strangway, D.W., 1983. 'A palaeomagnetic conglomerate test using the Abee E4 meteorite'. Earth Planet. Sci. Lett., 62, 169–179.

Wasilewski, P., 1974. 'Magnetic remanence mechanisms in iron and iron-nickel alloys, metallographic recognition criteria and implications for lunar science research'. The Moon, 9, 335–354.

Wasson, J.T., 1972. 'Formation of ordinary chondrites'. Rev. Geophys. Space Phys., 10, 711–759.

Wasson, J.T., 1974. 'Meteorites – Classification and Properties'. Springer Verlag, Berlin, 316 pp.

Wasson, J.T., 1985. 'Meteorites – their record of early Solar System history'. Freeman & Company, New York, 268 pp.

Westphal, M. and Whitechurch, H., 1983. 'Magnetic properties and palaeointensity determination of seven H-group chondrites'. Phys. Earth Planet. Inter., 31, 1-9.

Wood, C.A. and Ashwal, L.D., 1981. 'SNC meteorites: igneous rocks from Mars?'. Proc. 12th Lun. Planet. Sci. Conf., 12, 1359-1375.

GEODYNAMO THEORY

A.M. Soward
School of Mathematics
The University
Newcastle upon Tyne, NE1 7RU
U.K.

ABSTRACT.   Theoretical aspects of geodynamo models are outlined.   The importance of the α and ω-effects of mean field dynamo theory is explained.   The controversy concerning whether the Earth's dynamo is of α^2 or $\alpha\omega$-type is discussed.   Recent results point to an $\alpha\omega$-dynamo in which the primary force balance is between Coriolis, pressure and Lorentz forces.   Accordingly the magnetic field is strong of order $(\rho\Omega/\sigma)^{1/2}$ [fluid density ρ; electrical conductivity σ; rotation rate Ω], and, being large compared with the meridional magnetic field emerging at the Earth's surface, it is predominantly azimuthal.   If inertia and viscosity are ignored completely, the azimuthal component of the Lorentz force satisfies an integral constraint [Taylor's condition].   With an appropriate choice of geostrophic flow, the induced magnetic field might meet Taylor's condition but whether this is so for the Earth is a controversial matter yet to be resolved.

1.   INTRODUCTION

Dynamo theory has now developed to a high level of theoretical sophistication and many of the basic ideas are clearly explained in standard textbooks [see for example Moffatt, 1978 and Parker, 1979]. Motion in the Earth's core necessary to operate the dynamo is widely believed to be driven by buoyancy forces, and current opinion favours compositional effects for the energy source [Fearn, 1988].   Since compositional convection is complicated and not well understood, attempts at constructing consistent dynamic dynamo models have largely focused on the simpler mechanism of thermal convection.

Thermal convection in a rapidly rotating system like the Earth is asymmetric and well endowed with the helicity necessary for the dynamo process.   The three-dimensional complexity of the dynamo problem is usually simplified by dividing the problem into two parts.   First, equations governing the mean axisymmetric flow velocity and magnetic field are considered.   They involve terms which arise from the nonlinear interactions of the asymmetric motion and magnetic field averaged azimuthally.   This leads to mean-field dynamo theory [see for

F. J. Lowes et al. (eds.), Geomagnetism and Palaeomagnetism, 297–317.
© *1989 by Kluwer Academic Publishers.*

example Krause and Rädler, 1980] from which important concepts like
the α-effect emerge. Second, the dynamics is considered, to determine
the detailed nature of the asymmetric thermal convection. Often the
asymmetric motions occur on small scales and may be of a turbulent or
random wave character. For this reason, the structure of the
asymmetric fluctuating part is sometimes referred to as the
"microdynamics", while the large scale axisymmetric mean part is
called the "macrodynamics" [see for example Malkus and Proctor, 1975].
Of course, a completely self-consistent model of the geodynamo
involves the simultaneous solution of the equations governing both the
micro and macrodynamics. That requires large numerical computations
which are now just becoming feasible. Indeed Zhang and Busse (1988)
have recently made significant progress in that direction.

This paper outlines the development of geodynamo theory. In
Section 2, kinematic dynamo theory, which introduces the important
concepts of the α- and ω-effects, is summarised. Discussion of the
macrodynamics in Section 3 emphasises the contributions to the
ω-effect made by the thermal wind and geostrophic flow. The latter is
linked to an important [but controversial] constraint on the magnetic
field known as Taylor's (1963) condition [see (23) below]. Though
some models satisfy this condition, others, in particular Braginsky's
(1976) Model-Z, only satisfy a modified version [see (25) below]. The
macrodynamics discussed in Section 4 focusses attention on thermal
convection. The importance of the dynamical balance, in which
Coriolis, Lorentz and pressure forces are comparable dominating all
others, is stressed. The concluding Section 5 brings together the
ideas of the earlier sections to produce a picture of the complete
convectively driven dynamo. From our current understanding it is
tentatively concluded that the Earth is an $\alpha\omega$-dynamo with a strong
azimuthal magnetic field, large compared to the meridional magnetic
field emerging at the Earth's surface.

Many of the ideas summarised here are explained in greater depth
in Gubbins and Roberts (1988), Roberts and Gubbins (1988), Roberts
(1988a), and Fearn, Roberts and Soward (1988). They complement this
article. An appraisal of current results and possible future
directions of research are surveyed by Roberts (1988b).

The reader may find the following brief summary of subscript
notation used frequently in the following sections helpful for future
reference. We use B for Boundary, P for Poloidal [= meridional for
our axisymmetric fields], M for Magnetic, A[T] for Archimidean
[Thermal], C[G] for Coriolis [Geostrophic], E for Ekman, Ta for
Taylor, and c for critical. The sense of usage will become clear as
it is introduced, but, as an example, we have waves of frequency ω_{MAC}
[see (35)], and an angular velocity of magnitude $\omega_T + \omega_M + \omega_G$ [see
(18)].

2. KINEMATIC DYNAMOS

The generation of magnetic field, **B**, in an electrically conducting
fluid of magnetic diffusivity, $\eta [=(\sigma\mu)^{-1}]$, moving with velocity, **v**, is

governed by the magnetic induction equation,

$$\partial B/\partial t = \nabla \times (v \times B) + \eta \nabla^2 B, \qquad [\nabla \cdot B = 0]. \tag{1}$$

In the absence of motion $[v = 0]$, equation (1) reduces to a heat conduction equation and in the absence of any external electric current sources, the magnetic field decays. On the other hand, in a perfectly conducting fluid $[\eta = 0]$ magnetic field lines are material lines which move frozen to the fluid. Moreover, in an incompressible fluid $[\nabla \cdot v = 0]$, the magnetic field strength is proportional to the length of the material lines and so increases when two neighbouring fluid particles on that line separate. Intensification of magnetic field by this process is a simple matter. Nevertheless, in a fluid of finite electrical conductivity, Ohmic diffusion provides a competing process, by which the intensification of magnetic field is offset. The relative importance of the two competing effects in a fluid with typical velocity and length scales, U and L, is measured by the magnetic Reynolds number,

$$R_\omega = UL/\eta. \tag{2}$$

It might be anticipated that magnetic induction at sufficiently large R_ω would overcome Ohmic diffusion. Unfortunately that is not enough to ensure dynamo activity, as many of the antidynamo theorems show. Indeed the impossibility of an axisymmetric dynamo is now explained.

Consider an axisymmetric system in which z and s measure distance parallel to and normally from the axis of symmetry. This cylindrical polar coordinate system is completed by adding the azimuthal angle ϕ. The axisymmetric motion and magnetic field are expressed in terms of their azimuthal [V and B] and meridional $[v_p$ and $b_p]$ parts in the form,

$$\overline{V} = V\hat{\phi} + v_p \,, \qquad \overline{B} = B\hat{\phi} + b_p \,, \qquad [b_p = \nabla \times A\hat{\phi}], \tag{3}$$

where the hat is used here and subsequently to define unit vectors. From (1) the evolution of B and A [the magnetic potential for the meridional magnetic field] satisfy

$$sD(B/s)/Dt = sb_p \cdot \nabla \omega + \eta \Delta B, \qquad [\omega = V/s], \tag{4a}$$

$$s^{-1}D(sA)/Dt = \eta \Delta A, \tag{4b}$$

where

$$\Delta \equiv \nabla^2 - s^{-2}, \quad \text{and} \quad D/Dt \equiv \partial/\partial t + v_p \cdot \nabla \tag{4c}$$

is the material derivative. Remember that A and B are independent of ϕ and so cannot experience advection by the azimuthal motion, V. According to (4b), the total magnetic flux, $2\pi sA$, is conserved in a perfectly conducting fluid $[\eta = 0]$ but decays when the effects of finite diffusivity $[\eta \neq 0]$ are included. The resulting absence of meridional magnetic field, b_p, in (4a) heralds the collapse of B and the failure of the dynamo. When meridional magnetic field is present, however, deviations of the azimuthal motion from rigid rotation [i.e. $\nabla\omega \neq 0$] can pull that field out into the azimuthal direction. This is the well known ω-effect, which is a fundamental ingredient of the dynamo process.

The failure of the axisymmetric dynamo is a consequence of Cowling's (1957) well known antidynamo theorem and can be avoided when asymmetries are added. In that case the flow and magnetic field are expressed in the form,

$$V = V + v', \qquad B = \overline{B} + b', \tag{5}$$

where the bar denotes azimuthal average [as in (3)] and the prime identifies the remaining asymmetric part. Such a resolution of the dynamo problem was proposed by Braginsky (1964a,b) who argued that, if the azimuthal magnetic field and flow were large [$R_\omega \gg 1$], only a small asymmetry [$|v'/V| = O(R_\omega^{-1/2}) = |b'/B|$] would be necessary to complete the dynamo process. Specifically he showed that the mean electromotive force caused by the asymmetries is

$$\overline{v' \times b'} = \alpha B \hat{\phi}, \tag{6}$$

where $\alpha[= O(R_\omega^{-1}\eta/L)]$ is an axisymmetric function of position which measures the helicity of the flow [Soward, 1972]. The term αB is now added to the right of (4b) and, though small, is sufficient to generate meridional magnetic field, b_p, of order $R_\omega^{-1}B$. This is the α-effect which completes the links in the dynamo process. Indeed provided the meridional motion v_p is no larger than order $R_\omega^{-1}V$, the coupled equations (4a and b) are ordered consistently and provide an example of a nearly axially-symmetric $\alpha\omega$-dynamo.

An alternative basis for the α-effect emerges from two scale analysis [see for example Moffatt, 1978, and Krause and Radler, 1980]. Omitting details the idea is that small scale helical motions lead to the mean electromotive force

$$\overline{v' \times b'} = \alpha B \tag{7}$$

[similar to (6)], where the average is now taken over the short length scale, and α is closely related to the mean helicity, $\overline{v' \cdot (\nabla \times v')}$. The most striking feature of this isotropic α-effect is that the mean electromotive force is aligned parallel to $\overline{B}$, whereas $V \times B$ is necessarily perpendicular to B! When there is no mean motion the average of (1) becomes

$$\partial\overline{B}/\partial t = \nabla \times \alpha\overline{B} + \eta\nabla^2\overline{B}. \tag{8}$$

With axisymmetric meridional motion, v_p, added [but no differential rotation, $\omega = 0$], (8) yields the two equations,

$$sD(B/s)/Dt = -s^{-1}\nabla\alpha \cdot \nabla(sA) - \alpha\Delta A + \eta\Delta B, \tag{9a}$$

$$s^{-1}D(sA)/Dt = \alpha B + \eta \Delta A, \tag{9b}$$

analogous to (4a and b). Now, however, the α-effect provides the source for both the axisymmetric azimuthal and meridional magnetic fields, B and $\nabla \times A\hat{\phi}$. This α^2-dynamo will generally operate when the magnetic Reynolds number,

$$R_\alpha = \alpha L/\eta, \tag{10}$$

is of order unity. Then the meridional and azimuthal magnetic fields produced are of comparable size $[|\mathbf{b}_p| = O(B)]$.

When azimuthal flow, V, is included, the term $s\mathbf{b}_p \cdot \nabla \omega$ must be added to (9a). If the magnetic Reynolds number R_ω, based on V, is large, the α-effect in (9a) is negligible in comparison with the ω-effect and the dynamo is again of $\alpha\omega$-type. It operates when the Dynamo number,

$$D = R_\omega R_\alpha, \tag{11}$$

is of order unity. The distinction between α^2 and $\alpha\omega$-dynamos is central to our understanding of the geodynamo. Since only meridional magnetic field can be measured directly, the magnitude of the azimuthal magnetic field, B, in the core is linked to the magnitude of R_ω and/or R_α. So for an $\alpha\omega$-dynamo B is of order $R_\omega |\mathbf{b}_p|$.

3. MACRODYNAMICS

3.1. Braginsky's Model-Z

Motion in a rapidly rotating system is influenced greatly by the constraints of the Proudman-Taylor theorem. In particular when the Coriolis force balances the pressure gradient [geostrophy],

$$2\Omega \times \mathbf{V}_G = -\nabla p, \quad [\Omega = \Omega\hat{\mathbf{z}}], \tag{12}$$

the pressure p, and consequently the geostrophic velocity $\mathbf{V}_G$, are independent of z. [Note that $\Omega \cdot \nabla p = 0$]. So, if we model the Earth's core by a spherical cavity of radius L, we find that motion must follow geostrophic contours, s = constant. They, in turn, define geostrophic cylinders, $-z_B < z < z_B$, of constant height, $2z_B$, between the top and bottom boundaries of the cavity, $z = \pm z_B(s)$. This means that geostrophic motion in a sphere has to take the form

$$\mathbf{V}_G = s\omega_G\hat{\phi} = V_G\hat{\phi}, \tag{13}$$

in which the magnitude of $V_G(s)$ remains undetermined until secondary forces not included in the primary force balance (12) are considered.

We now build up a picture of the axisymmetric magnetic field and flow upon which the Braginsky (1976) $\alpha\omega$-dynamo model depends. We restrict attention to a Boussinesq fluid for which density variations are ignored except where they give rise to buoyancy forces. If, for simplicity, we suppose that the buoyancy force is thermal in origin and that $T(s,z)$ defines departures of the temperature distribution from its spherically symmetric state, then the motion $\mathbf{V}_T$ satisfies

$$2\boldsymbol{\Omega} \times \mathbf{V}_T = -\nabla(p/\rho) - \gamma T\mathbf{g}, \tag{14}$$

where $\mathbf{g}$ is gravity and γ is the coefficient of expansion. The curl of (14) eliminates the pressure and yields

$$2\Omega s(\partial\omega_T/\partial z) = \gamma(\mathbf{g} \times \nabla T)_\phi, \quad [\mathbf{V}_T = s\omega_T\hat{\boldsymbol{\phi}}]. \tag{15}$$

For given $T(s,z)$, the thermal wind, $\mathbf{V}_T$, is obtained by integrating (15) with respect to z. The constant of integration is simply the geostrophic flow (13).

When in addition the fluid is permeated by the strong azimuthal magnetic field, $B(s,z)\hat{\boldsymbol{\phi}}$, the associated Lorentz force is

$$\mathbf{j} \times \mathbf{B} = -\nabla p_M - (B^2/\mu s)\hat{\mathbf{s}}, \tag{16}$$

where $p_M = \frac{1}{2}B^2/\mu$ is the magnetic pressure. The radial magnetic force, $-B^2/\mu s$, emerges from the combined effects of tension and curvature in the magnetic field lines. Just as in the case of the thermal wind mentioned in the previous paragraph, this radial force drives a magnetic wind, $s\omega_M\hat{\boldsymbol{\phi}}$, of magnitude

$$\omega_M = (2\rho\mu\Omega)^{-1}(B/s)^2. \tag{17}$$

Together with the thermal wind, $s\omega_T$, and geostrophic flow, $s\omega_G$, the total contribution to the ω-effect in (4a) becomes

$$\omega = \omega_T + \omega_M + \omega_G. \tag{18}$$

Next we consider the influence of the weaker meridional magnetic field, $\mathbf{b}_p$, which drives meridional motion, $\mathbf{v}_{MP}$. The combination of the azimuthal and meridional magnetic fields leads to the ϕ-component of Lorentz force,

$$(\mathbf{j} \times \mathbf{B})_\phi = (\mu s)^{-1}(\mathbf{b}_p\cdot\nabla)(sB) = (\mu s)^{-1}\nabla\cdot(sBb_p), \tag{19}$$

which is balanced by the ϕ-component of the Coriolis force, $2\rho\Omega v_{Ms}$. Thus the radial velocity, v_{Ms}, is

$$v_{Ms} = (2\rho\Omega)^{-1}(\mathbf{j} \times \mathbf{B})_\phi. \tag{20}$$

Whence the axial velocity, v_{Mz}, is readily obtained from the continuity equation,

$$\partial v_{Mz}/\partial z = -s^{-1}\partial(sv_{Ms})/\partial s. \tag{21}$$

An interesting consequence of (20) emerges when the radial volume flux out across geostrophic cylinders,

$$J(s) = \int_{-z_B}^{z_B} \int_0^{2\pi} v_{Ms} s d\phi dz, \qquad (22)$$

is considered. Since this flux is zero by mass continuity, (20) yields the result,

$$\int_{-z_B}^{z_B} \int_0^{2\pi} (j \times B)_\phi \, d\phi dz \; [= (2\rho\Omega)J(s)] = 0, \qquad (23)$$

known as Taylor's condition [Taylor, 1963]. It says that the total azimuthal force on geostrophic cylinders vanishes. When the boundary of the cavity is an insulator, (19) and (23) yield, after an integration by parts, the alternative result,

$$0 = J = (\pi/\Omega s) \, \partial(s^2 M)/\partial s, \qquad (24a)$$

in which ρM is the integrated Maxwell stress,

$$\rho M = \mu^{-1} \int_{-z_B}^{z_B} Bb_s dz. \qquad (24b)$$

At this point in our discussion we recall that the geostrophic flow $V_G(s)$ is so far undetermined. Taylor's (1963) idea was that, by a suitable choice of $V_G(s)$, the magnetic fields produced by induction would be such that condition (23), or equivalently (24), is satisfied. Though Taylor's idea has had much support, it has also led to controversy. Specifically, do Taylor solutions [i.e. satisfying Taylor's condition] actually exist and do they pertain to the geodynamo? If not, what are the alternatives? Possibly the geostrophic flow is controlled by secondary effects not yet included. The most important of these emerges from core-mantle coupling. An electromagnetic couple arises from electric currents which leak into the weakly conducting mantle, while viscous coupling arises from Ekman suction into or out of the Ekman layer at the core/mantle boundary, CMB. Though the electromagnetic coupling is probably greater, the essential link between core and mantle can be appreciated from consideration of the viscous couple alone. If, in addition, fluid inertia is taken into account, the equation governing the geostrophic motion becomes

$$2z_B \frac{\partial \omega_G}{\partial t} + 2\left(\frac{L\nu\Omega}{z_B}\right)^{\frac{1}{2}} \omega_G = \frac{1}{s^3} \frac{\partial}{\partial s}(s^2 M), \qquad (25)$$

where ν is the kinematic viscosity, in place of the ideal Taylor condition (24a). Braginsky (1976), in the development of his $\alpha\omega$-dynamo called Model-Z, found that Taylor's condition was not satisfied. Instead large geostrophic flows emerged governed by the balance (25).

The energetics of Braginsky's (1976) Model-Z are of some

interest. Most of the energy [Kinetic + Magnetic] is contained in the strong azimuthal magnetic field, $B\hat{\phi}$. Nevertheless, since the inertia of the geostrophic flow, $V_G [= s\omega_G]$, may be significant [see (25)], it must be included also. The kinetic energy in the non-geostrophic velocity and the magnetic energy in the meridional magnetic field are relatively small and are ignored here. Thus the total energy,

$$\Phi = \int_V \tfrac{1}{2}(\mu^{-1}B^2 + \rho V_G^2)dV, \tag{26}$$

where the volume integration is taken over the fluid core, V, can be shown to satisfy

$$d\Phi/dt = S_T - (D_G + D_M). \tag{27a}$$

Here energy is lost by the viscous dissipation in thin Ekman layers adjacent to the CMB,

$$D_G = 4\pi(L\nu\Omega/z_B)^{\tfrac{1}{2}} \int_0^L \rho s V_G^2 ds \quad [>0], \tag{27b}$$

and by ohmic dissipation throughout the core,

$$D_M = (\eta/\mu) \int_V [|\nabla B|^2 + (B/s)^2]dV \quad [>0], \tag{27c}$$

and the only source of energy is the rate of working of the buoyancy forces,

$$S_T = -\int_V \mathbf{v}_P \cdot \mathbf{g} \; \gamma T dV \quad [= \mu^{-1} \int_V (\mathbf{b}_P \cdot \nabla)\omega_T dV]. \tag{27d}$$

The last term clearly identifies the thermal wind, $s\omega_T$, as the only part of the ω-effect which can be a genuine source of azimuthal magnetic field. It means that the magnetically driven motion (17) and (20) does not contribute to the energetics; the geostrophic flow, which could be large, can only take energy out of the system by viscous dissipation.

The $\alpha\omega$-dynamo solved numerically by Braginsky (1978) [also Braginsky and Roberts, 1987] is based on a prescribed thermal wind, $s\omega_T$. The geostrophic flow, $s\omega_G$, and the magnetically driven motion $s\omega_M\hat{\phi} + \mathbf{v}_{MP}$ are obtained as part of the solution together with the magnetic field, $B\hat{\phi} + \mathbf{b}_P$. Since viscous coupling with the CMB is important the Ekman number,

$$E = \nu/L^2\Omega, \tag{28a}$$

is central to the theory and appears in the expansion parameter,

$$\varepsilon_T = E^{\tfrac{1}{2}} R_T \; [<<1], \quad \bullet \quad [R_T = L^2\omega_T/\eta], \tag{28b}$$

where the magnetic Reynolds number based on the thermal wind velocity is large. [For comparison, note that Braginsky and Roberts (1987) use the magnetic Reynolds number, $\varepsilon_T^{1/2} R_T$, which is based on $\varepsilon_T^{1/2}\omega_T$, a hypothetical angular velocity. The parameter ε in their theory is $\varepsilon_T^{3/2}$.] The results of the numerical integrations suggest the ordering

$$b_z = O(\varepsilon_T^{-\frac{1}{2}} R_T^{-1} B), \qquad b_s = O(R_T^{-1} B), \tag{29a}$$

$$\omega_G = O(\varepsilon_T^{-\frac{1}{2}} \omega_T), \qquad \omega_M = O(\varepsilon_T^{\frac{1}{2}} \omega_T), \tag{29b}$$

$$v_M = O(R_T^{-1} L\omega_T), \qquad \alpha = O(\varepsilon^{-\frac{1}{2}} R_T^{-1} \eta/L). \tag{29c}$$

The magnitude of the azimuthal magnetic field, B, necessary to maintain the magnetically driven motion of the size advocated in (29b,c) is of order

$$\varepsilon_T^{\frac{1}{4}} R_T^{\frac{1}{2}} B_o, \tag{30a}$$

where

$$B_o = (\rho\mu\eta\Omega)^{\frac{1}{2}}. \tag{30b}$$

Two striking features of the solution have their origin in the modified Taylor's condition (25). Firstly, the meridional magnetic field is almost aligned with the z-axis [whence the name Model-Z] and is independent of z [$b_z \approx b_z(s)$]. Secondly, the geostrophic velocity is large. The former reduces the electromagnetic coupling between adjacent geostrophic cylinders. The latter increases the viscous coupling with the mantle so that a steady state balance in (25) can be achieved. The axial magnetic field adjusts to the external potential field in a thin magnetic layer close to the CMB.

3.2. The Malkus-Proctor dynamo

The failure of Braginsky's Model-Z to satisfy Taylor's condition (23) is puzzling. Another line of attack initiated by Malkus and Proctor (1975) focuses on α^2-dynamos. Such studies shed considerable light on the issues, and show that Taylor's condition can often be met. Many of the ideas developed for $\alpha\omega$-dynamos continue to hold for α^2-dynamos, except that now the meridional and azimuthal magnetic fields are of comparable size. This means that the magnetic energy of the meridional magnetic field, b_p, must be included in the total energy, Φ, of (26). Since there is no thermal wind [$\omega_T = 0$], the source S_T [see (27d)] is absent and is replaced by terms involving the α-effect. Here the α-effect is invoked to generate both meridional and azimuthal components of magnetic field as in (9). The ω-effect arising from the geostrophic flow is still present but cannot by itself generate magnetic field, as we explained in the previous subsection. The early numerical studies of Malkus and Proctor (1975) and Proctor (1977) were restricted to the case of finite Ekman number. Since the values of E used were never very small, they were not able to resolve the Taylor question completely. Nevertheless, for the small Ekman number limit

[E<<1] they proposed the scenario outlined below.

In the absence of any mean motion, the kinematic α^2-dynamo is characterised by the magnetic Reynolds number, R_α [see (10)]. For small R_α no dynamo solution exists. At some critical value $R_\alpha = R_{\alpha c}$, the zero solution bifurcates and, for $R_\alpha > R_{\alpha c}$, magnetic field is maintained which drives a geostrophic flow, $s\omega_G$. This influences the evolution of the magnetic field, when the magnetic Reynolds number $L^2\omega_G/\eta$, based on it is of order unity. The resulting Lorentz force is comparable to the viscous force in (25) when the magnetic field is weak of order

$$E^{\frac{1}{4}} B_o, \tag{31}$$

where B_o is defined by (30b). The corresponding magnetically driven motion is slow of order $E^{1/2}L\omega_G$ and has no inductive effect. The magnitude of the magnetic field increases with R_α. Simultaneously the geostrophic velocity, $s\omega_G$, adjusts its profile but its magnitude remains of order η/L. This means that the magnitude of the magnetic force in (25) remains of order $E^{1/2} B_o^2/\mu L$ whereas the magnitude of the magnetic field may increase dramatically. Indeed it is anticipated that a second critical value $R_\alpha = R_{\alpha Ta}$ is reached at which $|\mathbf{B}|/E^{1/4}B_o \to \infty$ while $\rho\mu L/M|\mathbf{B}|^2 \to 0$ [see (24b)]. This is the critical value for Taylor solutions at which Taylor's condition is met. Subsequently, as R_α continues to increase above $R_{\alpha Ta}$, finite amplitude Taylor solutions occur in which the magnetic field is of order B_o. This ensures a balance of Lorentz and Coriolis forces, and the ageostrophic magnetically driven flow $s\omega_M\hat{\boldsymbol{\phi}} + \mathbf{v}_{MP}$ now affects the magnetic field directly.

There is some evidence for the scenario just described from simplified models in which the spherical cavity is replaced by a plane layer of infinite horizontal extent and of constant width 2L. The system is referred to cartesian coordinates x,y,z. Electrically conducting fluid is contained inside the region, $|z| < L$, while the exterior region, $L < |z|$, is an insulator. Rotation remains in the z-direction. The horizontal x and y directions are identified with the radial and azimuthal directions respectively. Consequently the magnetic field is independent of y [corresponding to axisymmetry] and the geostrophic flow becomes $V_G(x)\hat{\mathbf{y}}$. To mimic the spherical geometry Soward and Jones (1983) sought steady solutions periodic in x with dipole symmetry. The geometrical simplifications enabled a wide class of weak field solutions [see (31)] to be obtained with relative ease. The results essentially confirmed the Malkus-Proctor scenario up to the critical magnetic Reynolds number $R_{\alpha Ta}$ for Taylor solutions. In some cases, however, the solution branch which emerged at $R_{\alpha c}$ did not connect continuously to the Taylor branch at $R_{\alpha Ta}$. Instead a finite jump was found to be necessary to make the connection. Ierley (1985) considered the corresponding model in a spherical cavity and obtained similar results.

Abdel-Aziz and Jones (1988) have also investigated $\alpha\omega$-dynamos in the plane layer which incorporates the thermal wind, $V_T(z)\hat{\mathbf{y}}$. These are properly characterised by the Dynamo number $D[= R_T R_\alpha,$ see (11)], but, if we keep R_T fixed, it is sufficient to consider variations of R_α as before. Abdel-Aziz and Jones (1988) found periodic travelling wave solutions which depend on x and t only through the combination, x-ct, where the constant c is the wave speed. They traced the evolution of these solutions through the complete Malkus-Proctor scenario and found finite amplitude Taylor solutions with magnetically driven ageostrophic flows when R_α exceed $R_{\alpha Ta}$. These travelling dynamo wave solutions have no direct analogue in a sphere because of the bounded geometry. The closest correspondence is oscillatory dynamo modes. If they exist, we might speculate that, like some of the α^2-dynamos mentioned above, they lie on solution branches totally disconnected from the branch which emerges from the bifurcation at $R_{\alpha c}$. That may explain why Braginsky (1978) does not find Taylor solutions, even though they might exist.

4. MICRODYNAMICS

4.1. The Governing Equations

In the previous Section we outlined the macrodynamics of the dynamo process. Now we will investigate the microdynamics necessary to complete the dynamo picture. Specifically we will discuss the buoyantly driven asymmetric motions which form the vital ingredient for the α-effect. As before, we consider a Boussinesq fluid governed by the equation of motion,

$$\partial \mathbf{v}/\partial t + \mathbf{v} \cdot \nabla \mathbf{v} + 2\Omega \times \mathbf{v} = -\nabla(p/\rho) + \rho^{-1}\mathbf{j} \times \mathbf{B} - \mathbf{g}\gamma T + \nu\nabla^2 \mathbf{v}, \quad (32a)$$

with

$$\nabla \cdot \mathbf{v} = 0. \tag{32b}$$

The basic temperature distribution, T_s, is assumed to be spherically symmetric so that the deviations, T, from that state are governed by the heat condution equation,

$$\partial T/\partial t + \mathbf{v} \cdot \nabla T = -\boldsymbol{\beta} \cdot \mathbf{v} + \kappa \nabla^2 T, \quad [\boldsymbol{\beta} \equiv \nabla T_s //g], \tag{33}$$

where κ is the thermal diffusivity. Throughout this Section, we will consider small disturbances to an equilibrium state [generally unstable], which includes a prescribed mean axisymmetric magnetic field.

4.2. Non-dissipative Waves

When there is no dissipation, $\nu = \kappa = \eta = 0$, and the basic state is uniform, plane wave solutions proportional to $e^{i(\mathbf{k} \cdot \mathbf{x} - \omega t)}$ exist, which satisfy the dispersion relation [see for example Acheson and Hide, 1973]

$$(\omega^2 - \omega_M^2)(\omega^2 - \omega_A^2 - \omega_M^2) = \omega^2\,\omega_C^2, \tag{34a}$$

where ω_C, ω_M and ω_A are the inertial, Alfvén and buoyancy frequencies,

$$\omega_C = 2\boldsymbol{\Omega}\cdot\hat{\mathbf{k}}, \qquad \omega_M = \mathbf{B}\cdot\mathbf{k}/(\mu\rho)^{\frac{1}{2}}, \qquad \omega_A = (-\gamma\mathbf{g}\cdot\boldsymbol{\beta})^{\frac{1}{2}}\,|\hat{\mathbf{g}}\times\hat{\mathbf{k}}|, \tag{34b}$$

[as before, the hat denotes unit vectors]. In the geodynamo context, we generally have $\omega_C \gg \omega_M$ and ω_A. The two distinct roots of the quadratic equation (34a) for ω^2 then define a fast inertial wave, $\omega[\approx \omega_C]$, and a slow hybrid wave,

$$\omega \approx \omega_{MAC} = (\omega_M/\omega_C)(\omega_M^2 + \omega_A^2)^{\frac{1}{2}}. \tag{35}$$

Braginsky (1967) remarks that this slow MAC wave is unstable when the fluid is sufficiently top heavy ($\mathbf{g}\cdot\boldsymbol{\beta} > 0$), specifically

$$-\omega_A^2 > \omega_M^2. \tag{36}$$

The fast inertial waves require large adverse density gradients to drive them and so they are unlikely to be relevant to the dynamo process.

4.3. Thermal Convection

When the effects of finite diffusivities are included, resistive instabilities are possible which are excited more easily than the non-dissipative waves discussed in the previous subsection. To appreciate their nature we must first introduce the dimensionless parameters which characterise our system. They are the Prandtl numbers,

$$P = \nu/\kappa, \qquad P_m = \nu/\eta, \qquad q = \kappa/\eta, \tag{37}$$

and the Ekman, Hartmann and Rayleigh numbers,

$$E = \nu/L^2\Omega, \qquad M = BL/\sqrt{\mu\rho\nu\eta}, \qquad Ra = g\gamma\beta L^4/\kappa\nu. \tag{38}$$

Since the Ekman number is small [$E \ll 1$] it is also useful to define the Elsasser and modified Rayleigh numbers,

$$\Lambda = EM^2 = B^2/\mu\rho\eta\Omega, \qquad Ra_E = ERa = g\gamma\beta L^2/\kappa\Omega, \tag{39}$$

which are independent of the viscosity, ν. The discussions which follow will indicate the importance of the various parameters.

A convective model, which has received much attention and closely resembles the situation in the Earth's core, concerns a self-gravitating spherical cavity rotating rapidly about the z-axis and containing a uniform distribution of heat sources. The non-magnetic problem was first considered by Roberts (1968) and Busse (1970). They noted that, in the small Ekman number limit, motion is strongly constrained to be two-dimensional by the Proudman-Taylor theorem. Consequently at the onset of instability columnar convection occurs close to the cylinder of radius $s_o [\approx \frac{1}{2}L]$, where as before L is the radius of the cavity. The azimuthal length scale of the convecting rolls, whose axes are aligned with the rotation axis, is short, of order

$$\ell = E^{\frac{1}{3}}L. \tag{40}$$

Though motion in the rolls necessarily varies with z to accommodate the top and bottom $[z = \pm z_B(s)]$ boundary conditions, it is, however, almost two-dimensional on the length scale, ℓ, of the convection. As a result of significant viscous and thermal dissipation on that short length scale, the critical value of the modified Rayleigh number required to drive the convection is large,

$$Ra_{Ec} = O(L/\ell) = O(E^{-\frac{1}{3}}). \tag{41}$$

Motion sets in as a travelling wave proportional to $e^{i(m\phi - \omega t)}$, in which m is of order L/ℓ $[= O(E^{-1/3})]$ and ω is closely related to the corresponding Rossby wave frequency of slow inertial waves of order $\Omega\ell/L$ $[= O(\Omega E^{1/3})$, see (34b)]. The realised values of m and ω are, of course, dependent on the Prandtl number P_m.

To determine the effect of a strong azimuthal magnetic field on the convection, Eltayeb and Kumar (1977), and Fearn (1979a,b) considered the special case of the magnetic field, $B = B_L(s/L)\hat{\phi}$, in which B_L is a constant. They found that, when the Elsasser number, Λ, is small, less than order $E^{1/3}$, the convection resembles the nonmagnetic case described in the previous paragraph. The presence of the magnetic field suppresses convection weakly so that, in general, the critical Rayleigh number, Ra_{Ec}, is an increasing function of Λ. When Λ is large compared to $E^{1/3}$ [but small compared to unity] the nature of the most unstable mode changes. Like the Rossby mode, it continues to be a travelling topographic wave localised close to some cylindrical surface, whose radius s_o depends on the value of Λ. On the other hand, inertia is no longer significant and is replaced by the Lorentz force, $[|j \times B| \gg \rho\partial u/\partial t]$, in the primary force balance. The most important characteristic of this convection from the dynamo point of view is that Ra_{Ec} is of order Λ^{-1} and so decreases dramatically with increasing magnetic field strength. It is often said that the rotational constraints, which lead to the short length scales are relaxed by the magnetic field. The effect is optimised when Λ and consequently Ra_{Ec} are of order unity. At the minimum value of Ra_{Ec} [at $\Lambda = \Lambda_o$, say] convection occurs on the length scale L of the sphere and m is typically unity. When Λ exceeds Λ_o, the magnetic field inhibits convection and Ra_{Ec} increases with Λ.

To investigate more detailed properties of the convection and dynamo applications, other simplified geometries have been employed. Busse (1976), for example, has considered convection in an annulus with tilted top and bottom boundaries. Such a model is appropriate in describing the localised topographic waves which occur at small Elsasser number, $\Lambda \ll 1$ [see also Soward, 1979a]. When Λ is of order unity, many important features of the convection also occur in the plane-layer model described in Subsection 3.2. Roberts and Stewartson (1974) considered the case, in which rotation and gravity are vertical [$\Omega = \Omega\hat{z}$ and $g = -g\hat{z}$], and the applied magnetic field is uniform and horizontal [$B = B_L\hat{y}$]. The effects of magnetic field line curvature [for example $B = B_L(s/L)\hat{\phi}$] which cause the convection pattern to propagate as a travelling wave, were investigated by Soward (1979b).

4.4. Other Instabilities

In addition to buoyancy driven convection, there are other instabilities which take energy from nonuniformities of the basic state. For example, when the applied azimuthal magnetic field, $B(s,z)\hat{\phi}$, is nonuniform, field gradient instabilities are possible. Stability criteria based on local analysis were developed by Acheson (1983) and confirmed by Fearn's (1983) numerical solution of the governing equations. On the other hand, differential rotation, $s\omega(s,z)\hat{\phi}$, is generally stabilizing for two reasons. Firstly, there is insufficient kinetic energy in the motion to promote shear flow instability against the constraints of rotation and the magnetic field. Secondly, the effect on asymmetric modes of convection [$m \neq 0$] is to shorten the radial length scale and to increase dissipation. The effect is minimised close to a resonant surface at which $\omega(s,z)$ equals the phase velocity, ω/m, of the unstable wave. Convection, therefore, is generally localised close to some circle, s = constant, z = constant, on the resonant surface in the neighbourhood of which conditions for convection are most favourable [Fearn and Proctor, 1983]. Finally we mention that other resistive instabilities have been studied [see for example Fearn, Roberts and Soward, 1988].

4.5. Taylor's Condition

The Roberts and Stewartson (1974) plane-layer model has also been employed to seek nonlinear marginal solutions of the convection problem satisfying Taylor's condition (23), in which the axisymmetric electric current and magnetic field are replaced by the asymmetric contributions, j' and b', induced by the convection. The nonlinearity stems from both Taylor's condition and the geostrophic velocity, $V_G(x)\hat{y}$, whose magnitude must be obtained as part of the solution. These solutions, nevertheless, are marginal because the mathematical problem is homogeneous in j' and b', with the consequence that no convective amplitude is determined. In this plane-layer geometry two distinct families of convection rolls are possible at the onset of instability [codimension 2 bifurcation at $Ra_E = Ra_{Ec}$], whose axes are inclined at angles of equal magnitude but of opposite sign to the applied uniform magnetic field $B_L\hat{y}$. Roberts and Stewartson (1975) regarded the addition of a small geostrophic velocity, V_G, as a third neutrally stable mode. They discussed the three-mode interaction within the framework of the modified Taylor's condition (25) and weakly nonlinear theory [see also Soward, 1980]. The existence of the two families of rolls at $Ra_E = Ra_{Ec}$ is peculiar to the system being

unbounded in the horizontal extent and permits the trivial solution
satisfying Taylor's condition (23) [$Ra_{ETa} = Ra_{Ec}$], which consists of
one family of rolls with $V_G = 0$. To remove the degeneracy, and to
more faithfully model the bounded geometry of the Earth's core, Soward
(1986) added vertical walls at $x = \pm \tilde{L}$ (say). In the bounded region
$|x| < \tilde{L}$, the two roll degeneracy is removed and, loosely speaking, the
amplitude of each family is linked through the boundary conditions.
Furthermore, these marginal Taylor solutions are not trivial because
they are no longer solutions of a linear problem and only exist when
$V_G(x) \not= 0$.

For the duct model just described, the small q limit,

$$q \ll 1, \tag{42}$$

appropriate to the Earth is particularly interesting. In that case
thermally driven convection is limited to a central region $|x| < \tilde{L}_0$,
where $\tilde{L}_0$ is a constant less than $\tilde{L}$, in which the Péclet number, $\tilde{L}V_G^0/\kappa$,
is of order unity. Outside, the geostrophic flow is an order of
magnitude, q^{-1}, larger and is characterised by a magnetic Reynolds
number of order unity. Here [$\tilde{L}_0 < |x| < \tilde{L}$] the large shear causes the
thermal disturbance to be strongly attenuated so that buoyancy forces
are negligible. The picture which emerges is consistent with Fearn
and Proctor's (1983) remarks about the stabilising role of
differential rotation. Now the line on the resonant surface mentioned
in Subsection 4.4 above has expanded to fill the entire region,
$|x| < \tilde{L}_0$. In this central region the geostrophic velocity is small
and the convection is steady with zero phase velocity.

The marginal Taylor solutions generally occur at a critical
Rayleigh number, Ra_{ETa}, which is in excess of the marginal value,
Ra_{Ec}, for the onset of convection with no geostrophic flow. The
scenario for convection is similar to that outlined for the α^2 and
$\alpha\omega$-dynamos in Subsection 3.2. Clearly the next step is to investigate
the fully nonlinear Taylor state which emerges when Ra_E exceeds Ra_{ETa},
but that is a task still to be undertaken. Currently progress is
being made to investigate the marginal Taylor problem in cylindrical
geometry [Skinner and Soward, 1988].

5. CONVECTION DRIVEN DYNAMOS

5.1. Weak Field Dynamos

Early attempts to model the geodynamo were based upon perturbation
methods which assumed that the magnetic field is weak. Accordingly
the onset of convection is not influenced by the magnetic field at
lowest order. Fortunately in a rapidly rotating system that motion is
well endowed with the helicity necessary to give an α-effect. At
higher order, when nonlinearities are taken into account, the
magnitude of the convective motions is determined partially by the
weak Lorentz force. So, for a given Rayleigh number, the amplitude of
the motion must lead to an α-effect which can drive a steady dynamo.

On the other hand, the magnitude of the magnetic field itself is fixed
by the requirement that the corresponding Lorentz force leads to the
appropriate equilibrium amplitude for the motion.

The weak field assumption forms the basis of Busse's (1975)
annulus model of the geodynamo and Childress and Soward's (1972) plane
layer model. For this equilibrium to be stable it is necessary that
the Rayleigh number increase with increasing field strength. For
small values of the Elsasser number, specifically $\Lambda \ll E^{1/3}$, thermal
convection in a sphere, as described in Subsection 4.3, has that
property. In this parameter regime we expect Busse's (1975) model to
be stable. This initial rise in the value of the critical Rayleigh
number, Ra_{Ec}, does not occur in a plane layer. That model appears to
be stable only for extremely weak magnetic fields, $\Lambda = O(E)$ [Soward,
1974]. For larger values of Λ, Fautrelle and Childress (1982) have
shown that the dynamo is unstable. Likewise we may expect the Busse
(1975) model to loose stability when $\Lambda \gg E^{1/3}$. This instability was
anticipated by Eltayeb and Roberts (1970). They pointed out that when
Ra_{Ec} decreases with Λ, it should be expected that the magnetic field
will increase. This phenomenon of "runaway field growth" persists
until Λ is of order unity, when the corresponding value of Ra_{Ec} is
minimised. For larger values of Λ, Ra_{Ec} again increases. Thus it is
to be expected that the dynamo stabilises close to the minimum of
Ra_{Ec}, where Λ is of order unity.

In both the spherical and plane-layer dynamo models multiple
length scale methods are employed which depend on the assumption that
the Ekman number is small. Recently Zhang and Busse (1988) [see also
Zhang, Busse and Hirsching, 1988] have investigated numerically
convection in a spherical shell at small but finite Ekman number.
Finite amplitude solutions are found which can generate magnetic
field. Their dynamo is subcritically unstable in the following sense.
As the strength of the magnetic field increases the amplitude of the
motion required to maintain it decreases, in accord with the Eltayeb
and Roberts (1970) scenario.

5.2. Strong Field Dynamos

Clearly the remarks of the previous subsection suggest that, in the
limit of small Ekman number, the geodynamo is of the strong field type
with Λ of order unity. The construction of such dynamos, which ignore
viscous effects and consequently satisfy Taylor's condition, has so
far proved elusive. One interesting approach has been initiated by
Fearn and Proctor (1984, 1987). Their idea is based on an iterative
approach. At the start of each iteration the axisymmetric part of the
magnetic field is assumed to be given. The critical Rayleigh number,
Ra_{Ec}, necessary to maintain marginal asymmetric convection is found.
The mean electromotive force, $\overline{\mathbf{v}' \times \mathbf{b}'}$, corresponding to the solution
is calculated and its magnitude is adjusted so that a steady dynamo
results. The axisymmetric part of the dynamo-produced magnetic field
is now used for the start of the next iteration. Unfortunately the
iterative scheme does not appear to converge to a dynamo solution.
The convergence difficulties appear to be linked with Taylor's

condition, and they disappear when Taylor's condition is ignored. Of course, the failure yet to find Taylor solutions is reminiscent of Braginsky's Model-Z, which only satisfies the modified condition (25). Indeed, the existence of strong field convective dynamos, which are independent of core-mantle coupling has yet to be demonstrated, and remains one of the central unanswered questions in geodynamo theory.

A separate issue which concerns strong field models is whether they are of α^2- or $\alpha\omega$-type. We can, however, make some tentative suggestions based upon mean field theory. Suppose convection occurs in our spherical cavity, radius L, on the short length scale, ℓ. If the asymmetric convective velocities are of order v' (say), then, in the notation of Sections 2 and 4.1, the perturbation temperature, T', and magnetic field, b', lead to mean heat fluxes and electromotive forces, which order of magnitude are

$$|\overline{v'T'}| = O(\ell^2 v'^2 \beta / \kappa), \quad |\overline{v' \times b'}| = O(\ell v'^2 B / \eta)$$ (43a,b)

respectively. The latter gives a magnetic Reynolds number based on the α-effect of magnitude

$$R_\alpha = O(L\ell v'^2 / \eta^2),$$ (44)

[see (10)]. The former, on the other hand, leads to the mean axisymmetric temperature perturbation

$$\overline{T} = O(L\ell^2 v'^2 \beta / \kappa^2).$$ (45)

By (14) this drives a thermal wind

$$s\omega_T = O(L\ell^2 v'^2 g\gamma\beta / \kappa^2 \Omega),$$ (46)

and leads to the magnetic Reynolds number,

$$R_T = O(\ell^2 v'^2 Ra_E / \eta\kappa),$$ (47)

[see (28b) and (39)]. Finally, from the estimates (44) and (47) we obtain

$$\frac{R_T}{R_\alpha} = O(q^{-1} \frac{\ell}{L} Ra_E).$$ (48)

The scale separation, $\ell \ll L$, employed in the above argument is applicable to thermal convection in a sphere when Λ is small. In that case marginal convection is characterised by

$$(\ell / L) Ra_{Ec} = O(1),$$ (49)

whenever q is not large [cf. (41)]. Once Λ is of order unity convection occurs on the length scale of the sphere, $\ell = O(L)$. Though the scale separation is no longer applicable, the estimates (43) to (49) should still apply with Ra_{Ec} of order unity, when $\Lambda = O(1)$, and large when $\Lambda \gg 1$. Consequently, when q is small, as it is for the

Earth, we see from (48) that

$$R_T \gg R_\alpha, \tag{50}$$

whatever the value of Λ. The above estimates are crude and must be interpreted cautiously. Indeed, if it is anticipated that turbulent values of κ and η are more appropriate, then q could be as large as order unity. In that case, R_α is of order R_T and the dynamo is no longer of $\alpha\omega$-type but may be more appropriately described as an $\alpha^2\omega$-dynamo.

   Finally we remark about the absence of the viscosity, ν, in the estimates we have just made. If Taylor's condition cannot be met, large geostrophic flows must ensue as in Braginsky's Model-Z. In that event our estimate must be revised and they will ultimately be viscosity dependent. Clearly, when the Ekman number is finite, as in Zhang and Busse's (1988) dynamo, we may expect the estimates to depend upon the Prandtl number, P_m. Indeed their dynamos generally exhibit $\alpha\omega$-behaviour but appear to be of α^2-type when P_m is sufficiently large.

6. ACKNOWLEDGEMENTS

This article was prepared while attending two SEDI (Study of the Earth's Deep Interior) meetings, namely the Symposium, "Structure and Dynamics of the Core and Adjacent Mantle", Blanes, Spain, 23-25 June, 1988, and the Workshop, "Earth's Core Boundary and Geodynamos", Liblice, Czechoslovakia, 27 June - 2 July, 1988. I have benefited from discussions with many of the participants but I mention particularly Drs. F.H. Busse, S. Childress, I.A. Eltayeb, D.R. Fearn, D. Gubbins, C.A. Jones, H.K. Moffatt and K.-H. Rädler. I also wish to thank Professor P.H. Roberts and Dr. M.R.E. Proctor for their useful comments.

REFERENCES

Abdel-Aziz, M.M. and Jones, C.A., 1988. '$\alpha\omega$-dynamos and Taylor's
      constraint', Geophys. Astrophys. Fluid Dynam., 44, to appear.
Acheson, D.J., 1983. 'Local analysis of thermal and magnetic
      instabilities in a rapidly rotating fluid', Geophys. Astrophys.
      Fluid Dynam., 27, 123-136.
Acheson, D.J. and Hide, R., 1973. 'Hydromagnetics of rotating
      fluids', Rep. Prog. Phys., 36, 159-221.
Braginsky, S.I., 1964a. 'Self excitation of a magnetic field during
      the motion of a highly conducting fluid', JETP, 47,
      1084-1098.
Braginsky, S.I., 1964b. 'Theory of the hydromagnetic dynamo', JETP,
      47, 2178-2193.
Braginsky, S.I., 1967. 'Magnetic waves in the Earth's core', Geomag.
      Aeron., 7, 851-859.

Braginsky, S.I., 1976. 'On the nearly axially-symmetric model of the
hydromagnetic dynamo of the Earth', Phys. Earth Planet. Inter.,
11, 191-199.

Braginsky, S.I., 1978. 'Nearly axially symmetric model of the
hydromagnetic dynamo of the Earth', Geomagn. Aeron., 18,
340-351.

Braginsky, S.I. and Roberts, P.H., 1987. 'A model-Z geodynamo',
Geophys. Astrophys. Fluid Dynam., 38, 327-349.

Busse, F.H., 1970. 'Thermal instabilities in rapidly rotating
systems', J. Fluid Mech., 44, 441-460.

Busse, F.H., 1975. 'A model of the geodynamo', Geophys. J. R. astr.
Soc., 42, 437-459.

Busse, F.H., 1976. 'Generation of planetary magnetism by convections,
Phys. Earth Planet. Inter., 12, 350-358.

Childress, S. and Soward, A.M., 1972. 'Convection-driven
hydromagnetic dynamos', Phys. Rev. Lett., 29, 837-839.

Cowling, T.G., 1957. 'The dynamo maintenance of steady magnetic
fields', Q.J. Mech. Appl. Math., 10, 129-136.

Eltayeb, I.A. and Kumar, S., 1977. 'Hydromagnetic convective
instabilities of a rotating self-gravitating fluid sphere
containing a uniform distribution of heat sources', Proc. R.
Soc. Lond., A353, 145-162.

Eltayeb, I.A. and Roberts, P.H., 1970. 'On the hydromagnetics of
rotating fluids', Astrophys. J., 162, 699-701.

Fautrelle, Y. and Childress, S., 1982. 'Convective dynamos with
intermediate and strong fields', Geophys. Astrophys. Fluid
Dynam., 22, 235-279.

Fearn, D.R., 1979a. 'Thermally driven hydromagnetic convection in a
rapidly rotating sphere', Proc. R. Soc. Lond., A369, 227-242.

Fearn, D.R., 1979b. 'Thermal and magnetically driven instabilities
in a rapidly rotating fluid sphere', Geophys. Astrophys. Fluid
Dynam., 14, 103-126.

Fearn, D.R., 1983. 'Hydromagnetic waves in a differentially rotating
annulus I. A test of local stability analysis', Geophys.
Astrophys. Fluid Dynam., 27, 137-162.

Fearn, D.R., 1988. 'Compositional convection and the Earth's core',
in these Proceedings.

Fearn, D.R. and Proctor, M.R.E., 1983. 'Hydromagnetic waves in a
differentially rotating sphere', J. Fluid Mech., 128 1-20.

Fearn, D.R. and Proctor, M.R.E., 1984. 'Self-consistent dynamo models
driven by hydromagnetic instabilities', Phys. Earth Planet.
Inter., 36, 78-84.

Fearn, D.R. and Proctor, M.R.E., 1987. 'On the computation of steady,
self-consistent, spherical dynamos', Geophys. Astrophys. Fluid
Dynam., 38, 293-325.

Fearn, D.R., Roberts, P.H. and Soward, A.M., 1988. 'Convection,
stability and the dynamo', in Energy Stability and Convection
(G.P. Galdi and B. Straughan, eds.), Longman, Harlow, 60-324.

Gubbins, D. and Roberts, P.H., 1988. 'Magnetohydrodynamics of the
Earth's core', in Geomagnetism Vol. 2 (J.A. Jacobs, ed.),
Academic Press, London, 1-183.

Ierley, G.R., 1985. 'Macrodynamics of α^2-dynamos', Geophys. Astrophys. Fluid Dynam., 34, 143-173.

Krause, F. and Rädler, K.-H., 1980. Mean-Field Magnetohydrodynamics and Dynamo Theory, Pergamon, Oxford.

Malkus, M.V.R. and Proctor, M.R.E., 1975. 'The macrodynamics of -effect dynamos in rotating fluids', J. Fluid Mech., 67, 417-443.

Moffatt, H.K., 1978. Magnetic Field Generation in Electrically Conducting Fluids, Cambridge University Press.

Parker, E.N., 1979. Cosmical Magnetic Fields, their origin and their activity, Clarendon Press, Oxford.

Proctor, M.R.E., 1977. 'Numerical solutions of the non-linear -effect dynamo equations', J. Fluid Mech., 80, 769-784.

Roberts, P.H., 1968. 'On the thermal instability of a rotating fluid sphere containing heat sources', Phil. Trans. R. Soc. Lond., A263, 93-117.

Roberts, P.H., 1988a. 'Origin of the main field: Dynamics', in Geomagnetism Vol. 2 (J.A. Jacobs, Ed.), Academic Press, London, 251-306.

Roberts, P.H., 1988b. 'Future of geodynamo theory', Geophys. Astrophys. Fluid Dynam., 44, to appear.

Roberts, P.H. and Gubbins, D., 1988. 'Origin of the main field: kinematics', in Geomagnetism Vol. 2 (J.A. Jacobs, ed.) Academic Press, London, 185-249.

Roberts, P.H. and Stewartson, K., 1974. 'On finite amplitude convection in a rotating magnetic system', Phil. Trans. R. Soc. Lond., A277, 287-315.

Roberts, P.H. and Stewartson, K., 1975. 'Double roll convection in a rotating magnetic system', J. Fluid Mech., 68, 447-466.

Skinner, P.H. and Soward, A.M., 1988. 'Convection in a rotating magnetic system and Taylor's constraint', Geophys. Astrophys. Fluid Dynam., 44, to appear.

Soward, A.M., 1972. 'A kinematic theory of large magnetic Reynolds number dynamos', Phil. Trans. R. Soc. Lond., A272, 431-462.

Soward, A.M., 1974. 'A convection-driven dynamo 1. The weak field case', Phil. Trans. R. Soc. Lond., A275, 611-651.

Soward, A.M., 1979a. 'Convection driven dynamos', Phys. Earth Planet. Inter., 20, 134-151.

Soward, A.M., 1979b. 'Thermal and magnetically driven convection in a rapidly rotating fluid layer', J. Fluid Mech., 90, 669-684.

Soward, A.M., 1980. 'Finite amplitude thermal convection and geostrophic flow in a rotating magnetic system', J. Fluid Mech., 98, 449-471.

Soward, A.M., 1986. 'Non-linear marginal convection in a rotating magnetic system', Geophys. Astrophys. Fluid Dynam., 35, 329-371.

Soward, A.M. and Jones, C.A., 1983. 'α^2-dynamos and Taylor's constraint', Geophys. Astrophys. Fluid Dynam., 27, 87-122.

Taylor, J.B., 1963. 'The magneto-hydrodynamics of a rotating fluid and the Earth's dynamo problem', <u>Proc. R. Soc. Lond.</u>, A**274**, 274-283.

Zhang, K.K. and Busse, F.H., 1988. 'Finite amplitude convection and magnetic field generation in a rotating spherical shell', <u>Geophys. Astrophys. Fluid Dynam</u>, **44**, to appear.

Zhang, K.K., Busse, F.H. and Hirsching, N., 1988. 'Numerical models in the theory of geomagnetism', in these Proceedings.

LONG-TERM PALAEOFIELD VARIATIONS AND THE GEOMAGNETIC DYNAMO

Poorna C. Pal
Instituto Astronomico e Geofisico
Universidade de Sao Paulo
Caixa Postal 30,627
01051 Sao Paulo, SP, Brazil

ABSTRACT. The geomagnetic field and its reversals are usually
attributed to magnetohydrodynamic (MHD) processes in the Earth's fluid
and metallic outer core. But changes in the reversal frequency
perhaps result from thermal or velocity perturbations at the
core-mantle boundary (CMB). Assuming that the reversals are endemic
to the geomagnetic dynamo behaviour, it is argued here that while the
chronological coincidence of reversal spurts and catastrophic episodes
does not necessarily imply their causal association, the observed
paucity of reversals and stronger fields during the ~35-60 Myr long
superchrons of fixed polarity state, require sources extrinsic to the
geomagnetic dynamo mechanism. Two likely sources, thermal cataclysms
at the CMB and impulsive changes in the tidal torque, are examined in
this context. The former seems more likely to have affected the
field's hemispheric asymmetry than its polarity stability or strength,
however, while the latter favours a causal link between the fixed
polarity superchrons and impulsive changes in the Earth's rotational
velocity.

1. INTRODUCTION

The palaeomagnetic conclusion that the geomagnetic field has reversed
its polarity often in the geological past (Jacobs, 1984) and hence
results through a process that can enable its intermittent destruction
and rejuvenation, has provided the strongest basis yet to postulate
the geomagnetic origin in magnetohydrodynamic (MHD) processes (Gubbins
and Roberts, 1987) in the Earth's fluid and metallic outer core. For
a given velocity field $\mathbf{v}$, the generation of magnetic field B here
obeys the induction equation which can be written as

$$\dot{\mathbf{B}} = R_m \, \nabla\times(\mathbf{v}\times\mathbf{B}) + \nabla^2\mathbf{B} \tag{1}$$

where R_m (= $\mu\sigma Uc$) is the magnetic Reynolds number associated with
zonal nongeostrophic shear, while μ, σ, U and c denote the magnetic
permeability, electrical conductivity, zonal flow speed of order

319

F. J. Lowes et al. (eds.), Geomagnetism and Palaeomagnetism, 319–334.
© 1989 by Kluwer Academic Publishers.

1-10 km/year, and the core radius, respectively, and dot denotes time-derivative.

The motions **v** that generate **B** have to be sufficiently complex, so as to circumvent Cowling's antidynamo theorem and its corollaries, and also vigorous enough to resist the back reaction, through Lenz's law, of the Lorentz force $(\nabla \times B) \times B$ which rises quadratically with B. Also, a given **v** can support **B** of either polarity, normal (N) or reversed (R). Indeed, the disc-dynamo analogue for the geomagnetic field (Bullard, 1955) (Figure 1) readily mimics the geomagnetic reversals, either when two dynamos are coupled (Rikitake, 1958) or when slight adjustments are made in the single dynamo itself (Robbins, 1977). The reversals in these models (Krause and Roberts, 1980) occur on a time scale consistent with the electromagnetic time scale $t_n = L^2 \mu \sigma$, where L ⩽ c, since turbulence can enhance the effective magnetic diffusivity of a fluid. For $\mu = 4\pi \times 10^{-7}$ H/m, $\sigma = (3-5) \times 10^5$ S/m and L = $(1-3.48) \times 10^6$ m, t_n works out to $10^4 - 10^{5.5}$ years. This compares well with the durations of individual N and R polarity epochs in the observed record, particularly if the progressive lengthening of these epochs towards a fixed polarity superchron reflects an extrinsic restraint, and suggests that **reversals are endemic to the geomagnetic dynamo behaviour.**

The presence of ~35-60 Myr long fixed polarity superchrons in the palaeomagnetic record (Harland et al., 1982) presents a problem

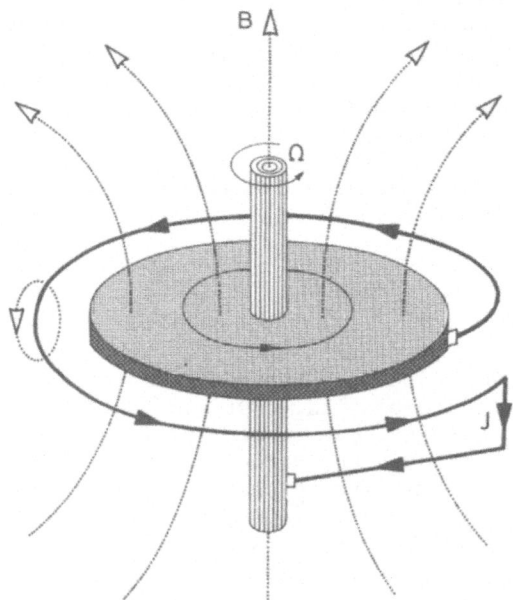

Figure 1. Faraday's homopolar dynamo explains the features of the Earth's magnetic field. The magnetic field **B** is created by a current of density J flowing in the circular loop round the periphery of the disc rotating with an angular velocity Ω. Modified from Bullard (1955) and Roberts (1987).

however. The models show no comparable periods of protracted polarity
stability (Crossley et al., 1986), when the dynamo's auto-reversing
mechanism is completely dormant, while the statistical (McFadden and
Merrill, 1986) and palaeointensity (Pal and Roberts, 1988) arguments
suggest that the dynamo regime changes significantly during such
superchrons. Indeed, the dynamo's sensitivity to perturbations at the
CMB (Braginskii, 1983) makes it plausible that such superchron-related
changes arise from sources extrinsic to the dynamo mechanism. An
extrinsic source, either extraterrestrial altogether or from within
the planetary interior, is also sometimes invoked to explain a ~30 Myr
harmonic in the superchrons of frequent reversals.

These two aspects of the long-term palaeofield behaviour, and
their implications for understanding the geomagnetic dynamo mechanism,
are examined here.

2. REVERSAL SPURTS AND EXTRATERRESTRIAL CATASTROPHISM

While several earlier studies had proposed ~30-700 Myr harmonics in
the chronostratigraphically poorly controlled palaeomagnetic reversal
data then available, the ~30 Myr harmonic has aroused considerable
interest as the likely geomagnetic signature of extraterrestrial
catastrophism. This harmonic has been recently reiterated from an
analysis (Stothers, 1986) of the magnetostratigraphically calibrated
Cox-82 reversal record (Harland et al., 1982) for the past ~165 Myr.
But the length of the Cretaceous N superchron (~35 Myr) is too close
to this harmonic for it to have been left uncontaminated. As Pal and
Creer (1986) have argued, it therefore appears more likely that,
rather than a harmonic, what can be asserted is the presence of
reversal spurts, spaced ~30 Myr apart, in the post-Santonian and
Barremian/Oxfordian segments of frequency reversals. Figure 2A
illustrates this by comparing the reversal frequency in the Cox-82
scale for a 15 Myr window with that computed by filtering the
harmonics ⩽18 Myr in the data obtained using a 5 Myr window. The
Oxfordian, late Valanginian, late Maastrichtian, late Eocene/early
Oligocene and middle/late Miocene reversal spurts are conspicuous here
and occur superposed over linearly decreasing trends towards the
Cretaceous N superchron. The argument (McFadden and Merrill, 1984),
that almost a third of the "true" reversals are perhaps missing from
the observed record, suggests that these spurts may well be stronger
than they appear here.

As the apparent non-stationarity of the reversal record makes
using a fixed-time window seem inappropriate here, the reversal
frequency in the Cox-82 scale is estimated in Figure 2B as the
reciprocal of mean interval length computed for a sliding window of 16
consecutive N and R intervals. While the increasingly wider
time-windows towards the Cretaceous N superchron may have subdued the
reversal spurts closer to it, the spurts seen here clearly corroborate
those in Figure 2A.

These reversal spurts chronologically coincide with the
extinction events (Raup, 1987), making it tempting to ascribe them to

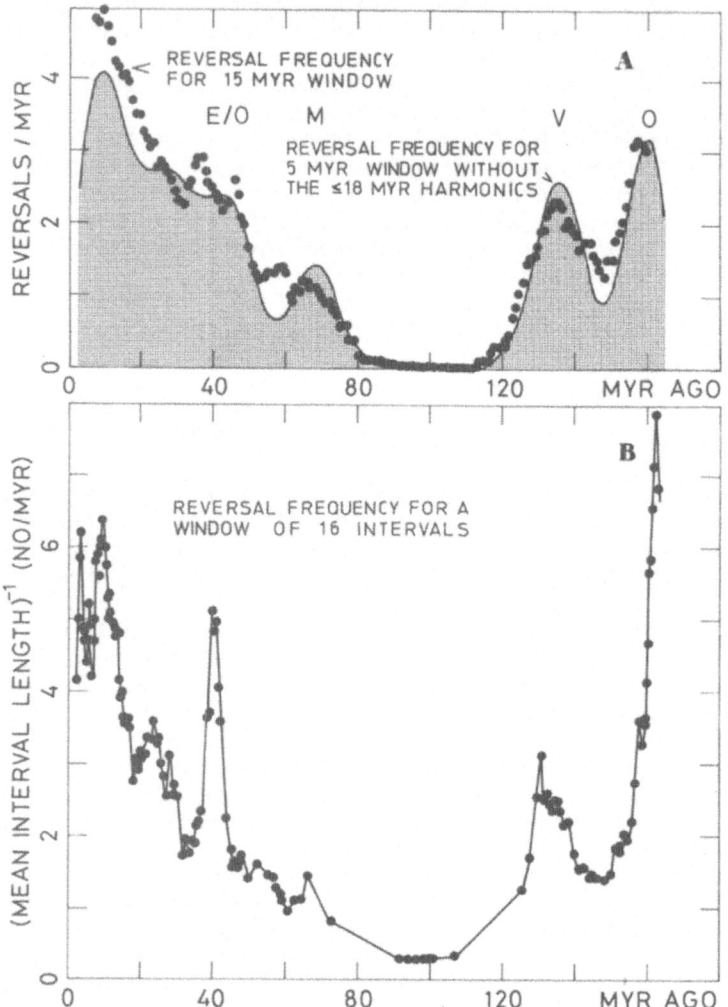

Figure 2. Reversal frequency in the Cox-82 (Harland et al., 1982) scale seen through rectangular windows having (A) fixed-time and (B) fixed-intervals.

extraterrestrial catastrophism. Note that a bolide impact can conceivably produce (Muller and Morris, 1986) an equatorial shear of $\sim 10^{-5}$ m/s at the core surface. The Proudman-Taylor constraint encourages core-convection in steady rolls aligned to the spin axis. For a convective cell of radius 10^6 m, the impact-induced shear ($\dot{e}$) suggests 10^{-2} W $\lesssim \nu (\dot{e})^2 V \leqslant 10^3$ W of viscous power dissipation in the core (volume V, viscosity $\nu = 10^{-2}$-10^3 kg^{2}m /s), a minuscule amount compared to the ohmic dissipation of 10^{12} W. But the impact-induced shear, being comparable to the zonal flow speed in eq. (1), can double R_m. In view of the inferred (Pal and Roberts, 1988) increase in R_m

during a superchron of frequent reversals, then impacts could perhaps
be taken to encourage the already occurring reversals, if the problem,
that the core-mantle coupling would restore the torque balance too
rapidly for a reversal to occur, is solved by assuming impact showers.

The thermal cataclysmic model (Loper and McCartney, 1986) offers
an attractive alternative here. But the reason why the same source,
viz. a thermal destabilization of D" layer, can cause reversals in one
case (Loper and McCartney, 1986) but a cessation of reversal activity
in the other (McFadden and Merrill, 1986), remains inexplicable.

The imperative of invoking an extra-dynamo source for the
reversal spurts in Figures 2A and 2B may seem arguable however. If
the dynamo's auto-reversals possess a Markovian memory, for instance,
then successive intervals will carry the cumulative effects of that
memory until it lasts. The periodicity apparent in Figure 3A, where
the reversal frequency variation of Figure 2B is shown as a function
of interval number rather than time, corroborates this. The
corresponding periodogram in Figure 3B suggests that the span of such
a memory could be ~40 intervals. But whether this reflects an
intrinsic periodicity in the dynamo's auto-reversals, or a flicker
noise input (Crossley et al., 1986), or the evolution of a
deterministic chaos (Grebogi et al., 1987), is difficult to discern.

3. THE FIXED POLARITY SUPERCHRONS

Turning now to the question of fixed polarity superchrons, note that
the best documented of them (Harland et al., 1982) occurred during
Cretaceous (N polarity: 83-118 Myr) and Permian (R polarity: 260-320
Myr) times. Although some reversals within these superchrons have
been reported, the overwhelming evidence is that of a stable polarity
of the geomagnetic field during these two superchrons. A similar
superchron is also likely to have characterized Cambrian times (R
polarity: 535-590 Myr) (Khramov and Rodionov, 1981).

Based on the magnetostratigraphically calibrated reversal record,
a gradual growth of reversal frequency following a fixed polarity
superchron is an important aspect of the long-term reversal behaviour.
No comparable time-dependent behaviour is seen in the gamma
distribution parameter however, suggesting (McFadden and Merrill,
1986) a gradual relaxation of the extrinsic restraining of the dyamo's
reversal mechanism that produced the Cretaceous N superchron. The
statistical similarity of the declining trends towards this
superchron, apparent in Figures 2A and 2B, is however sensitive to
chronostratigraphic scaling. The recent changes in the Jurassic/early
Cretaceous segment of the reversal record (Lowrie and Ogg, 1986) thus
make the decay of reversal rate before the Cretaceous superchron
appear far more rapid than its rise since then. This is also seen in
Figure 3A.

In terms of dynamo behaviour, perhaps an even more significant
aspect of the palaeomagnetic record is the strengthening of the
geomagnetic field with its protracted polarity stability during a
fixed polarity superchron. Note that although the palaeointensity

record is replete with uncertainties, Pal and Roberts (1988) have also marshalled marine and satellite magnetic evidence to support this postulate. Pronounced MAGSAT anomalies are indeed observed over the Cretaceous "Quiet Zones" in the central Pacific, north-central Atlantic and southwestern Indian Oceans, for instance, suggesting appreciably stronger magnetization of the ocean floor formed during the Cretaceous N superchron. This could be ascribed (LaBrecque and Raymond, 1985) to either a stronger magnetizing field in the middle Cretaceous or a stronger viscous remanence acquired during the Cretaceous N superchron itself, of which the former is consistent with the land-based palaeointensity data. As against this, the model (Johnson, 1985) ascribing the strong magnetization of Cretaceous ocean floor to a viscous imprint of the present field has, in the absence of

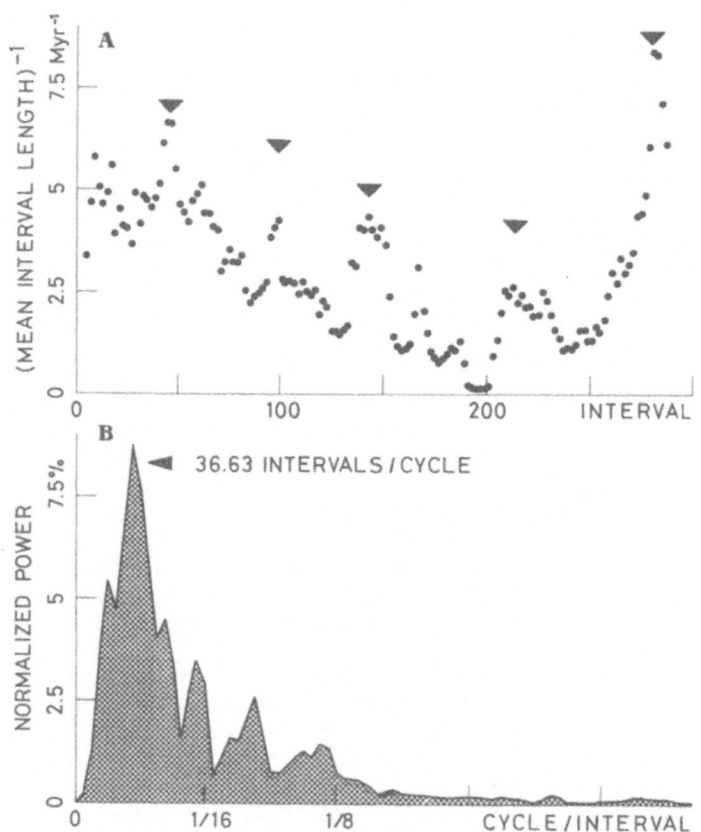

Figure 3. (A) the data of Figure 2(B) shown as the reversal frequency versus interval number suggest a possible Markovian memory. (B) The periodogram for this plot suggests that this Markovian memory may have a life of ~40 intervals.

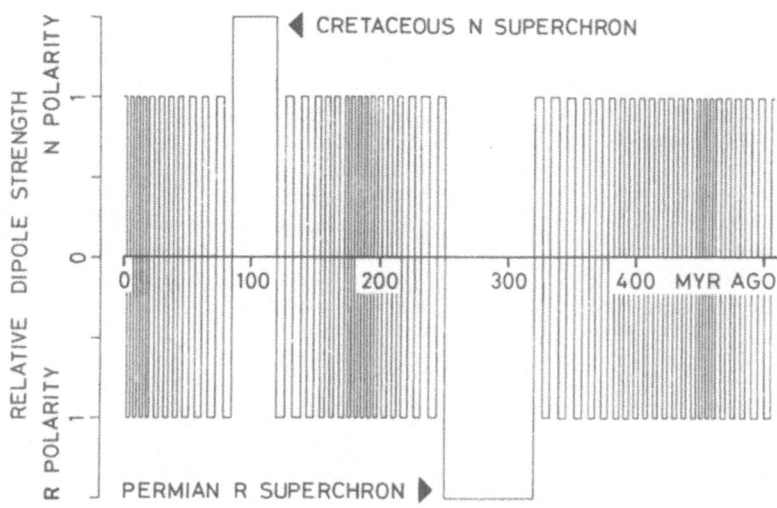

Figure 4. A schematic plot of the long-term palaeofield variations, suggesting alternating superchrons of frequent reversals and fixed polarity state, and appreciable strengthening of the field during the latter.

any significant rock magnetic peculiarities of this oceanic crust, rather untenable implications.

The tentative picture that emerges (Figure 4) is that the long-term palaeofield variations occur as superchrons of frequent reversals punctuated by those of fixed polarity state and appreciably stronger field. Within the former, the reversal frequency grows progressively following a fixed polarity superchron.

What specific peculiarities of the geodynamo behaviour are manifest here? The spherically-bounded core geometry allows a geophysically convenient decomposition of the axisymmetric parts of divergenceless v and B into their toroidal (v_T, B_T) and poloidal (v_p, B_p) parts. The field B_p is then repeatedly wrapped round the core by the zonal flow v_T possessing a non-zero shear (the differential motion or ω-effect), to produce a strong B_T which then reinforces B_p through the α-effect, a twisting of the field lines due to the effects of the waves (the asymmetric part v' of v) that drift longitudinally round the core. This is the strong-field or αω-dynamo (Braginskii, 1976). Since the α-effect can also produce B_T from B_p, although not as efficiently as the ω-effect does, a weak-field or α^2-dynamo (Busse, 1975) is also possible. Because of the problems of runaway growth and unrealistically weak fields but high dissipations in the α^2-dynamo (Soward, 1979; Roberts, 1987), and the evidences of sunspot-type flux patches of radial field (Gubbins and Bloxham, 1987) and fluid flow (Jault et al., 1988) on the core surface, we will focus on the αω-dynamo however. In this case, Hide (1982) has shown that

$$B_T = B_s(R_m)^{\frac{1}{2}} \qquad \text{and} \qquad B_P = B_s(R_m)^{-\frac{1}{2}} \qquad (2)$$

where $B_s = (\rho\Omega/\sigma)^{\frac{1}{2}}$ is the scale magnetic field strength, ρ is density and Ω is the rotational angular velocity. Note that since B_T does not escape through the mantle, the observed palaeointensity data pertain to B_P. The stronger palaeofield therefore signifies stronger B_P and weaker R_m during a fixed polarity superchron. Since an $\alpha\omega$-dynamo becomes regenerative when the dynamo number $D = R_\omega R_\alpha = O(1)$ (Parker, 1979), where $R_\omega \sim R_m$ and $R_\alpha = \alpha L \mu \sigma$ is the magnetic Reynolds number associated with the α-effect, this implies increases in R_α and α-effect during a fixed polarity superchron. The meridional flow v_P, which can change an oscillatory $\alpha\omega$-dynamo to a stationary one (Roberts, 1972), also becomes stronger during such times.

4. FIXED POLARITY SUPERCHRONS AND EXTRA-DYNAMO SOURCES

The possibility that a stronger α-effect can efficiently increase B_P now necessitates evidence of a relatively vigorous convective regime in the core, such as would enhance α, during a fixed polarity superchron. A thermal cataclysm at the CMB, say the destabilization of mantle's basal (D") layer, can conceivably affect this. Note the sensitive role that all kinds of core-mantle coupling play in Braginskii's (1976) model-Z dynamo, for instance. This justifies the invoking of such a mechanism by McFadden and Merrill (1986), to explain the Cretacous N superchron and the progressive decrease in reversal frequency towards it, even though their argument, based on Olson's (1983) model, does not explain why the two sources of helicity fluctuation in the core, thermal convection and segregation of the inner core, should compete rather than cooperate. As mentioned before, for instance, Loper and McCartney (1986) have ascribed spurts in the reversal activity to this mechanism.

If a thermal cataclysm at the CMB indeed produced the Cretaceous N superchron, then the resulting plume may well have eventually warmed up the mantle and thus enhanced the geothermal flux during the late Cretaceous/Eocene interval (Sprague and Pollack, 1980). Since the mantle's large thermal inertia would inhibit an explicit correlation of geomagnetic and geodynamic activities, this approximate correspondence should certainly suffice. But then, as shown in Figure 5A, this interval of high geothermal flux coincided with the sign reversal of the quadrupole field (Lee and Lilley, 1986). Perhaps the plume produced by this cataclysm itself had a hemispheric bias, depending on the latitude on the CMB at which the cataclysm occurred, as the geodynamic record indicates. The Phanerozoic profile of fractional heat flow variations (Turcotte and Schubert, 1982) also shows no comparable thermal surge during the Permian R superchron (Figure 5B).

It is not clear, therefore, whether a thermal cataclysm at the CMB promotes the field's polarity stability and strength or merely affects its hemispheric bias. Interestingly, Gubbins (1987) has recently sought an association between the growth of a hot patch at

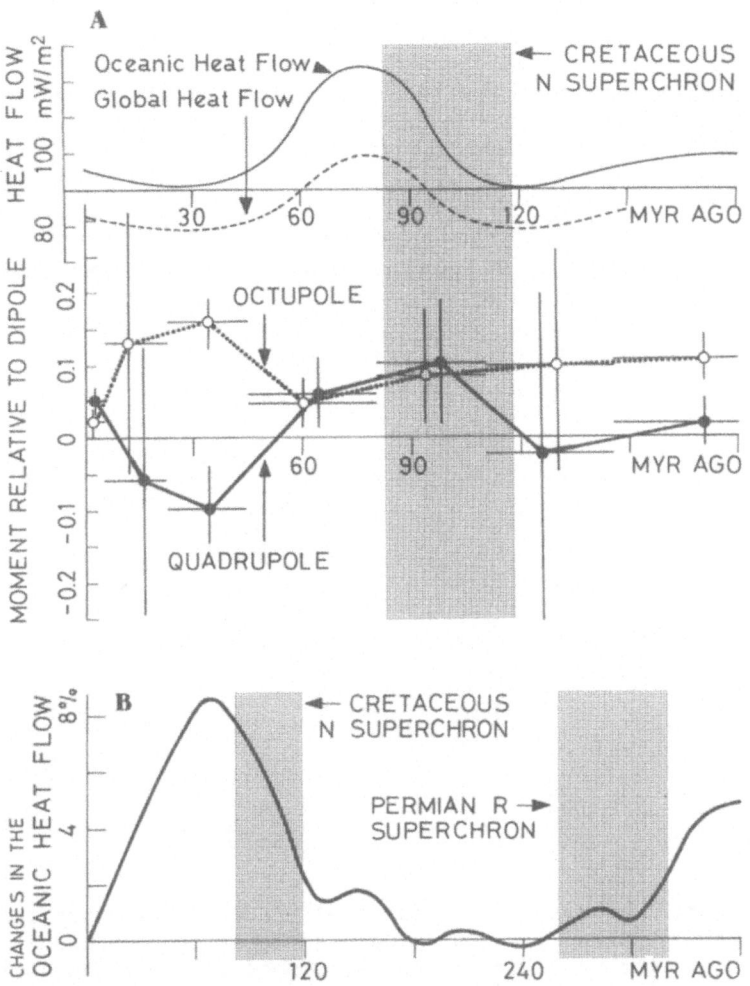

Figure 5. (A) The increased geothermal flux during late Cretaceous and Eocene (Sprague and Pollack, 1980) coincided with sign-reversal of the quadrupole field (Lee and Lilley, 1986). (B) The Cretaceous N superchron can be related to this geothermal surge, with a time-lag of ~30 Myr, but the changes in the mean oceanic heat flow since the Palaeozoic (Turcotte and Schubert, 1982) show no corresponding geothermal anomaly for the Permian R superchron.

the CMB beneath Africa and the increased reversal frequency since the late Cretaceous. But Pal's (1983) correlation of long wavelength geoid anomalies and Phanerozoic palaeocontinental configurations favours a longevity of such patches extending to the late Palaeozoic.

On the other hand, the long-term changes in the geodynamo behaviour show a better correspondence with the planetary rotational history. The astronomical, historical, and Phanerozoic palaeontological records show a ~5×10^{-22} rad/s² deceleration of Earth's rotation (Lambeck, 1980), for instance. This largely reflects lunar tidal interaction with the oceans and corresponds to a secular increase in the Moon's geocentric distance R.

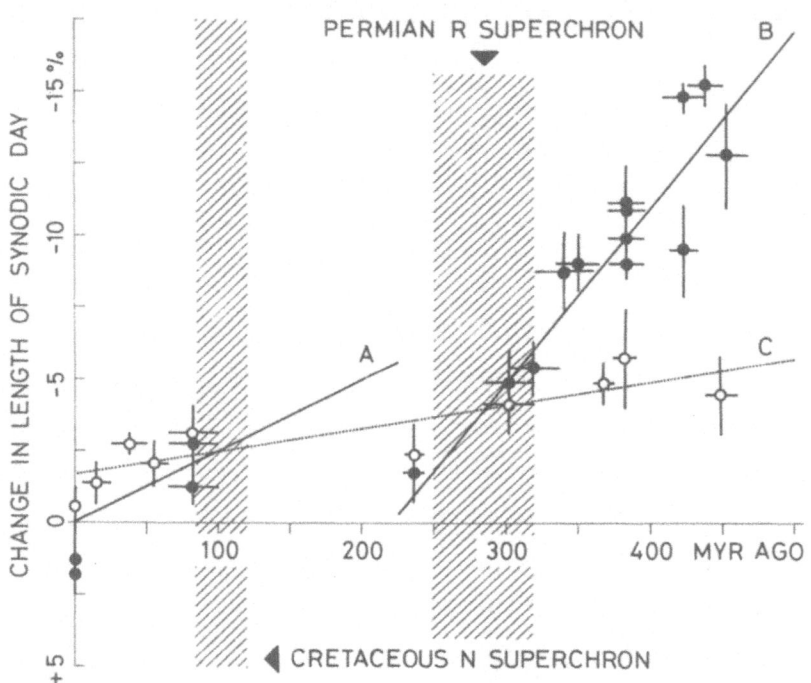

Figure 6. Earth's Phanerozoic rotational history shows that impulsive changes in the dominantly steady tidal deceleration coincided with the Permian R and Cretaceous N superchrons. Based on the palaeontological data summarized by Lambeck (1980). Open circles denote relative changes in the number of solar days per synodic (lunar) month, closed circles those in the number of solar days per year. Lines A, B and C are explained in the text.

Earth's tidal deceleration hardly remained constant throughout the geologic past however. Fluctuations in the Earth's rotation through the Phanerozoic are evident in Figure 6 which shows, relative to their present values, the changes in the planetary spin, inferred from the number of days per year, and those in the lunar orbital velocity, as inferred from the number of days per lunar month, based on the Phanerozoic palaeontological data summarized by Lambeck (1980). It is apparent here that the presently observed deceleration can be only extrapolated backwards (line A) until the late Mesozoic. The linear fit for the Palaeozoic data (line B) suggests a faster though steady deceleration during that era. As for the early and middle

Mesozoic, the data are sparse although a slower but steady deceleration, slightly above that given by the relative changes in the lunar orbital velocity (line C), is consistent with the available data for the middle Triassic. Sudden or impulsive changes in the planetary spin, perhaps attendant on enhanced dissipations from more extensive shallow seas (Krohn and Sundermann, 1982), thus coincided with the Permian and Cretaceous fixed polarity superchrons, whereas steady tidal torques characterized the superchrons of frequent reversals.

5. THE COUPLING OF ROTATIONAL AND GEODYNAMO CHANGES

The implicit assumption of Earth's non-stationary tidal deceleration is reasonable indeed - neither the terrestrial nor the lunar histories reveal any signs of the Gerstenkorn event that an Earth-Moon proximity within the Roche limit (Stacey, 1977; Lambeck, 1980) implies. A correlation between the rotational changes and the palaeomagnetic record has been attempted before (Creer, 1975). What is new here, however, is the correlation of impulsive changes in the Earth's tidal deceleration with the occurrences of fixed polarity superchrons.

What does this correlation connote? First of all, it suggests that the motions of the core and the mantle are coupled. Now, if Ω_M and Ω_c are the spin angular velocities of the mantle and the core, and I_M and I_c the corresponding inertial momenta, then the conservation of the whole Earth's angular momentum requires that

$$\hat{\Omega}_c = -(I_M/I_c)\hat{\Omega}_M = -8.5 \; \hat{\Omega}_M \tag{3}$$

where $I_M = 7.2 \times 10^{37}$ kg m^2 and $I_c = 8.5 \times 10^{36}$ kg m^2, i.e., a deceleration of the mantle's spin produces a nearly ten-fold spin-up of the core which, for the time scale of interest here, perhaps gets accomplished almost instantaneously as the core's spin-up time t_s ~4×10^{11} s (Gubbins and Roberts, 1987). But irrespective of whether the core-mantle coupling is dominantly electromagnetic or not (Lambeck, 1980; Runcorn, 1982; Courtillot and Le Mouel, 1984; Stix and Roberts, 1984), when the relative motion of the core increases so does the restoring couple, produced by the Faraday tension that opposes the tendency of the zonal flow associated with Ω_c to bow the entrained lines of B_P into the azimuthal direction, tending to reduce that relative motion.

What if changes in the fluid flow v indeed occurred consequent to these rotational changes? After all, even the zonal flow v_T, which is the dominant component of $v = (v_T + v' + v_P)$ in Braginskii's (1976) expansion scheme $(v_T/v') \sim (v_T/v_P) \sim (R_\omega)^{1/2}$ where R_ω is of order 10^2, barely corresponds to $\omega \sim 10^{-11}$ rad/s (Jault et al., 1988), i.e. ~$10^{-7} \; \Omega_c$. But even if the violations of Lenz's law occurred, the average tidal deceleration of ~5×10^{-22} rad/s^2 (Lambeck, 1980), with a doubling time for Ω_c and v comparable to the length of the Phanerozoic, is too weak to have produced any palaeomagnetically observable changes within the intervals of steady tidal torque.

The effects of the Permian and middle Cretaceous impulsive

changes in tidal torque on the asymmetric ($\mathbf{v}'$) and meridional ($\mathbf{v_p}$) components of $\mathbf{v}$ would hardly contravene Lenz's law on the other hand, since these flows are unaffected by the Coriolis forces and are therefore sensitive to external forcing. The large geostrophic shears created by the sudden changes in the tidal torque should then enhance both these flows immediately. As for the former, note that in Braginskii's (1976) model-Z dynamo, $\mathbf{v}'$ induces the asymmetric field $\mathbf{B}'$ from $\mathbf{B_T}$ which then induces through $\mathbf{v}'$ an electromotive force whose axisymmetric part is parallel to $\mathbf{B_T}$, thus generating the poloidal field $\mathbf{B_p}$ through the α-effect. Even a subtle rise in $\mathbf{v}'$ can thus enhance $\mathbf{B_p}$ substantially. The latter, viz. a concomitant enhancement in $\mathbf{v_p}$ through the Ekman suction, is also immediately relevant here. An efficient functioning of the $\alpha\omega$-dynamo requires (Steenbeck and Krause, 1969) a physical separation of the regions of greatest shear from those where the α-effect is maximum. Since the indirect but crucial role that $\mathbf{v_p}$ plays in the process of field generation is to advect $\mathbf{B_T}$ to regions of high ω-effect, and $\mathbf{B_p}$ to those of high α-effect, a large $\mathbf{v_p}$ can modify the field and lower the critical D for self-excitation (Braginskii, 1976, 1983). Indeed, even in an α^2-dynamo, a non-zero meridional circulation is known (Krause and Radler, 1980) to excite the axisymmetric field more readily. But what is most interesting here is that a suppression of the dynamo's auto-reversing mechanism, as during a fixed polarity superchron, becomes a natural corollary of a stronger $\mathbf{v_p}$ when we note, following Roberts (1972), that in the case of a kinematic $\alpha\omega$-dynamo, the field most readily excited to dynamo action is a stationary dipole if meridional circulation is sufficiently large and $\alpha\omega > 0$ but an oscillating dipolar solution is preferred when it is weak (or absent) and $\alpha\omega < 0$.

Although speculative, perhaps this is what the observed data (e.g. Figure 2A), which show a rapid decrease in the reversal rate before the Cretaceous N superchron but a gentler rise since then, and far more pronounced spurts prior to this superchron than since then, indeed imply. After all, increases in the meridional flow, attendant on the impulsive changes in rotation during Permian and middle Cretaceous times, must have been sudden while the restoration of the pre-existing steady state must have taken longer.

6. DISCUSSION

Two inferences therefore appear appropriate. One is that, although the irregularly occurring palaeogeomagnetic reversals can be considered endemic to the geodynamo behaviour, the fixed polarity superchrons reflect basic changes in the core's MHD state. A decomposition of the palaeomagnetic reversal record into alternate superchrons of fixed polarity state and frequent reversals (Figure 4) is thus justified. The other inference is that such changes in the core's MHD state probably result from sudden changes in the planetary deceleration, as the resulting geostrophic shears enhance the meridional and asymmetric flows which then strengthen the observed field and also suppress its reversal activity during a fixed polarity superchron.

The inferred importance of asymmetric flow during a fixed polarity superchron need not translate into any palaeomagnetically observable increased separation between the geomagnetic and geographic axes. Note that although $B'/B_p \sim R_\omega^{\frac{1}{2}}$ within the dynamo region (Braginskii, 1976), this relation is reversed in the relatively insulating mantle. Thus, if R_ω remains steady, as would result from a steady v_T, then so will the tilt of the dipole axis. The palaeosecular variations too may then remain largely unaffected, or may be slightly reduced at best (Merrill and McElhinny, 1983), if an enhanced geostrophic shear is assumed to discourage the dipole's wobble. Perhaps such a shear also retards any slippage of the mantle past the core, thus explaining the observation (Courtillot and Besse, 1987; Gordon, 1987) of a reduced polar wander during the fixed polarity superchrons.

The rather speculative model proposed here also suggests that the reversal behaviour may not be as sensitive to thermal changes at the CMB as is sometimes suspected. Perhaps this is the reason why it is not clear whether such changes inhibit the reversals (McFadden and Merrill, 1986) or promote them (Loper and McCartney, 1986). It may well be that changes in the thermal regime at the CMB affect only the nondipole field. Another interesting corollary of this model is that if spasmodic rotational changes can affect the geodynamo so dramatically as to cause the fixed polarity superchrons then the Poincaré force (Malkus, 1968) probably plays a greater role in the workings of the core MHD than Rochester et al. (1975) have argued. The most intriguing, however, is the possibility that, by recording a relatively minor part of the geodynamo mechanism, viz., the role of the asymmetric and meridional flows and fields, the palaeomagnetic relevance to this mechanism critically hinges on the latter indeed being such that, to quote Roberts (1987), "the tail can wag the dog"!

ACKNOWLEDGEMENTS

I thank Professors Paul Roberts, Kenneth Creer and Igor Pacca for their critical comments and helpful discussions.

REFERENCES

Braginskii, S.I., 1976. 'On the nearly axially-symmetrical model of the hydromagnetic dynamo of the Earth'. Phys. Earth Planet. Int., 11, 191-199.

Braginskii, S.I., 1983. 'On the theories of the geomagnetic field and its secular variations' in Magnetic Field and the Processes in the Earth's Interior (Ed: V. Bucha et al.), 308-332 (Academia, Prague).

Bullard, E.C., 1955. 'The stability of a homopolar dynamo', Proc. Camb. Phil. Soc., 51, 744-760.

Busse, F.H., 1975. 'A model of the geodynamo', Geophys. Jour. R. Astr. Soc., 42, 437-459.

Courtillot, V. and Le Mouel, J.L., 1984. 'Geomagnetic secular variation impulses', Nature, 311, 709-716.

Courtillot, V. and Besse, J., 1987. 'Magnetic field reversals, polar wander, and core-mantle coupling', Science, 237, 1140-1147.

Creer, K.M., 1975. 'On a tentative correlation between changes in the geomagnetic polarity bias and reversal frequency and the Earth's rotation through Phanerozoic time' in Growth Rhythms and the History of Earth's Rotation (Ed: G.D. Resenberg and S.K. Runcorn), 293-317 (Wiley, London).

Crossley, D., Jensen, O. and Jacobs, J., 1986. 'The stochastic excitation of reversals in simple dynamos', Phys. Earth Planet. Int., 42, 143-153.

Gordon, R.G., 1987. 'Polar wandering and paleomagnetism', Ann. Rev. Earth Planet. Sci., 15, 567-593.

Grebogi, C., Ott, E. and Yorke, J.A., 1987. 'Chaos, strange attractors, and fractal basin boundaries in nonlinear dynamics', Science, 238, 632-638.

Gubbins, D., 1987. 'Mechanism for geomagnetic polarity reversals', Nature, 326, 167-169.

Gubbins, D. and Bloxham, J., 1987. 'Morphology of the geomagnetic field and implications for the geodynamo', Nature, 325, 509-511.

Gubbins, D. and Roberts, P.H., 1987. 'Magnetohydrodynamics of the Earth's core' in Geomagnetism (Ed: J.A. Jacobs), 2, 1-183 (Academic Press, London).

Harland, W.B., Cox, A., Llewellyn, P.G., Pickton, C.A.G., Smith, A.G. and Walters, R., 1982. A Geologic Time Scale (Cambridge Univ. Press, Cambridge).

Hide, R., 1982. 'On the role of rotation in the generation of magnetic fields by fluid motions', Phil. Trans. R. Soc. Lond., A306, 223-234.

Jacobs, J.A., 1984. Reversals of the Earth's Magnetic Field (Adam Hilger, Bristol).

Jault, D., Gire, C. and Le Mouel, J.L., 1988. 'Westward drift, core motions and exchanges of angular momentum between core and mantle', Nature, 333, 353-356.

Johnson, B.D., 1985. 'Viscous remanent magnetization model for the Broken Ridge satellite magnetic anomaly', Jour. Geophys. Res., 90, 2640-2646.

Khramov, A.N. and Rodionov, V.P., 1981. 'The geomagnetic field during Palaeozoic time' in Global Reconstructions and the Geomagnetic Field during the Palaeozoic (Ed: M.W. McElhinny, A.N. Kramov, M. Ozima and D.A. Valencio), 99-115 (D. Reidel, Dordrecht).

Krause, F. and Radler, K.H., 1980. Mean-Field Magnetohydrodynamics and Dynamo Theory (Pergamon, Oxford).

Krause, F. and Roberts, P.H., 1980. 'Strange attractor character of large-scale non-linear dynamos', Adv. Space Res., 1, 231-240.

Krohn, J. and Sundermann, J., 1982. 'Paleotides before the Persian' in Tidal Friction and the Earth's Rotation (Ed: P. Brosche and J. Sundermann), 190–209 (Springer-Verlag, Berlin).

LaBrecque, J.L. and Raymond, C.A., 1985. 'Seafloor spreading anomalies in the Magsat field of the North Atlantic', Jour. Geophys. Res., 90, 2565–2575.

Lambeck, K., 1980. The Earth's Variable Rotation: Geophysical Causes and Consequences (Cambridge Univ. Press, Cambridge).

Lee, S. and Lilley, F.E.M., 1986. 'On paleomagnetic data and dynamo theory', Jour. Geomag. Geoelectr., 38, 797–806.

Loper, D.E. and McCartney, K., 1986. 'Mantle plumes and the periodicity of magnetic field reversals', Geophys. Res. Lett., 13, 1525–1528.

Lowrie, W. and Ogg, J.G., 1986. 'A magnetic polarity time scale for the early Cretaceous and late Jurassic', Earth Planet. Sci. Lett., 76, 341–349.

Malkus, W.V.R., 1968. 'Precession of the Earth as the cause of geomagnetism', Science, 160, 259–264.

McFadden, P.L. and Merrill, R.T., 1984. 'Lower mantle convection and geomagnetism', Jour. Geophys. Res., 89, 3354–3362.

McFadden, P.L. and Merrill, R.T., 1986. 'Geodynamo energy source constraints from palaeomagnetic data', Phys. Earth Planet. Int., 43, 22–33.

Merrill, R.T. and McElhinny, M.W., 1983. The Earth's Magnetic Field: Its History, Origin and Planetary Perspective (Academic Press, London).

Muller, R.A. and Morris, D.E., 1986. 'Geomagnetic reversals from impacts on the Earth', Geophys. Res. Lett., 13, 1177–1180.

Olson, P., 1983. 'Geomagnetic polarity reversals in the turbulant core', Phys. Earth Planet. Int., 33, 260–274.

Pal, P.C., 1983. 'Palaeocontinental configurations and geoid anomalies', Nature, 303, 513–516.

Pal, P.C. and Creer, K.M., 1986. 'Geomagnetic reversal spurts and episodes of extraterrestrial catastrophism', Nature, 320, 148–150.

Pal, P.C. and Roberts, P.H., 1988. 'Long-term polarity stability and strength of the geomagnetic dipole', Nature, 331, 702–705.

Parker, E.N., 1979. Cosmical Magnetic Fields: Their Origin and Their Activity (Clarendon, Oxford).

Raup, D.M., 1987. 'Mass extinction: a commentary', Palaeontology, 30, 1–13.

Rikitake, T., 1958. 'Oscillations of a system of disk dynamos', Proc. Camb. Phil. Soc., 54, 89–105.

Robbins, K.A., 1977. 'A new approach to subcritical instability and turbulent transitions in a simple dynamo', Math. Proc. Camb. Phil. Soc., 82, 309–325.

Roberts, P.H., 1972. 'Kinematic dynamo models', Phil. Trans. Soc. Lond., A272, 663–703.

Roberts, P.H., 1987. 'Origin of the Main Field: Dynamics' in Geomagnetism (Ed: J.A. Jacobs), 2, 251–306 (Academic Press, London).

Rochester, M.G., Jacobs, J.A., Smylie, D.E. and Chong, K.F., 1975. 'Can precession power the geomagnetic dynamo?', Geophys. Jour. R. Astr. Soc., **43**, 661–678.

Runcorn, S.K., 1982. 'The role of the core in irregular fluctuations of the Earth's rotation and the excitation of the Chandler wobble', Phil. Trans. R. Soc. Lond., **A306**, 261–270.

Soward, A.M., 1979. 'Convection driven dynamos', Phys. Earth Planet. Int., **20**, 134–151.

Sprague, D. and Pollack, H.N., 1980. 'Heat flow in the Mesozoic and Cenozoic', Nature, **285**, 393–395.

Stacey, F.D., 1977. Physics of the Earth (John Wiley, New York).

Steenbeck, M. and Krause, F., 1969. 'Zur dynamotheorie stellarer und planetarer magnetfelder. I. Berechnung sonnenahnlicher wecheselfeldgeneratoren', Astron. Nachr., **291**, 49–84.

Stix, M. and Roberts, P.H., 1984. 'Time-dependent electromagnetic core-mantle coupling', Phys. Earth Planet. Int., **36**, 49–60.

Stothers, R.B., 1986. 'Periodicity of the Earth's magnetic reversals', Nature, **322**, 444–446.

Turcotte, D.L. and Schubert, G., 1982. Geodynamics: Application of Continuum Physics to Geological Problems (John Wiley, New York).

COMPOSITIONAL CONVECTION AND THE EARTH'S CORE

David R. Fearn
Department of Mathematics
University of Glasgow
University Gardens
Glasgow, G12 8QW
U.K.

ABSTRACT. The Earth is cooling and the solid inner core is growing as
the outer core freezes. The outer core is composed primarily of iron
but it contains a small fraction of some lighter constituent.
Freezing such a mixture leaves most of the light constituent in the
remaining fluid (the outer core), the frozen solid (the inner core)
being nearly pure iron. The fluid close to the freezing interface
[the inner-core boundary (ICB)] is therefore rich in the light
constituent. This effect together with the latent heat released by
the freezing process produces buoyant fluid at the ICB. This fluid
rises, stirring the outer core and so is the source of the energy
required to maintain the geomagnetic field against ohmic losses. This
picture of the evolution of the Earth's core is now widely accepted,
but the consequences of the core being a freezing mixture have yet to
be fully explored. Here we review the background to this theory and
discuss its consequences for the structure of the inner core, the
structure of the complex freezing region at the ICB, the flow in this
region and the flow in the outer core.

1. INTRODUCTION

Typical strengths of the Earth's magnetic field measured today are
comparable with palaeomagnetic determinations of the field strength
over the past 3×10^9 years [see for example Stacey (1977)]. When
this observation is compared with the ohmic decay timescale (the
e-folding time) which is of the order of 10^5 years (based on the
radius of the core), it is immediately obvious that the field cannot
be a fossil field. Some mechanism must continually be regenerating
the field, the power source being just sufficient to balance the ohmic
losses. The most plausible mechanism is that provided by dynamo
theory [see for example Moffatt (1978), and Fearn, Roberts and Soward
(1988)]. The motion of a conducting fluid in a magnetic field causes
electrical currents to flow. Under certain conditions, these induced
currents produce a magnetic field that reinforces the original field
and in this manner the field may be maintained against ohmic losses.

335

F. J. Lowes et al. (eds.), Geomagnetism and Palaeomagnetism, 335–346.
© *1989 by Kluwer Academic Publishers.*

A key problem in dynamo theory is the source of the fluid motions in the outer core, i.e. what is the power source for the geodynamo?

For many years the favoured power source was thermal convection due to radioactive heat sources distributed throughout the core, but more recently this has fallen out of favour; for two reasons. First, it remains a matter of speculation how much, if any, Potassium 40 is dissolved in the outer core [see Brett (1976), Verhoogen (1980), and Liu (1986)], and second, such a heat source is an inefficient means of driving fluid motions in the core. This is principally because a large amount of heat can be <u>conducted</u> down the adiabatic temperature gradient. Typical estimates of the adiabatic heat flux at the core–mantle boundary (CMB) are in the range $2-4 \times 10^{12}$ W [see Loper and Roberts (1983)]. Only radiogenic heat in excess of this amount can contribute to convection in the outer core. Precessional torques acting at the CMB have been suggested as an alternative source of fluid motions in the core. However, the studies of Loper (1975) and Rochester et al. (1975) predict that precession is an inefficient means of stirring the core, most energy being dissipated in a thin boundary layer at the CMB.

More recently, it has become widely accepted that the Earth is cooling [see for example Stacey (1980)], and the solid inner core has grown through the freezing of the outer core. The freezing releases latent heat and (as described in Section 3) compositionally buoyant fluid. Together these may provide an adequate power source for the geodynamo (see Section 5).

Verhoogen (1961) was the first to associate freezing in the core with the dynamo power source. He discounted the contribution of the chemical segregation associated with the freezing of a mixture, preferring convection driven by the latent heat released during the crystallization of the inner core and the specific heat given out by the cooling core. Braginsky (1963) was the first to recognise the importance of compositional effects. His study was motivated by the belief that seismological observations required the presence of a layer (known as the F-layer) in the outer core, adjacent to the ICB in which the concentration of Fe increased with depth. Braginsky showed how the freezing of an iron–poor core (a core containing a <u>smaller</u> mass fraction of Fe than a eutectic mixture) could explain the presence of the F-layer. In recent Earth models, the evidence for an F-layer is much weaker [see for example Bolt (1982)], and it seems unlikely that an iron–poor core is consistent with the presence of the inner core [Fearn and Loper (1983, 85)]. However, Braginsky's (1963) work is important in being the first to propose compositionally driven convection as the dynamo power source.

More than a decade later, as the inadequacy of radiogenic heating and precession was realised, Braginsky's idea was revived [see Gubbins (1976, 77), Loper (1978, a,b), and Gubbins et al. (1979)]. These studies demonstrated that large quantities of energy are available through the continual segregation caused by the freezing of the inner core. The dense iron is concentrated towards the centre, the light constituent(s) in the shrinking outer core, releasing gravitational potential energy. The dynamo powered by this process is often

therefore called the "gravitationally powered dynamo". Estimates of
the power available vary (see Section 5) but conductive losses are
much smaller than for thermal buoyancy and it seems likely that a
substantial magnetic field can be generated by this mechanism.

2. COMPOSITION OF THE EARTH'S CORE

The Earth's core cannot be directly sampled, so our knowledge of its
composition is all based on indirect information. Earth models based
on seismic travel time and free oscillation data impose strong
constraints upon the density distribution [see for example the review
by Bolt (1982)]. The relative abundances of the various constituents
of the Earth are inferred from those of chondritic meteorites [see for
example Brett (1976)], and it is concluded that the primary
constituent of the core is Fe. Comparison of the density of Fe at
core pressures [see for example Ahrens (1980), and Anderson (1986)]
with the seismically determined core density shows the outer core to
be some 10% lighter than pure Fe. The inner core density is
consistent with it being pure Fe [Anderson (1986)], but may contain
some light constituent [Jephcoat and Olson (1987)]. There is a
density jump $\Delta\rho$ at the ICB. Its magnitude is not well determined; for
example, Masters (1979) finds $\Delta\rho = 0.87 \pm 0.32$ g cm^{-3}, the model PREM
[Dziewonski and Anderson (1981)] has $\Delta\rho = 0.63$ g cm^{-3}, while CAL8
[Bolt (1982)] has 1.17 g cm^{-3} [see also Anderson (1986)]. What is
clear is that this density jump is larger than can be due to the
difference in density between solid and liquid Fe. Estimates for the
latter range from 0.05 g cm^{-3} [Masters (1979)], through 0.12 g cm^{-3}
[Poirier (1986)], to 0.17 g cm^{-3} [Spiliopoulos and Stacey (1984)].
This all leads to the conclusion that the outer core is not pure iron
but an alloy of iron and some lighter constituent.

Many elements have been suggested as the light constituent; S,
Si, O and others [see Brett (1976), Stevenson (1981)]. The most
popular candidate is Sulphur, with 9-13 wt % S required to explain the
outer core density [Ahrens and Jeanloz (1987)]. McCammon et al.
(1983) argue in favour of an Fe-FeO core [see also Knittle and Jeanloz
(1986)], with some 8% O required [Ahrens (1980)]. This concentration
of O may be larger than that required for a eutectic Fe-FeO mixture
[McCammon et al. (1983)] and this is unlikely to be consistent with
the presence of the inner core [Fearn and Loper (1983, 85)]. We shall
not concern ourselves further with the identity of the light
constituent (which may be a mixture of the candidates together with
traces of other elements); it is sufficient that the iron is mixed
with some lighter constituent X. In the following section we discuss
the freezing of such a mixture (which is a much more complicated
process than the freezing of a pure substance) and show that the
differences in composition and density of the inner and outer cores
arise as a natural consequence of the cooling of the core.

3. FREEZING OF THE INNER CORE

Two features characterize the freezing of a mixture. First, the
temperature below which freezing occurs (the liquidus temperature T_L)
depends on the composition of the fluid. Second, the composition of
the solid is, in general, different from that of the fluid from which
it has frozen [see for example, Chalmers (1964)]. A familiar example
is that of salt water. Its freezing temperature falls as the
concentration of salt is increased and the solid that freezes when
salt water is cooled is almost pure H_2O, the salt being rejected into
the remaining fluid. Most authors model the core as a binary Fe-X
alloy. This is almost certainly a gross oversimplification but the
model is the simplest that includes the essential physics; the two
features discussed above. The model is far from simple, though, and
many details of the freezing process are still poorly understood [see
the work by Hills and Roberts (1987a, b, 88)].

The principal features of the core are easily explained by the
freezing of an Fe-X mixture. Starting with a completely molten core
containing a mass fraction ξ of X, the temperature falls as the Earth
cools until, somewhere in the core, the liquidus temperature is
reached. The temperature T in the core varies with depth, and hence
with pressure p, so $T = T(p)$. The liquidus temperature is a function
of pressure (as well as of composition), so $T_L = T_L(p, \xi)$. The
temperature gradient dT/dp cannot be steeper than the adiabatic
gradient dT_A/dp (since steeper gradients are convectively unstable),
so provided $dT_L/dp > dT_A/dp$ the liquidus temperature is reached first
at the centre of the core, with $T > T_L$ elsewhere. It is generally
accepted that this is the case [see for example Verhoogen (1980),
Stevenson (1981), Loper and Roberts (1983)], so freezing takes place
first at the centre of the core. The solid that freezes contains a
smaller mass fraction ξ_S of X than the fluid outer core ($\xi_S < \xi$). (If
the inner core is pure iron, then $\xi_S = 0$.) As cooling continues,
freezing proceeds, resulting in the growth of a dense solid inner
core. The density difference between the inner and outer cores is
explained by the larger concentration of the light constituent in the
fluid outer core.

The difference in composition between fluid and solid means that
if in a time dt, a mass dM of solid freezes (containing mass ξ_S dM of
X) then a mass $(\xi - \xi_S)$dM of X must be rejected into the fluid above.
The upward flux I of X at the ICB is therefore proportional to the
rate of growth of the inner core; $I = (\xi - \xi_S)dM/dt$. The heat
released by the freezing inner core is $Q = LdM/dt$ where L is latent
heat of fusion. Both this heat and the excess light constituent must
be removed from the ICB into the outer core if freezing is to
continue. There can be no fluid motion normal to the boundary so both
the heat and the excess X must be removed from the immediate vicinity
of the ICB by conduction.

Let us first consider a flat ICB. The heat flux Q is
proportional to the temperature gradient, so

$$Q = \rho gkA \; dT/dp, \tag{1}$$

and the upward flux of X is given by

$$I = \rho^2 gDA \; [d\xi/dp + (\partial\rho^{-1}/\partial\xi)_{p,T}/(\partial\mu/\partial\xi)_{p,T}], \tag{2}$$

where ρ is the density, g the gravitational acceleration, k the thermal conductivity, A the surface area, D the material diffusion coefficient and μ the chemical potential, [neglecting the Soret effect, see Loper and Roberts (1983), and Landau and Lifshitz (1959)]. The second term on the right hand of (2) is known as the barodiffusive flux, and is positive. From (2) it is clear that for small I, $d\xi/dp < 0$, i.e. the concentration of the light constituent decreases with depth, a stable configuration. It is only if I exceeds the barodiffusive flux that the core is compositionally unstable (light fluid below heavy) and compositionally driven convection ensues. The barodiffusive flux of X is thus a source of inefficiency for the dynamo in a manner similar to the conduction of heat down the adiabatic gradient. However, because of the small diffusivity D of X in Fe [Loper and Roberts (1983), Poirier (1988)], the barodiffusive flux is a much less serious source of energy loss than the adiabatic heat flux [Gubbins et al (1979), Loper and Roberts (1983)].

As the rate of growth (dM/dt) of the inner core is increased, the fluxes Q of heat and I of light constituent increase in proportion. From (1) and (2) this requires that dT/dp and $d\xi/dp$ increase. The ICB is at the liquidus temperature T_L and

$$dT_L/dp = (\partial T_L/\partial p)_\xi + (\partial T_L/\partial\xi)_p \; d\xi/dp. \tag{3}$$

The liquidus temperature T_L decreases when ξ increases, so $(\partial T_L/\partial\xi)_p < 0$. As dM/dt increases, $d\xi/dp$ increases, and hence [from (3)] dT_L/dp decreases. Since dT/dp increases as the growth rate increases, there must be some critical growth rate for which $dT_L/dp = dT/dp$. For larger growth rates $dT_L/dp < dT/dp$ at the ICB. Since $T = T_L$ at the ICB, this requires $T < T_L$ above the ICB, i.e. the fluid above the ICB is <u>below</u> its freezing temperature. This situation is known as constitutional supercooling and will occur if the rate of crystallisation of the inner core is sufficiently high. Loper and Roberts (1981) estimate the inner-core growth rate to be about 500 times the critical growth rate.

4. STRUCTURE OF THE INNER CORE AND INNER CORE BOUNDARY

In practice, a state with fluid above a flat interface being below its freezing temperature is unstable to the growth of solid, usually in the form of dendrites. In this manner the system greatly increases the surface area over which freezing takes place, permitting much larger diffusive fluxes of heat and light constituent. Once sufficiently far away from the freezing interface, the excess heat and light constituent can be convected into the outer core. From Loper and Roberts (1981) estimate of the critical growth rate, it seems very likely that the ICB is not flat, but consists of a region, partly fluid and partly solid, in which freezing is taking place. In the metallurgical literature, such a region is called a mushy zone.

The depth of a mushy zone is dependent on the temperature gradient in the fluid and in a laboratory is typically of the order of a few centimetres [see for example Copley et al. (1970)]. In the Earth, temperature gradients are much smaller, and in addition there is the effect of pressure on the liquidus temperature. Fearn et al. (1981) estimate the mushy zone depth to be of the order of thousands of kilometres. In other words, the mushy zone extends all the way from the ICB to the centre of the inner core. What does this mean in practice?

The bottom on the mushy zone is determined by the liquidus temperature $T_e(p)$ of a eutectic mixture. The statement that the mushy zone extends to the centre of the core means that the temperature at the centre of the core exceeds $T_e(p_c)$ where p_c is the pressure at the centre of the core. We envisage the freezing process taking place as follows. Fluid from the outer core flows into a porous dendritic region at the ICB. As it descends, a fluid parcel with composition ξ reaches its liquidus temperature. Freezing commences and the dendrites become coated in solid of composition ξ_s. The excess light constituent and latent heat rejected by the freezing process are transported by the fluid parcel. As freezing proceeds, the temperature of the parcel and the mass fraction ξ of light constituent in the parcel increase. What remains of the parcel becomes buoyant and eventually rises back into the outer core. The details of this freezing process are very complicated. The first attempts to describe the process [Roberts and Loper (1983), and Hills et al. (1983)] use equilibrium thermodynamics, an approximation valid only if melting and freezing take place rapidly in response to changes in temperature, pressure and composition. More recent work is aimed at understanding the effect of finite melting/freezing times, though only at present for a pure material, not a mixture, see Hills and Roberts (1987a, b). Even if melting and freezing are rapid, equilibrium thermodynamics may still not be valid because negligible diffusion of light constituent in the solid phase means that ξ_s cannot adjust to changing p and T. Hills and Roberts (1988) have constructed a "quasi-equilibrium" theory which takes this into account, but limit themselves to freezing only. Melting is more complicated, requiring a knowledge of the previous history of the system.

The depth of the region in which there is a significant fraction of fluid is not known. Loper (1983) estimates it to be of the order of one kilometre. Going deeper in the mushy zone, the fraction of solid increases. To escape to the outer core, fluid faces an increasingly tortuous journey. As freezing proceeds, there is the possibility of parcels of fluid becoming trapped. As the Earth continues to cool, freezing will take place within such fluid inclusions. In a similar manner to the fluid parcel described above, the mass fraction ξ of the light constituent in the inclusion will increase, and since $\partial T_L/\partial \xi < 0$, the liquidus temperature of the

remaining fluid in the inclusion will fall. This process will
continue until the fluid reaches is eutectic composition. It is only
if T falls below T_e that all the fluid will freeze. The work of Fearn
et al. (1981) (described earlier) shows that it is likely that $T > T_e$
at the centre of the inner core and hence there is the possibility
that the inner core is not completely solid but contains fluid
inclusions throughout its interior. The size, composition and
distribution of such inclusions must depend on the detailed structure
of the freezing region at the ICB. At present, no estimates are
available, but seismological observations may provide a clue. The
absorption of seismic waves in the inner core is anomalously high
[Doornbos (1974), Cormier (1981)]. This may be explained by
absorption associated with melting/freezing and diffusion in fluid
inclusions in the inner core [Loper and Fearn (1983)]. The absorption
is dependent on the size, density and distribution of the inclusions,
so seismic observations may indirectly provide constraints on these
quantities.

Recent analyses of travel time and free oscillation data have
provided evidence for the inner core being anisotropic, with the
rotation axis being an axis of symmetry [see Morelli et al. (1986),
Woodhouse et al. (1986), and Cormier (1986)]. If the inner core has
grown through crystallization of Fe from the outer core, then these
observations suggest that the Earth's rotation and perhaps
differential rotation in the outer core play an important role in the
freezing process. So far, no studies have incorporated these effects
into models of the freezing inner core.

5. FLOW IN THE OUTER CORE AND THE GEODYNAMO

The nature of the convection in the outer core is of great interest
for constructing realistic dynamo models. Our only clues to the flow
above the mushy zone come from laboratory experiments. Copley et al.
(1970) find that upward flow out of the mushy zone is concentrated in
narrow "chimneys". The upward flow of buoyant (hot and light) fluid
is fast, in contrast to the slow downward flow into the mushy zone
elsewhere [see also Roberts and Loper (1983), Loper (1983)]. In the
experiments, the plume of buoyant fluid may rise all the way to the
top of the fluid layer. In the Earth it is possible that the same may
happen (creating a stably stratified layer beneath the CMB), though
the plume is subject to many influences in the outer core; Lorentz
forces, Coriolis forces and strong differential rotation, for example,
Moffatt (1988) has made the first attempt to incorporate some of these
effects into a model of compositional convection. He envisages the
buoyant material rising in distinct "bubbles" rather than in plumes.
A great deal of work remains to be done before these effects are
understood and compositional convection can be incorporated in
theoretical dynamo models.

Some information on the fluid motions in the outer core at the
CMB can be inferred from observations of the geomagnetic secular
variation, and the possible existence of a shallow, stably stratified

layer beneath the CMB has been investigated [Whaler (1980, 86)]. Fearn and Loper (1981) have shown that if the excess light constituent released by the freezing of the inner core is evenly distributed throughout the outer core, then only very close to the CMB can this redistribution be achieved by diffusion (the barodiffusive flux) in a stable layer. They estimate the depth of this region to be about 70 km (but make no claims that they believe such a stable layer exists).

The gravitational potential energy available from the segregation associated with the freezing of the inner core depends on the difference in density $\Delta\rho$ between the inner and outer cores. To estimate the rate at which this energy is released, most authors assume a steady growth rate for the inner core over the age of the Earth. Using values of $\Delta\rho$ between 0.2 and 1.25 g cm^{-3}, the gravitational energy released is estimated to be between 1.2×10^{11} and 10^{12}W [Gubbins et al. (1979), Loper and Roberts (1983), and Loper (1984)]. A typical estimate of the latent heat released is about 10^{12}W (taking L = 10^6 J kg^{-1}), see Gubbins et al. (1979) and Anderson and Young (1987). The latent heat released at the ICB is greater than that which can be conducted down the adiabat at the ICB [which is $O(10^{11}W)$], the excess light constituent released is greater than the barodiffusive flux, so convection close to the ICB is driven by both thermal and compositional effects. The total heat flux includes latent heat released at the ICB, specific heat released throughout the core as it cools, ohmic dissipation and any radiogenic heating. At the MCB the adiabatic heat flux is $2-4 \times 10^{12}$W [Loper and Roberts (1983)]. Whether the total heat flux exceeds this or not is unclear. If it does, then convection throughout the core is driven by thermal and compositional buoyancy. If not, then, in the upper part of the core convection carries heat downward, compositional effects driving convection against a stable temperature distribution [Gubbins et al. (1979), Loper and Roberts (1983)].

The efficiency with which gravitational and thermal energy can be converted into fluid motions and subsequently the efficiency with which these motions can be converted into magnetic field energy are considered in detail by Gubbins et al. (1979) and Verhoogen (1980). The major energy loss is the heat conducted down the adiabatic gradient, and only a fraction of any remaining heat can be converted into work because of heat engine efficiency limitations. Gravitational energy is more efficiently converted into fluid motions since the barodiffusive flux is relatively small and there are no Carnot type efficiency limits. Most of the convective energy is converted into magnetic energy and then into ohmic heat. The useful thermal and gravitational energy available enables the ohmic heating to be estimated. This then leaves the problem of what strength of magnetic field would produce this heating. To answer this question we require a detailed knowledge of the field structure in the core, which is not available. Only estimates of the relationship between ohmic heating and field strength can be made. They range from $10^{14} - 10^{16}$ WT^{-2} [Loper and Roberts (1983)]. Lower estimates are based on scaling arguments, larger ones on specific kinematic dynamo models. The uncertainty in this value produces an order of magnitude uncertainty in the field strength corresponding to a given ohmic heating.

A cooling core means that there is a gravitational contribution to the dynamo. Verhoogen's (1980) view is that this contribution is very uncertain because of our poor knowledge of the physical properties of the core. Gubbins et al. (1979) give estimates for the magnetic field strength of up to about 2×10^{-2}T, comparable with the estimates of Loper and Roberts (1983). Our knowledge of the physical properties of the core is improving, but the uncertainties are still large, so any field strength estimate should be taken only as a rough guide. What seems clear, though, is that the convection associated with the freezing inner core is capable of maintaining a toroidal magnetic field in the core that is substantially larger than the observed poloidal field.

6. CONCLUDING REMARKS

We have reviewed the consequences for the core of a cooling Earth. It has not been our intention to repeat detailed calculations, particularly relating to the power available to drive the geodynamo. For more details, the reader is referred to the comprehensive reviews by Gubbins et al. (1979), Verhoogen (1980), Loper and Roberts (1983) and Loper (1984). Rather, it has been our aim to emphasise the wealth of physical phenomena introduced by recognising the core as a freezing mixture. The structure of the freezing interface at the ICB, the flow through it and throughout the core are major problems which are only beginning to be tackled.

Finally a few remarks about the temperature in the core. This is strongly dependent on the concentration of the light constituent. An upper bound on the temperature at the ICB is the melting temperature T_m of pure iron at the ICB pressure. The most recent estimates for this are based on new high pressure experimental results, giving T_m = 7600 ± 500K [Williams et al. (1987)], significantly higher than earlier estimates of $T_m \simeq$ 6200K [Anderson (1986), Poirier (1986)]. The light constituent in the outer core depresses the freezing temperature. Taking this into account, Williams et al. (1987) estimate the temperature at the ICB to be 6600K (assuming a 1000K depression of the melting point). Poirier (1986) estimates a depression of 1150K with 8% S. Ahrens and Jeanloz (1987) give the ICB temperature to be about 4400K, based on an 11% concentration of S.

REFERENCES

Ahrens, T.J., 1980. 'Dynamic compression of Earth materials', Science, 207, 1035-1044.

Ahrens, T.J. and Jeanloz, R., 1987. 'Pyrite: shock compression, isentropic release, and composition of the Earth's core', J. Geophys. Res., 92, 10363-10375.

Anderson, O.L., 1986. 'Properties of iron at the Earth's core conditions', Geophys. J. R. Astr. Soc., 84, 561-579.

Anderson, O.L. and Young, D., 1987. 'Crystallization of the core and flow of heat in the Earth', abstract U2-17, IUGG XIX General Assembly, Vancouver, Canada.

Bolt, B.A., 1982. 'The constitution of the core: Seismological evidence', Phil. Trans. R. Soc. London A, 306, 11-20.

Braginsky, S.I., 1963. 'Structure of the F-layer and reasons for convection in the Earth's core', Doklady Acad. Nauk SSSR, 149, 8-10.

Brett, R., 1976. 'The current status of speculations on the composition of the core of the Earth', Rev. Geophys. Space Phys., 14, 375-383.

Chalmers, B., 1964. Principles of Solidification, Wiley, New York.

Copley, S.M., Giamei, A.F., Johnson, S.M. and Hornbecker, M.F., 1970. 'The origin of freckles in unidirectionally solidified castings', Met. Trans., 1, 2193-2204.

Cormier, V.F., 1981. 'Short-period PKP phases and the anelasticity mechanism of the inner core', Phys. Earth Planet. Inter., 24, 291-301.

Cormier, V.F., 1986. 'A search for lateral heterogeneity in the inner core from differential travel times near PKP-D and PKP-C', Geophys. Res. Lett., 13, 1553-1556.

Doornbos, D.J., 1974. 'The anelasticity of the inner core', Geophys. J. R. Astr. Soc., 38, 397-415.

Dziewonski, A.M. and Anderson, D.L., 1981. 'Preliminary reference Earth model', Phys. Earth Planet Inter., 25, 297-356.

Fearn, D.R. and Loper, D.E., 1981. 'Compositional convection and stratification of the Earth's core', Nature, 289, 393-394.

Fearn, D.R. and Loper, D.E., 1983. 'The evolution of an iron-poor core I. Constraints on the growth of the inner core', in Stellar and Planetary Magnetism (A.M. Soward, ed.), Gordon and Breach, London, 351-370.

Fearn, D.R. and Loper, D.E., 1985. 'Pressure freezing of the Earth's inner core', Phys. Earth Planet. Inter., 39, 5-13.

Fearn, D.R., Loper, D.E. and Roberts, P.H., 1981. 'Structure of the Earth's inner core', Nature, 292, 232-233.

Fearn, D.R., Roberts, P.H. and Soward, A.M., 1988. 'Convection, stability and the dynamo', in Energy Stability and Convection (G.P. Galdi and B. Straughan, eds.), Longman, Harlow, 60-324.

Gubbins, D., 1976. 'Observational constraints on the generation process of the Earth's magnetic field', Geophys. J. R. Astr. Soc., 47, 19-39.

Gubbins, D., 1977. 'Energetics of the Earth's core', J. Geophys., 43, 453-464.

Gubbins, D., Masters, T.G. and Jacobs, J.A., 1979. 'Thermal evolution of the Earth's core', Geophys. J. R. Astr. Soc., 59, 57-99.

Hills, R.N., Loper, D.E. and Roberts, P.H., 1983. 'A thermodynamically consistent model of a mushy zone', Q. J. Mech. appl. Math., 36, 505-539.

Hills, R.N. and Roberts, P.H., 1987a. 'Relaxation effects in a mixed phase region I: General theory', J. Non-equilib. Thermodyn., 12, 169-181.

Hills, R.N. and Roberts, P.H., 1987B. 'Relaxation effects in a mixed phase region II: Illustrative examples', J. Non-equilib. Thermodyn., 12, 183-195.

Hills, R.N. and Roberts, P.H., 1988. 'A generalised Scheil-Pfann theory for a dynamical theory of a mushy zone', submitted.

Jephcoat, A. and Olson, P., 1987. 'Is the inner core of the Earth pure iron?', Nature, 325, 332-335.

Knittle, E. and Jeanloz, R., 1986. 'High pressure metallization of FeO and implications for the Earth's core', Geophys. Res. Lett., 13, 1541-1544.

Liu, L., 1986. 'Potassium and the Earth's core', Geophys. Res. Lett., 13, 1145-1148.

Landau, L.D. and Lifshitz, E.M., 1959. Fluid Mechanics, Pergamon Press, Oxford.

Loper, D.E., 1975. 'Torque balance and energy budget for the precessionally driven dynamo', Phys. Earth Planet. Inter., 11, 43-60.

Loper, D.E., 1978a. 'The gravitationally powered dynamo', Geophys. J. R. Astr. Soc., 54, 389-404.

Loper, D.E., 1978b. 'Some thermal consequences of a gravitationally powered dynamo', J. Geophys. Res., 83, 5961-5970.

Loper, D.E., 1983. 'Structure of the inner core boundary', Geophys. Astrophys. Fluid Dynam., 25, 139-155.

Loper, D.E., 1984. 'Structure of the core and lower mantle', Adv. Geophys., 26, 1-34.

Loper, D.E. and Fearn, D.R., 1983. 'A seismic model of a partially molten inner core', J. Geophys. Res., 88, 1235-1242.

Loper, D.E. and Roberts, P.H., 1981. 'A study of conditions at the inner core boundary of the Earth', Phys. Earth Planet. Inter., 24, 302-307.

Loper, D.E. and Roberts, P.H., 1983. 'Compositional convection and the gravitationally powered dynamo', in Stellar and Planetary Magnetism, (A.M. Soward, ed.), Gordon and Breach, London, 297-327.

McCammon, C.A., Ringwood, A.E. and Jackson, I., 1983. 'Thermodynamics of the system Fe-FeO-MgO at high pressure and temperature and a model for formation of the Earth's core', Geophys. J. R. Astro. Soc., 72, 577-595.

Masters, G., 1979. 'Observational constraints on the chemical and thermal structure of the Earth's deep interior', Geophys. J. R. astr. Soc., 57, 507-534.

Moffatt, H.K., 1978. Magnetic Field Generation in Electrically
    Conducting Fluids, Cambridge University Press.
Moffatt, H.K., 1988. 'Compositional convection and the dynamo
    problem', submitted.
Morelli, A., Dziewonski, A.M. and Woodhouse, J.H., 1986. 'Anisotropy
    of the inner core inferred from PKIKP travel times', Geophys.
    Res. Lett., 13, 1545-1548.
Poirer, J.P., 1986. 'Dislocation - mediated melting of iron and the
    temperature of the Earth's core', Geophys. J. R. Astr. Soc.,
    85, 315-328.
Poirer, J.P., 1988. 'Transport properties of liquid metals and
    viscosity of the Earth's core', Geophys. J., 92, 99-105.
Roberts, P.H. and Loper, D.E., 1983. 'Towards a theory of the
    structure and evolution of a dendrite layer', in Stellar and
    Planetary Magnetism (A.M. Soward, ed.), Gordon and Breach,
    London, 329-349.
Rochester, M.G., Jacobs, J.A., Smylie, D.E. and Chong, K.F., 1975.
    'Can precession power the geomagnetic dynamo?', Geophys. J. R.
    Astr. Soc., 43, 661-678.
Spiliopoulos, S. and Stacey, F.D., 1984. 'The Earth's thermal
    profile: is there a mid-mantle boundary layer?', Geodynamics,
    1, 61-77.
Stacey, F.D., 1977. Physics of the Earth, Wiley, New York.
Stacey, F.D., 1980. 'The cooling Earth; a reappraisal', Phys. Earth
    Planet. Inter., 22, 89-96.
Stevenson, D.J., 1981. 'Models of the Earth's core', Science, 214,
    611-619.
Verhoogen, J., 1961. 'Heat balance of the Earth's core', Geophys. J.
    R. Astr. Soc., 4, 276-281.
Verhoogen, J., 1980. Energetics of the Earth, National Academy of
    Sciences, Washington D.C.
Whaler, K.A., 1980. 'Does the whole of the Earth's core convect?',
    Nature, 287, 528-530.
Whaler, K.A., 1986. 'Geomagnetic evidence for fluid upwelling at the
    core-mantle boundary', Geophys. J. R. Astr. Soc., 86,
    563-588.
Williams, Q., Jeanloz, R., Bass, J., Svendsen, B. and Ahrens, T.J.,
    1987. 'The melting curve of iron to 250 gigapascals: a
    constraint on the temperature at the Earth's center', Science
    236, 181-182.
Woodhouse, J.H., Giardini, D. and Li, X.-D., 1986. 'Evidence for
    inner core anisotropy from free oscillations', Geophys. Res.
    Lett., 13, 1549-1552.

NUMERICAL MODELS IN THE THEORY OF GEOMAGNETISM

K.-K. Zhang
Dept. of Earth & Space Science
University of California
Los Angeles
CA 90024, USA

F.H. Busse and W. Hirsching
Physikalisches Institut
Universitat Bayreuth
Postfach 101251
8580 Bayreuth, Germany

ABSTRACT. The potential of numerical models for the theoretical understanding of the dynamics of the Earth's core and the origin of the geomagnetic field is discussed. Some initial results of the analysis of convection driven dynamos in rotating spherical fluid shells are presented. As the parameter regime of the problem is more fully explored, the numerical simulation of processes such as geomagnetic secular variation and geomagnetic reversals becomes feasible. Because the physical conditions of the Earth's core are not sufficiently well understood, the interaction between numerical modelling and geomagnetic observations will become increasingly important. The decade fluctuations in the length of the day may provide additional constraints on models of the geodynamo if the process of electromagnetic core-mantle coupling is sufficiently strong.

1. INTRODUCTION

The progress in the theoretical understanding of the dynamo mechanism of the generation of magnetic fields in electrically conducting fluids and the increase in the speed and capacity of modern electronic computers have brought us to the stage where the numerical modelling of the geodynamo has become feasible. The word dynamo encompasses the large scale fluid dynamics of the Earth's core, the process of magnetic field generation by the fluid motions and the feedback on the fluid flow provided by the Lorentz forces. The difficulties in the understanding of this highly nonlinear system arise from the fact that important parameters have not yet become accessible to measurements or to reliable theoretical inference. The strength of the zonal component of the magnetic field in the core is not measurable at the Earth's surface, for example, because of the low electrical conductivity of the bulk of the Earth's mantle. Theoretical estimates are thus ranging over two orders of magnitude. The viscosity of the liquid iron alloy occupying the outer core of the Earth is not known within an order of magnitude and the value of 10^{-2} cm^2/sec used for

347

F. J. Lowes et al. (eds.), Geomagnetism and Palaeomagnetism, 347–358.
© 1989 by Kluwer Academic Publishers.

the kinematic viscosity in most papers on the subject has been chosen
for reasons of a convenient convention rather than on the basis of a
reliable theory. Numerical modellers of the geodynamo are thus at a
disadvantage in comparison with meteorologists who have succeeded
reasonably well in their numerical simulations of the dynamics of the
atmosphere. Even solar physicists modelling the magnetic field of the
Sun have direct observational access to all components of the magnetic
fields at the surface of the sun, and the dynamics of the solar
convection zone can be seen in the form of the solar granulation and
supergranulation.

In spite of these difficulties there are good reasons for the
expectation that numerical models may be ultimately successful in
providing a detailed understanding of the dynamo. The fluid
velocities in the outer core of the Earth are only fractions of
mm/sec, the magnetic Reynolds number is of the order 10^3 at most, and
the level of turbulence is presumably low in comparison to that of the
Earth's atmosphere or that of the sun. The constraint imposed by the
Coriolis force on the motions in the rapidly rotating core fluid is
released to some extent by the action of the Lorentz force. But this
coupling with the magnetic field implies that small scale motions are
quickly dissipated by Ohmic heating. These effects suggest that the
Earth's core may be more readily accessible to numerical modelling
than other geophysical and astrophysical systems to which numerical
flow simulations have been applied. At the present time we are still
far from achieving the goal of realistic simulations as will become
evident in the following. The rapid progress in numerical modelling
suggests, however, that major advances can be expected within the next
decade.

The numerical analysis of convection driven dynamos in rotating
spherical shells that is outlined in this paper has the purpose of
providing a foundation for later more detailed simulations. Only the
most important physical effects are thus taken into account, but an
attempt is made to explore a wide range of the parameter space. Using
the Galerkin method for the numerical approximation of the solutions
of the basic equations we keep the computational work relatively
lucid. There is no need at the current level of understanding of the
problem to resort to extensive computations on supercomputers. As in
other problems of fluid turbulence, it is illuminating to analyze the
bifurcation structure of solutions of the basic equations which
reveals the ways in which complexity in the spatial and time
dependence is introduced into the problem. This bifurcation structure
can be investigated best when one starts with a most symmetric
solution as the basic state of the problem. Subsequent bifurcations
from this basic solution are typically characterized by broken
symmetries. Many of the physical phenomena associated with the broken
symmetries can thus be traced to particular bifurcations or
instabilities. As in the case of convection in plane layers (see, for
example, Busse, 1981) the physical mechanisms leading to complex forms
of flow can thus be isolated.

The plan of this paper is as follows. In Section 2 the basic
equations are presented and the bifurcation structure of solutions is

outlined. In Section 3 examples of the different solutions are
presented and in Section 4 further developments of the numerical
approach are described. For more extensive discussions of the
analysis and its results we shall refer to the papers of Zhang and
Busse (1987, 1988, 1989) now referred to as ZB87, ZB88, ZB89. In
Section 5 the problem of core-mantle coupling is addressed and some
concluding remarks are made in the final section of the paper.

2. THE MATHEMATICAL PROBLEMS

In order to describe the problem of convection flow and magnetic field
generation in a rotating spherical fluid shell we use the equations of
motion in the Boussinesq approximation (i.e. material properties are
assumed to be constant except for the temperature dependence of the
density in the gravity force term), the heat equation for the
deviation θ of the temperature from its static distribution T_s, and
the equation of induction for the magnetic flux density $\mathbf{B}$,

$$\left[\frac{\partial}{\partial t} + \mathbf{u} \cdot \nabla\right] \mathbf{u} + 2\Omega \times \mathbf{u} =$$

$$-\nabla\pi - \alpha g\theta + \nu\nabla^2 \mathbf{u} + \frac{1}{\rho_0\mu}(\nabla \times \mathbf{B}) \times \mathbf{B} \tag{1a}$$

$$\nabla \cdot \mathbf{u} = 0 \tag{1b}$$

$$\left[\frac{\partial}{\partial t} + \mathbf{u} \cdot \nabla\right]\theta + \mathbf{u} \cdot \nabla T_s = \kappa\nabla^2\theta \tag{1c}$$

$$\left[\frac{\partial}{\partial t} + \mathbf{u} \cdot \nabla\right]\mathbf{B} - \mathbf{B} \cdot \nabla\mathbf{u} = \lambda\nabla^2\mathbf{B} \tag{1d}$$

where Ω denotes the angular velocity vector of the rotating system, ν,
κ and λ are the viscous, thermal and magnetic diffusivities, α is the
thermal expansivity of the density ρ_0 and $\mathbf{g}$ is the gravity force. All
terms that can be written as gradients in the equation of motion (1a)
have been combined into $\nabla\pi$, and μ denotes the magnetic permeability
which can also be expressed through λ and the electrical conductivity
σ, $\mu = (\lambda\sigma)^{-1}$. For simplicity we assume that thermal buoyancy
provides the driving force for the motions. Alternatively θ could be
interpreted as the concentration of light elements in which case κ
would denote an average diffusivity of these elements. Homogeneous
boundary conditions for the dependent variables $\mathbf{u}$, $\mathbf{B}$, θ can be applied
at the inner and outer radii, r_i and r_o, of the spherical fluid shell.
For details we refer to ZB87, ZB88.

Important dimensionless parameters of the problem are the
Rayleigh number R, the Taylor number τ^2, the Prandtl number P and the
magnetic Prandtl number P_m which are defined by

$$R = \frac{\alpha\Delta T g_0 (r_o - r_i)^3}{\nu\kappa}, \qquad \tau = 2|\Omega|(r_o - r_i)^2/\nu, \qquad P = \frac{\nu}{\kappa}, \qquad P_m = \frac{\nu}{\lambda},$$

where ΔT is the temperature difference across the spherical shell, which in the Earth's core would correspond to the superadiabatic part of the temperature difference between inner-outer core boundary and core-mantle boundary. The static solution of the problem, corresponding to $\mathbf{u} = \mathbf{B} \equiv 0$, $\theta \equiv 0$, exist for all values of these parameters, but becomes unstable when R reaches sufficiently high values. Infinitesimal disturbances $\mathbf{u}$, θ grow when the Rayleigh number exceeds a critical value R_c which depends in general on P and τ^2 (for detailed results see ZB87). These solutions reach finite amplitudes at supercritical Rayleigh numbers, but subcritical branches are also possible, especially at low Prandtl numbers. The preferred convection solutions corresponding to the lowest values of R_c are characterized by a periodic structure in the azimuthal direction (with respect to the axis of rotation) which is described by a wavenumber m_o. An example for $m_o = 4$ is shown in Figure 1. The convection pattern is time dependent in the form of a azimuthally travelling wave such that the pattern becomes steady with respect to a system moving relative to the rotating system with the phase velocity c. At sufficiently high values of τ the phase velocity is positive, i.e. the waves travel in the eastward direction.

The simplicity of the drifting convection solution permits a considerable saving of computational expenses in the numerical approximation of this solution. The symmetry of this type of convection is broken when the next bifurcation takes place. As the amplitude of convection increases two types of instability can occur. A purely hydrodynamic instability may cause a transition to a more complex spatial structure or a magnetic disturbance may grow without changing the velocity field in the immediate neighbourhood of the bifurcation point. The former kind of bifurcation is familiar from the work on the transition to turbulence in a convection layer (Busse, 1981) or similar fluid systems. The latter type of bifurcation is the dynamo process which represents the basis of the theory of the origin of planetary magnetism.

In both cases the structure of the bifurcating solution can be characterized according to the particular symmetries that are broken. Because of the periodic structure of the convection solution from which the tertiary solutions bifurcate, Floquet's theory is an indispensible tool of the analysis. Restricting the attention to the dynamo process we first distinguish those bifurcations that tend to change the azimuthal periodicity of the convection pattern from those that preserve it. In the mathematical parlance we may say that the Floquet exponent is vanishing for the latter. They are also of primary physical interest because they exhibit an axisymmetric component of the magnetic field. In restricting the attention to the periodicity preserving bifurcations we observe that the imaginary part of the growth rate can either be finite, indicating an oscillatory dynamo, or vanish as in the case of a steady dynamo. Again, geophysical motivations and matters of computational convenience impel us to focus the attention on the latter case. Because of the symmetry with respect to the equatorial plane there is still another possibility for a distinction between bifurcating solutions namely the quadrupolar class for which the radial component of the magnetic field

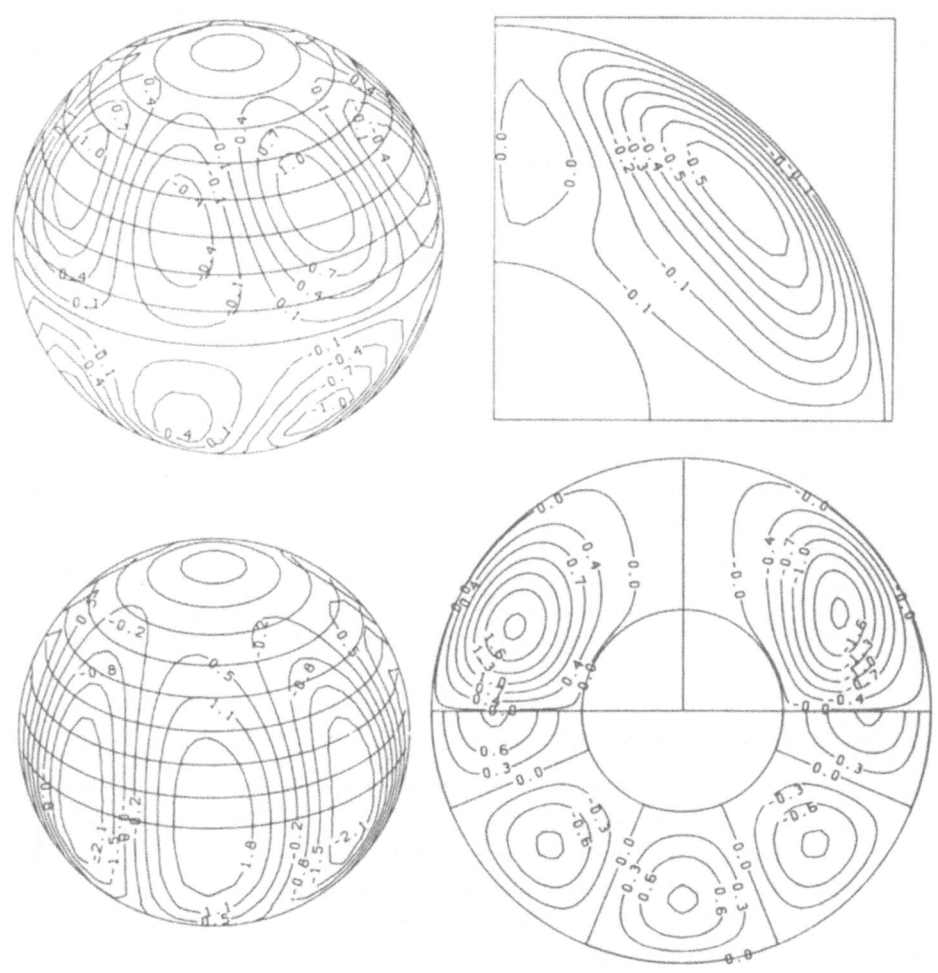

Figure 1. Convection flow with m = 4 in a rotating spherical shell with radius ratio η = 0.4 at R = 3300, τ² = 10⁵, P = 1. Upper left picture shows the streamlines of the toroidal component of the velocity at the outer boundary; the lines of constant radial valocity at the mid–radius are shown in the lower left. The upper right picture shows the meridional streamlines of the axisymmetric component of the velocity field and the streamlines of the fluctuating poloidal component in two meridional sections (90° apart in phase) and in the equatorial plane are shown in the lower right.

is symmetric with respect to the equatorial plane and the dipolar
class for which it is antisymmetric. As summarized in Figure 2 there
are thus eight different types of magnetic solutions which can be
separated according to their symmetry properties.

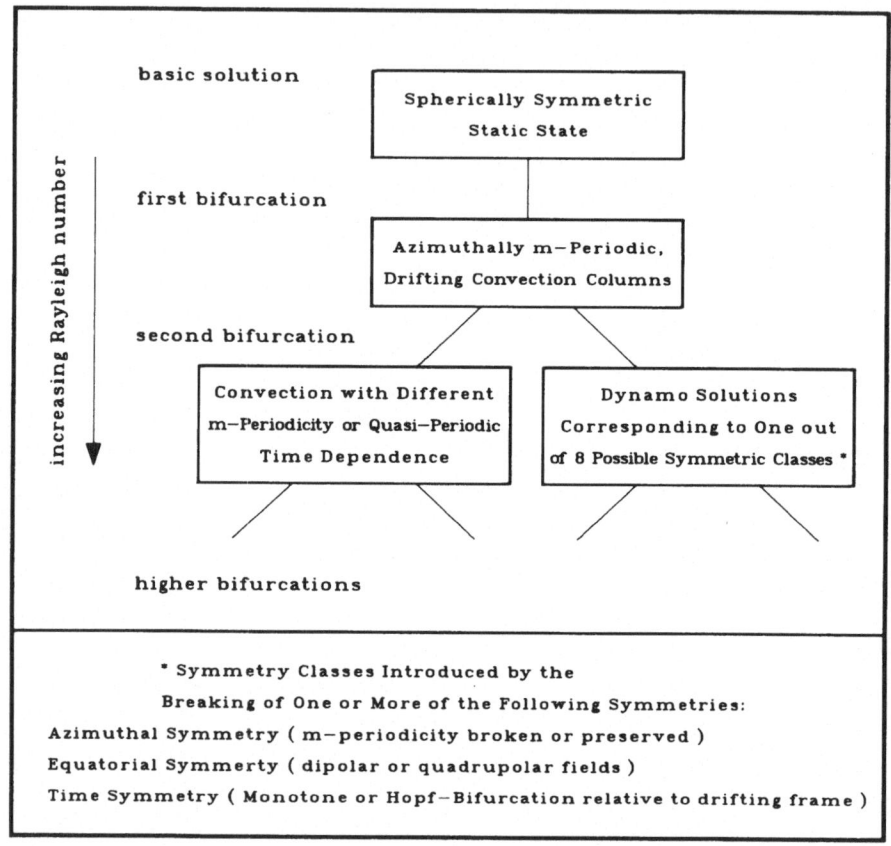

Figure 2. Bifurcation-Diagram.

3. SPHERICAL DYNAMOS

In contrast to the first bifurcation leading to the preferred
solutions in the form of convection columns that are symmetric with
respect to the equatorial plane, there is usually no single symmetry
class among the magnetic solutions which is preferred over a wide
region of the parameter space.   The critical magnetic Prandtl number
for the onset of oscillatory or steady dynamos in the form of either
quadrupolar or dipolar type fields often lie fairly close for a given

convection velocity field as shown in ZB88. The Lorentz force which enters the force balance as the strength of the magnetic field grows appears to be more important in selecting preferred configurations than the linear equation of induction.

Since the differential rotation tends to be nearly constant on coaxial cylinders at high rotation rates there is a qualitative difference between solutions of the dipolar and quadrupolar type. The poloidal component of the latter necessarily intersects the surfaces of constant differential rotation and the Lorentz force thus exerts a strong braking effect on the zonal flow. A dipolar field, on the other hand, can adjust itself in such a way that the field lines of the axisymmetric component become nearly parallel to the surfaces of constant angular velocity. An example for this behaviour is seen in the two upper pictures of Figure 3.

It is generally found that the scale of convection increases with increasing magnetic field strength and that the spatial dependence of various components of the convection field becomes smoother. There is thus hope that solutions can be extended to much larger Taylor numbers than have been accessible in the nonmagnetic case if a sufficiently strongly magnetic field is present. A difficulty in pursuing this idea in the numerical analysis arises from the fact that magnetic fields often tend to decay when the Rayleigh number or Taylor number is increased. Apparently a particular dynamo mechanism is most efficient only in a restricted region of the parameter space and becomes superseded by a different mechanism if the system is moved to a different parameter regime. The "island"-structure evident in the plot of the critical magnetic Prandtl numbers in ZB88 is a typical consequence of this property.

As an example of a quadrupolar dynamo we show in Figure 4 the magnetic field which starts to grow in the case of the convection velocity field of Figure 1 when P_m exceeds the value 89.3. Because the wavenumber m is higher than in Figure 3, more small scale structure is noticeable in the magnetic field and a higher value of P_m is required for onset of dynamo action.

4. COMPLEX DYNAMOS

The numerical approximations for spherical dynamos that have been obtained until now by the method outlined in Section 2 are but the simplest examples of a wide variety of possible solutions. A common property of all dynamo solutions corresponding to the second bifurcation in the diagram of Figure 2 is the symmetry of the Lorentz force with respect to the equatorial plane. Thus the symmetry of the convection velocity field with respect to the equatorial plane is not affected by the growth to finite amplitudes of the magnetic field. This situation will change when a higher bifurcation introduces a magnetic field component with the other equatorial symmetry, opposite to the one realized in the second bifurcation.

The fact that magnetic field components with quadrupolar symmetry make a significant contribution to the observed geomagnetic field

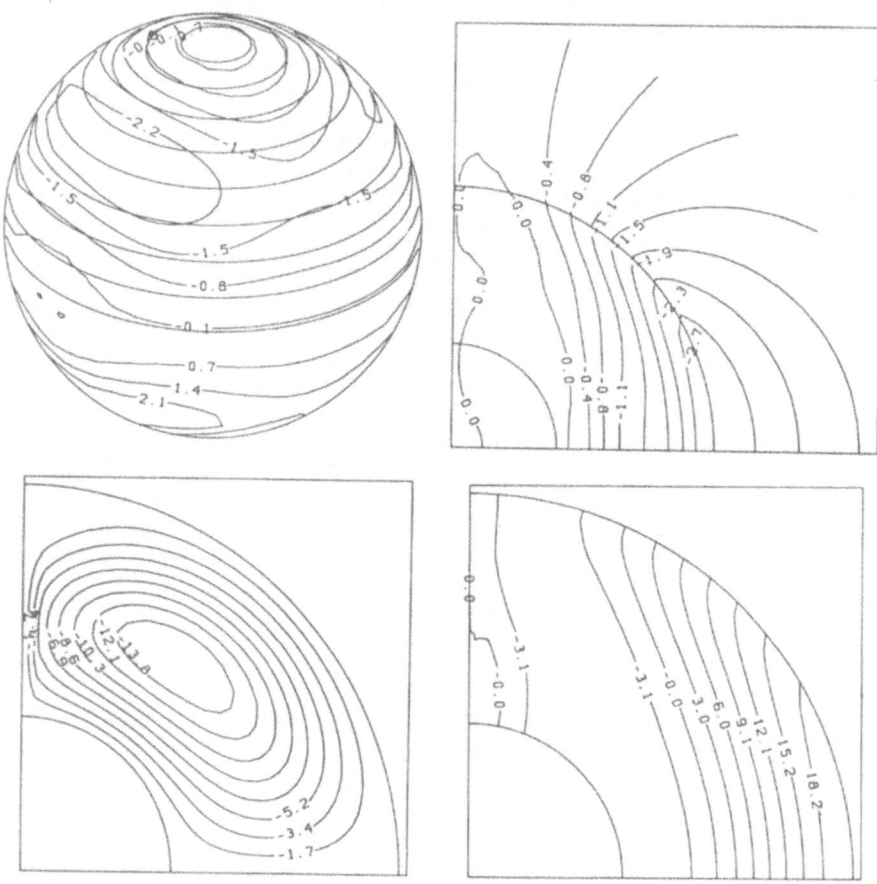

Figure 3. Dipolar magnetic field generated by convection (m$_0$ = 2) in a spherical shell with η = 0.4 at R = 3000, τ² = 587000, P =0.1, P$_m$ = 25. The upper left picture shows the lines of constant radial component of the magnetic field at mid-radius, the lines of constant axisymmetric zonal field are shown in the lower left. The field lines of the axisymmetric poloidal component are shown in the upper right, while lines of constant differential rotation are shown in the lower right for comparison.

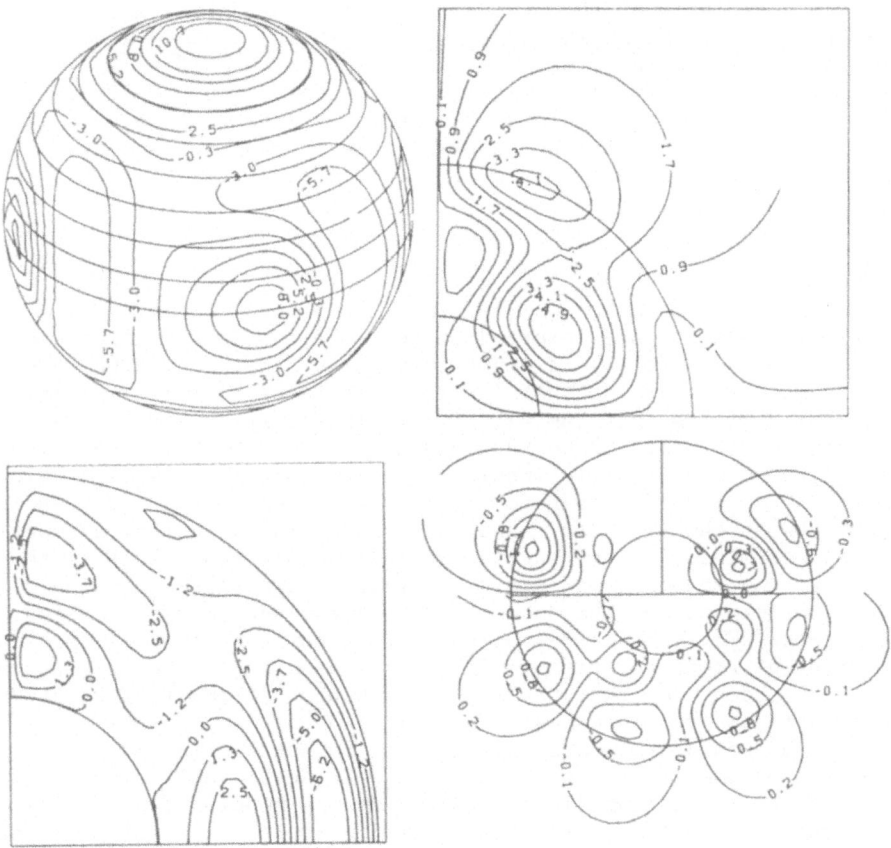

Figure 4. Quadrupolar magnetic field generated by the convection flow of figure 1 with P_m = 89.3. In contrast to Figure 3 the field has infinitesimal amplitude. The pictures show the same features as in Figure 3 except for the lower right picture which shows the field lines of the fluctuating poloidal component of the magnetic field in two meridional planes (90° apart in phase) and in the equatorial plane.

indicates that asymmetric dynamos are not atypical. It will be of interest to explore the higher bifurcations of dipolar dynamos in order to analyse the changes in the dynamics which a quadrupolar magnetic field may introduce. In general any higher bifurcation will introduce a time dependence that can not be eliminated by the introduction of a suitably drifting frame of reference. The extension of the analysis to the higher bifurcations thus not only encounters the loss of spatial symmetry but the introduction of time as additional variable of the problem as well. The Galerkin method can still be used if the coefficients are no longer regarded as constants but are treated as functions of time. With the use of a semi-implicit Crank-Nicholson scheme the system of nonlinear ordinary differential equations from the coefficients can be integrated in time. It is expected that phenomena like secular variation and perhaps even reversals of the magnetic field can be modelled eventually in this way.

5. CORE-MANTLE COUPLING

Because viscous stresses exerted at the core-mantle interface are very small and because it is difficult to resolve numerically the thin Ekman boundary layers at that interface it is usually assumed in the numerical analysis that the velocity field satisfies stress-free boundary conditions. Since the outside of the spherical shell is assumed to be electrically insulating no Lorentz force exists beyond the boundary. As a consequence no angular momentum transfer can take place between the fluid core and the outside in the numerical model. This indeterminacy of the angular momentum expresses itself in the fact that the component of rigid rotation in the representation of the velocity field can not be determined from the equations of motion. In the papers ZB87, ZB88, ZB89 the convention has been adopted that the component of rigid rotation vanishes. Alternatively the assumption could have been made that the angular momentum of the flow relative to the rotating system vanishes.

The motion of the core relative to the Earth's mantle is an important parameter of the problem, however, especially since it may offer eventually an additional constraint for models of a realistic geodynamo. A simple way for the determination of this relative motion opens up when the effect of a small, but finite electrical conductivity in the region outside the spherical fluid shell is considered. Introducing the ratio ε between the outer and the inner electrical conductivities as a perturbation parameter, we can determine the toroidal magnetic field in the outside region which is introduced by the leaking currents of the order ε. Another contribution to the Lorentz force action in the outside region arises from the time dependence and from the advection by the as yet unspecified relative motion between mantle and core. The condition that the torque of the Lorentz force exerted in the mantle about the axis of rotation vanishes in the stationary state can be used to determine the relative rate of rigid rotation between core and mantle.

This rate is independent of ε, but is sensitive to the conductivity distribution in the lower mantle. For a more detailed analysis we refer to Busse (1988).

The computations of the relative angular velocity that have been done so far on a relatively small sample of dynamo solutions show that the core moves retrograde relative to the mantle. But this motion is usually smaller than prograde drift of the convection pattern. Only in cases when the relative angular velocity exceeds the drift rate in magnitude may this effect be interpreted as a possible explanation of the westward drift of the non-dipole part of the geomagnetic field. There are other possible explanations, however, and a wider range of the parameter space needs to be explored before the geophysical relevance of the various effects can be estimated.

6. CONCLUDING REMARKS

The approach towards a theory of the geodynamo outlined in this paper is based on the strategy that the most simple models must be understood first before more complicated processes in the core can be simulated successfully. Some complications rise from the nonuniform boundary conditions at the core-mantle interface. Seismic evidence indicates strong lateral temperature variations and bumps at the core-mantle boundary (Hager et al., 1985; Morelli and Dziewonski, 1987). Bloxham and Gubbins (1987) have pointed out that effects of those inhomogeneities may be noticeable in the observed secular variation. In order to separate those regional influences from the global properties of the geodynamo more detailed data from archaeomagnetic sources on the long-time behaviour of the secular variation will be desirable.

In contrast to other bifurcation problems in fluid dynamics which usually exhibit a well defined preferred solution, the onset of dynamo action in convecting rotating spherical shells occurs in a number of different ways. In addition to the possibilities mentioned in Section 2, there are more complicated types of dynamos which bifurcate after purely hydrodynamic higher bifurcations have occured. A major open problem are the conditions leading to the preference of steady, dipolar dynamos rather than oscillatory or quadrupolar dynamos. It is not clear whether this question can be fully resolved on the basis of numerical models. Indeed, the close competition between dipolar and quadrupolar and between steady and oscillatory dynamos could be the basic physical reason for the occurence of reversals and excursions in the palaeomagnetic history of the Earth's magnetic field.

7. ACKNOWLEDGEMENTS

The authors' research on the origin of geomagnetism has been supported by the Deutsche Forschungsgemeinschaft.

358

8. REFERENCES

Bloxham, J. and Gubbins, D., 1987. 'Thermal core-mantle
    interactions'. <u>Nature</u> **325**, 511-513.
Busse, F.H., 1981. 'Transition to Turbulence in Rayleigh-Bénard
    Convection'. Chapter 5 in "Hydrodynamic Instabilities and the
    Transition to Turbulence", H.L. Swinney and J.P. Gollub, Editors,
    Springer-Verlag.
Busse, F.H., 1988. 'Core-mantle coupling and the geodynamo', to
    appear in the Proceedings of the I. Workshop on "Interaction of
    the Solid Planet with the Atmosphere and Climate".
Hager, B.H., Clayton, R.W., Richards, M.A., Comer, R.P. and
    Dziewonski, A.M., 1985. 'Lower mantle heterogeneity, dynamic
    topography and the geoid'. <u>Nature,</u> **313**, 541-545.
Morelli, A., and Dziewonski, A.M., 1987. 'Topography of the core-
    mantle boundary and lateral homogeneity of the liquid core'.
    <u>Nature,</u> **325**, 678-683.
Zhang, K.-K. and Busse, F.H., 1987. 'On the Onset of Convection in
    Rotating Spherical Shells'. <u>Geophy. Astrophys. Fluid Dyn.</u>, **39**,
    119-147.
Zhang, K.-K. and Busse, F.H., 1988. 'Finite amplitude convection and
    magnetic field generation in a rotating spherical shell'.
    <u>Geophys. Astrophys. Fluid Dyn.</u>, in press.
Zhang, K.-K. and Busse, F.H., 1989. Magnetohydrodynamic convection
    driven dynamos in rotating spherical shells, to be submitted to
    Geophys. Astrophys. Fluid Dyn.

INDEX